GUIDELINES FOR DRINKING-WATER QUALITY

Volume 2

Health Criteria and Other Supporting Information

World Health Organization

Geneva 1984

ISBN 92 4 154169 5

TYPESET IN INDIA

PRINTED IN BELGIUM

83/5957 – Macmillan/Ceuterick – 8000

CONTENTS

PART III. HEALTH-RELATED INORGANIC CONSTITUENTS

PART IV. HEALTH-RELATED ORGANIC CONSTITUENTS

PART V.   AESTHETIC CONSTITUENTS AND CHARACTERISTICS

PART VI.   RADIOACTIVE MATERIALS

PREFACE

The last edition of the *International standards for drinking-water[a]* was issued in 1971 and that of the *European standards for drinking-water[b]* in 1970. These standards have now been reviewed, revised, and combined, the resulting publication being issued in three separate volumes under the title *Guidelines for drinking-water quality*. In brief, the three volumes contain the following:

Volume 1: Recommendations. This volume presents recommended guideline values *per se*, together with the information essential to understanding the basis for the recommended guideline values as well as information on monitoring requirements. Where possible, suggestions are included regarding remedial measures to ensure compliance with the guideline values. The guidelines cover the microbiological, biological, chemical, organoleptic, and radiological quality of drinking-water.

Volume 2: Health criteria and other supporting information. The second volume sets out the health criteria for those drinking-water pollutants and other constituents that were examined with a view to recommending guideline values. In addition, it provides information regarding the detection of contaminants in water and measures for their control. It contains a review of the toxicological, epidemiological, and clinical evidence that was available and used in deriving the recommended guideline values.

Volume 3: Guidelines for drinking-water quality control for small-community supplies. The final volume deals specifically with the problem of small communities mainly located in rural areas. It contains information on techniques for the assessment and control of contamination of water supplies to these communities, including simple methods for sampling and analysis, sanitary surveys, and other means of investigating and controlling drinking-water quality in these areas. This volume is concerned essentially with the bacteriological safety of water.

Volume 2 thus reviews the evidence for recommending guideline values, summarizes and evaluates the available information on the health and sensory effects of drinking-water constituents, and provides a convenient reference source for those involved in developing and implementing national standards, as well as for those engaged in research work. This volume elaborates greatly on the health risk information presented in volume 1 and should be considered as a vital companion document to it.

---

[a] *International standards for drinking-water*, 3rd ed. Geneva, World Health Organization, 1971.
[b] *European standards for drinking-water*, 2nd ed. Geneva, World Health Organization, 1970.

vii

The microbiological and biological aspects of drinking-water quality are covered in Parts I and II of this volume. In addition to a description of the waterborne bacterial pathogens, including the rationale for using indicator organisms, detailed information is provided on the surveillance requirements and possibilities for safeguarding the bacteriological quality of drinking-water supplies. Subsections deal with the collection, storage, and transport of water samples, recommended methods for the detection of various microorganisms, and disinfection practice. A brief account of viruses in drinking-water is included as well.

While no guideline values as such are put forward for biological contaminants, the presence of pathogenic protozoans and helminths is of concern. Free-living organisms that may occur in water supplies are also discussed. No guideline values are recommended since standard methodology for surveillance is lacking in this field. Therefore, emphasis is placed on general source protection to minimize health problems from biological agents in drinking-water supplies.

A wide range of health-related inorganic and organic chemicals were considered by the task groups that met in the process of preparing the guidelines. Of the 37 inorganic and the 46 organic chemicals they reviewed in detail, guideline values were set for 9 inorganic and 15 organic chemicals, while tentative guideline values were set for 3 organic chemicals. The information on health effects considered in the development of the guideline values, together with other supporting data, is summarized in Parts III and IV of this volume. In addition, summaries are also included for some of the other chemicals and constituents that were reviewed but for which, for various reasons, it was not considered appropriate to recommend health-based guideline values. These summaries relate to asbestos, barium, beryllium, hardness, nickel, nitrite, silver, sodium, vinyl chloride, and certain chlorophenols and chlorobenzenes.

The information contained in these summaries usually includes the following: (a) a general description of the chemical, its major sources and concentrations in water; (b) information on routes of human exposure (drinking-water, food, air, etc.) including their relative significance; (c) metabolism (absorption, distribution, retention, elimination, and biotransformation); and (d) evidence of health effects, which includes a description of the adverse biological effects observed and an assessment of their health significance, identification of the most sensitive population groups at risk, relationships between dose and effects, and the prevalence of the effects. Each summary contains a list of the relevant references.

A total of 22 constituents and characteristics of drinking-water were carefully examined to ascertain their influence on its aesthetic and organoleptic qualities. Guideline values were recommended for 15 of them. The basic information used in arriving at the guideline values is summarized in this volume with reference to the sources, occurrence, routes of exposure, and the health and other effects. These summaries will be found in Part V, which also provides information on the influence of temperature, dissolved oxygen, and pH on drinking-water quality.

Part VI deals with radioactive materials in drinking-water and was prepared in close cooperation with the International Commission on Radiological Protection (ICRP). It explains the basis for the guideline values for gross alpha and gross beta activity and gives guidance with respect to the application of these guideline values in practice.

The preparation of volumes 1 and 2 of the new guidelines took over three years and involved the active participation of nearly 30 WHO Member States, scores of scientists, and meetings of 10 task groups. The work of these institutions and scientists, whose names appear in Annex 1 to volume 1, was central to the successful completion of the guidelines and is much appreciated. The collaboration of the national focal points for the WHO Environmental Health Criteria Programme, various international organizations, and individual experts was most helpful and their continuing participation contributed effectively to the work. The coordinators for this work were Dr H. Galal Gorchev of WHO headquarters and Mr W. Lewis of the WHO Regional Office for Europe.

The preparation of these guidelines was made possible by the financial support afforded to WHO by the Danish International Development Agency (DANIDA) and by the United Nations Environment Programme (UNEP), which provided funds to offset the costs of publishing this volume. Their contribution is most gratefully acknowledged. Appreciation is also expressed to the Environmental Protection Agency (EPA) in the USA which supported this effort by secondment of Dr Galal Gorchev for a period of two years.

It is recognized that as new information becomes available, the basis of the recommended guideline values will need to be reviewed and revised and new or changed guideline values may have to be recommended. As regards chemicals in drinking-water supplies, such reviews will be facilitated in the future through the International Programme on Chemical Safety (IPCS), which is a cooperative venture of the United Nations Environment Programme, the International Labour Organisation (ILO), and WHO. Two of its main objectives are (a) the evaluation of the effects of chemicals on human health and the environment, and (b) the development of guidelines for exposure limits (such as acceptable daily intakes and maximum permissible or desirable levels in air, water, food, and the working environment) for various classes of chemicals, including food additives, industrial chemicals, toxic substances of natural origin, plastics, packing materials, and pesticides. In particular, it was already recognized by the final task group meeting that new evidence regarding the potential health hazards from asbestos, sodium, nitrate/nitrite, nickel, chloroform, other trihalomethanes, trichloroethene, tetrachloroethene, and carbon tetrachloride was likely to become available over the coming years and would necessitate a reconsideration of these substances.

PART I. MICROBIOLOGICAL ASPECTS

1. THE BACTERIOLOGICAL QUALITY OF DRINKING-WATER

The most common and widespread danger associated with drinking-water is contamination, either directly or indirectly, by sewage, by other wastes, or by human or animal excrement. If such contamination is recent, and if among the contributors there are carriers of communicable enteric diseases, some of the living causal agents may be present. The drinking of water so contaminated or its use in the preparation of certain foods may result in further cases of infection.

1.1 Waterborne bacterial pathogens

Faecal pollution of drinking-water may introduce a variety of intestinal pathogens—bacterial, viral, and parasitic—their presence being related to microbial diseases and carriers, present at that moment in the community. Intestinal bacterial pathogens are widely distributed throughout the world. Those known to have occurred in contaminated drinking-water include strains of *Salmonella, Shigella*, enterotoxigenic *Escherichia coli, Vibrio cholerae, Yersinia enterocolitica*, and *Campylobacter fetus*. These organisms may cause diseases that vary in severity from mild gastroenteritis to severe and sometimes fatal dysentery, cholera, or typhoid.

Other organisms, naturally present in the environment and not regarded as pathogens, may also cause occasional opportunist disease. Such organisms in drinking-water may cause infection predominantly among people whose local or general natural defence mechanisms are impaired; this is most likely to be the case in the very old, the very young, and patients in hospitals, for example, with burns or on immunosuppressive therapy. Potable water used by patients for drinking and bathing, if it contains excessive numbers of organisms such as *Pseudomonas, Flavobacterium, Acinetobacter, Klebsiella*, and *Serratia*, may produce a variety of infections involving the skin and mucous membranes of the eye, ear, nose, and throat.

The modes of transmission of bacterial pathogens include ingestion of contaminated water and food, contact with infected persons or animals, and exposure to aerosols. The significance of the water route in the spread of intestinal bacterial infections varies considerably, both with the disease and with local circumstances. Although *Shigella* may be waterborne, water is not usually the main route for the spread of

3

shigellosis, but rather person-to-person contact in crowded living conditions; in contrast, cholera is usually waterborne and salmonellosis food-borne.

Among the various waterborne pathogens, there exists a wide range of minimum infectious dose levels necessary to cause a human infection. With *Salmonella typhi*, ingestion of relatively few organisms can cause disease; with *Shigella flexneri*, several hundred cells may be needed, whereas many millions of cells of *Salmonella* serotypes are usually required to cause gastroenteritis. Similarly, with toxigenic organisms such as enteropathogenic *E. coli* and *V. cholerae*, as many as 10^8 organisms may be necessary to cause illness. The size of the infective dose also varies in different persons with age, nutritional status, and general health at the time of exposure. The significance of routes of transmission other than drinking-water should not be underestimated as the provision of a safe potable supply by itself will not necessarily prevent infection without accompanying improvements in sanitation and personal habits. Education in simple applied hygiene is essential.

1.2 Rationale for the use of indicator organisms

The recognition that microbial infections can be waterborne has led to the development of methods for routine examination to ensure that water intended for human consumption is free from excremental pollution. Although it is now possible to detect the presence of many pathogens in water, the methods of isolation and enumeration are often complex and time-consuming. It is therefore impracticable to monitor drinking-water for every possible microbial pathogen that might occur with contamination. A more logical approach is the detection of organisms normally present in the faeces of man and other warm-blooded animals as indicators of excremental pollution, as well as of the efficacy of water treatment and disinfection. The presence of such organisms indicates the presence of faecal material, and thus that intestinal pathogens could be present. Conversely, the absence of faecal commensal organisms indicates that pathogens are probably also absent. Search for such indicators of faecal pollution thus provides a means of quality control. Surveillance of the bacterial quality of raw water is also important, not only in the assessment of the degree of pollution, but also in the choice of the best source and the treatment needed.

Bacteriological examination offers the most sensitive test for the detection of recent and therefore potentially dangerous faecal pollution, thereby providing a hygienic assessment of water quality with a sensitivity and specificity that is absent from routine chemical analyses. It is essential that water is examined regularly and frequently as contamination may be intermittent and may not be detected by the examination of a single sample. For this reason, it is important that drinking-water is examined frequently by a simple test rather than infrequently by a more complicated test or series of tests. Priority must always be given to ensuring that routine bacterial examination is maintained whenever manpower and facilities are limited.

It must be appreciated that all a bacteriological analysis can prove is that, at the time of examination, contamination, or bacteria indicative of faecal pollution, could or could not be demonstrated in a given sample of water using specified culture methods. In addition, the results of routine bacteriological examination must always be interpreted in the light of a thorough knowledge of the water supplies, including their source, treatment, and distribution. Whenever changes in conditions lead to deterioration in the quality of the water supplied, or even if they should suggest an increased possibility of contamination, the frequency of bacteriological examination should be increased, so that a series of samples from well chosen locations may identify the hazard and allow remedial action to be taken. Whenever a sanitary survey, including visual inspection, indicates that a water supply is obviously subject to pollution, remedial action must be taken, irrespective of the results of bacteriological examination. For unpiped rural supplies, sanitary surveys may often be the only form of examination that can be undertaken regularly.

1.2.1 Organisms indicative of faecal pollution

The use of normal intestinal organisms as indicators of faecal pollution rather than the pathogens themselves is a universally accepted principle for monitoring and assessing the microbial safety of water supplies (*1*). Ideally, the finding of such indicator bacteria should denote the possible presence of all relevant pathogens. Indicator organisms should be abundant in excrement but absent, or present only in small numbers, in other sources; they should be easily isolated, identified and enumerated and should be unable to grow in water. They should also survive longer than pathogens in water and be more resistant to disinfectants, such as chlorine. In practice, these criteria cannot all be met by any one organism, although many of them are fulfilled by coliform organisms, especially *Escherichia coli* as the essential indicator of pollution by faecal material of human or animal origin. Other microorganisms that satisfy some of these criteria, though not to the same extent as coliform organisms, can also be used as supplementary indicators of faecal pollution in certain circumstances. The significance that can be attached to the presence or absence of particular faecal indicators varies with each organism and especially with the degree to which that organism can be specifically associated with faeces.

Organisms used as bacterial indicators of faecal pollution include the coliform group of organisms as a whole, *E. coli* and coliform organisms that have been described as "faecal coliforms", faecal streptococci, and sulfite-reducing clostridia, especially *Clostridium perfringens*. Anaerobic bacteria, such as bifidobacteria and *Bacteroides*, are more abundant than coliform organisms in faeces, but routine methods for their detection and enumeration are not yet available. The other groups of organisms also have non-faecal sources in the environment, and may even grow in the aquatic environment, thus decreasing the confidence with which their presence may be associated with excremental pollution. Full identification of these indicator organisms would require an extensive series of tests, which would be impracticable in routine monitoring. Water bacteriologists have therefore evolved definitions of indicator species and groups that are practical rather than taxonomic and are based largely on detection and enumeration in water, usually by either multiple-tube methods or membrane-filtration techniques.

1.2.1.1 *Coliform organisms (total coliforms)*

Coliform organisms have long been recognized as a suitable microbial indicator of drinking-water quality, largely because these organisms are easy to detect and enumerate in water. They are characterized broadly by their ability to ferment lactose in culture at 35 °C or 37 °C, and include *E. coli, Citrobacter, Enterobacter,* and *Klebsiella* species. Coliform bacteria should not be detectable in treated water supplies, and if found, suggest inadequate treatment or post-treatment contamination (*2*). In this sense, the coliform test is used as an indicator of treatment efficiency. Although coliform organisms may not be directly related to the presence of viruses in drinking-water, the use of the coliform test is still essential for monitoring the microbial quality of public water supplies (*3*). The cysts of some parasites are known to be more resistant than coliform organisms to disinfection. The absence of coliform organisms in surface water that has only been disinfected will not necessarily indicate freedom from the cysts of *Giardia*, amoebae, and other parasites. Furthermore, coliform bacteria are derived not only from the faeces of warm-blooded animals but also from vegetation and soil (*4–6*). Under certain conditions, coliform organisms may also persist on nutrients derived from non-metallic construction materials. For these reasons, the presence of small numbers of coliform organisms (1–10 organisms per 100 ml), particularly in untreated groundwater, may be of limited sanitary significance provided faecal coliform organisms are absent.

1.2.1.2 *Faecal (thermotolerant) coliform organisms*

These are coliform organisms that are able to ferment lactose at 44.0 °C or 44.5 °C; they comprise the genus *Escherichia* and to a lesser extent occasional strains of *Enterobacter, Citrobacter,* and *Klebsiella*. Of these organisms, only *E. coli* is specifically of faecal origin, being always present in the faeces of man, animals, and birds in large numbers, and rarely found in water or soil that has not been subject to faecal pollution. Complete identification of *E. coli* in terms of modern taxonomy would require an extensive series of tests, which would be impracticable for routine water examination. Therefore, detection and identification of these organisms as faecal organisms or presumptive *E. coli* is considered to provide sufficient information to assess the faecal nature of pollution. Regrowth of faecal coliform organisms (*7*) in the distribution system is unlikely unless sufficient bacterial nutrients are present (biochemical oxygen demand (BOD) greater than 14 mg/l), water temperature is above 13 °C, and there is no free chlorine residual.

1.2.2 Other indicators of faecal pollution

If there is any doubt, especially when coliform organisms are found in the absence of faecal coliform organisms and *E. coli*, other indicator organisms may be used to confirm the excremental nature of the contamination. These secondary indicator organisms include the faecal streptococci and sulfite-reducing clostridia, especially *C. perfringens*.

1.2.2.1 *Faecal streptococci*

The occurrence of faecal streptococci in water generally indicates faecal pollution (*8, 9*). This term refers to those streptococci normally present in the faeces of man and animals. It includes *S. faecalis, S. faecium, S. durans, S. bovis*, and *S. avium*, as well as strains with properties intermediate between them. These organisms rarely multiply in polluted water and they may be slightly more resistant to disinfection than coliform organisms. However, this indicator group has rarely been recommended for control of drinking-water quality because of their persistence in water with moderate salt concentration (*10*), such as might occur with blended water supplies. Furthermore, the widespread occurrence of *S. faecalis* var. *liquifaciens* may detract from the significance of numbers of faecal streptococci less than 100 per 100 ml in drinking-water, unless strain identification is part of the routine procedure. When used as a supplementary bacterial indicator, the ratio of faecal coliform organisms to faecal streptococci (> 3:1 for human wastes; < 0.7:1 for other animal wastes) may be useful in locating the origin of faecal pollution in heavily contaminated sources of raw water, provided sufficient data are collected. In addition, these organisms can be used to assess the significance of doubtful results with the coliform test, particularly if coliform organisms are found in the absence of faecal coliform organisms. They can also be of value in checking water in the distribution system following repairs to mains.

1.2.2.2 *Sulfite-reducing clostridia*

These are anaerobic spore-forming organisms, of which the most characteristic, *C. perfringens (C. welchii)*, is normally present in faeces though in much smaller numbers than *E. coli*. Clostridial spores can survive in water longer than organisms of the coliform group and they can resist disinfection if the concentration, contact time, or pH is unsatisfactory. Their persistence in disinfected waters may thus indicate deficiencies in treatment (11).

However, it would not be desirable to consider these organisms for the routine monitoring of distribution systems since they tend to survive and accumulate; they may thus be detected remote in both time and place from the original source of pollution and consequently give rise to false alarms.

1.2.3 Indicators of water quality

Apart from the colony counts, the use of other microorganisms, including *Pseudomonas aeruginosa*, has been advocated to assess the hygienic quality of drinking-water (12, 13). However, neither examination for these organisms, nor colony counts are essential for the routine monitoring of hygienic quality. They are of value in certain circumstances in giving an indication of the general cleanliness of the distribution system and in assessing the quality of bottled water.

1.2.3.1 *Pseudomonas aeruginosa*

This organism often occurs in the faeces of man but in much lower numbers than coliform organisms. It is an opportunist pathogen of the very young and old and those already debilitated by disease, being frequently isolated from persons with urinary tract infections and skin burns (14). The organism occurs in raw water, usually in the presence of coliform organisms. However, in drinking-water it may occur in the absence of coliform organisms (15, 16) and the ability of some materials used in the construction of distribution and plumbing systems to support the growth of the organisms may account, in part, for this (17). Although the presence of the organism in potable water should not be ignored, it should not be used for the routine examination of water for the presence of faecal pollution. Examination for *P. aeruginosa* may be of value in certain circumstances, as for example, in the reconstitution of rehydration mixtures, baby foods, and pharmaceutical preparations, as well as in the surveillance of hospital water supplies and bottled water (18).

1.2.3.2 *Colony counts*

Colony counts may be used to assess the general bacterial content of water. They do not represent the total number of microorganisms present in water, but simply those that are able to form visible colonies in nutrient media under specified culture conditions. They are of little value in detecting the presence of faecal pollution and should not be considered essential for assessing the safety of potable water supplies, although a sudden increase in the colony count from a groundwater source may be an early sign of pollution of the aquifer (19). They are useful in assessing the efficiency of water-treatment processes, specifically coagulation, filtration, and disinfection, the objective being to maintain as low a density as possible in the treated water. They may also be used to assess the cleanliness and integrity of the distribution system and the suitability of water for use in the manufacture of food and drink where, to minimize the risk of spoilage, numbers should be low. The main value of colony counts lies in the comparison of results obtained from regular samples from the same supply so that any significant change from the normal range in a particular location can be detected.

1.3 Nuisance organisms

These constitute a morphologically and physiologically diverse group of organisms, which includes planktonic and sessile algae, fungi, crustacea, and protozoa, as well as actinomycetes, and iron and sulfur bacteria. These organisms can cause objectionable tastes, colour, odour, and turbidity and may interfere with treatment processes by blocking strainers and filters. In addition, certain planktonic organisms, although not themselves harmful, may harbour pathogens and protect them against disinfection by chlorine. Most of these nuisance organisms can be controlled relatively easily by the usual water-treatment processes. Furthermore, problems of taste, odour, colour, and turbidity, which may be caused by them, are covered indirectly by the guidelines for the aesthetic characteristics of water given in Part V. The presence of certain organisms in water may be an indication of corrosion of cast iron or of biodeterioration of construction materials that support the growth of microorganisms (17). These include nonmetallic materials, such as plastics, rubbers, jointing compounds, and pipe-lining materials, which provide organic nutrients and thus encourage the growth of micro-organisms including sometimes coliform organisms and *P. aeruginosa*. Although treatment will normally remove such organisms, they may establish themselves in sediment or slime and on materials within the distribution system, where their presence may support the growth of *Asellus* and other animal populations. Residual disinfection will help control such infestations, but occasional flushing and physical cleaning of pipe sections with polyurethane foam plugs may be needed.

Nuisance organisms may also cause problems with groundwater sources by encrusting well-screens, thus causing loss of yield and impairing the aesthetic quality of the supply. Indeed, their presence may indicate organic pollution of the aquifer (20).

Routine monitoring of such nuisance organisms cannot be recommended because of their diverse nature and unpredictable occurrence, although bacteriologists should be aware that they can impair water quality. It is not practicable to specify any quantitative limit for nuisance organisms.

1.4 Disinfection

The principal reasons for disinfecting drinking-water are to ensure the destruction of pathogens, to maintain a protective barrier against pathogens entering the distribution system, and to suppress bacterial regrowth in the pipe environment. Because of the importance of disinfection in safeguarding the hygienic quality of potable supplies, it is essential that the concentration of disinfectant should be measured frequently and preferably recorded continuously. For small supplies, especially those known to be at risk, simple disinfection and measuring devices should be available.

1.4.1 Disinfection efficiency

The comparative efficiency of disinfectants may be expressed in terms of either the relative concentrations needed to attain the same rate of disinfection or the relative rates of disinfection produced by the same concentration of disinfecting agent. However, because of the different nature of the microorganisms and the difficulty of standardizing test conditions, such as pH, temperature, and the chemical characteristics of the water, only generalized statements can be made about the comparative efficiencies of different disinfectants. Within these limitations, disinfecting agents may be grouped according to their efficacy. Thus, the use of chlorine, chlorine dioxide, or ozone is preferable, although for chlorine the pH should be less than 8.0. Because chloramines are only slowly biocidal, their use as primary disinfecting agents for water-treatment purposes is not recommended, although they may be used for the maintenance of residuals in distribution systems where the contact time is longer.

Similarly, in decreasing order, the relative resistance of different types of microorganisms and their probable survival may be listed as follows: protozoan cysts, enteroviruses, enterobacteria. Although there are distinct differences in the time required to inactivate enteroviruses as compared with enterobacteria, the minimum conditions of disinfectant-residual and contact-time required to ensure a microbiologically safe water supply can be achieved readily. It is, therefore, recommended that water from potentially polluted sources should always be disinfected. This would ensure inactivation of certain organisms, including some viruses, which may be relatively more resistant than faecal indicator bacteria.

1.4.2 Disinfectant residuals

In addition to disinfection efficiency, another important consideration is the ability of these agents to remain as residual disinfectants during the storage and distribution of potable water. Except for ozone, all of the other practicable disinfectants (chlorine, chlorine dioxide, and chloramines) can provide a persistent residual for continued microbial control once the finished water enters the distribution network. However, chloramines are such slow biocides that any decision as to their use should be evaluated carefully with sufficient bacteriological data collected throughout the distribution system to demonstrate effectiveness in controlling microbial regrowth and providing protection against a moderate degree (1% sewage) of contamination from cross-connections (21).

All supplies obtained from surface sources should be provided with disinfection as minimum treatment. Where disinfection is practised, a measurable residual should be maintained throughout the distribution system. Maintenance and monitoring of a chlorine residual offers two benefits. A chlorine residual will suppress the growth of organisms within the system and may afford some protection against contamination entering through cross-connection or leakage. The sudden disappearance of the residual provides an immediate indication of the entry of oxidizable matter into the system or of a malfunction of the treatment process. If chlorine is employed, it is desirable that a free chlorine residual of 0.2 to 0.5 mg/litre be maintained and monitored daily throughout the entire system. When the residual in the supply is less than that routinely expected at a particular point, then remedial action, including increased chlorination, flushing and a sanitary survey, should be considered as the loss of residual may indicate the entry of pollution into the pipework. Booster or relay chlorination may be needed to ensure that this residual is maintained throughout the system. It is recognized that excessive levels of free chlorine may react with organic matter to produce tastes and odours in some waters. In these instances, the control agency or the medical officer of health should encourage necessary improvements in treatment or distribution and as a temporary measure establish a suitable concentration of chlorine residual to ensure microbiologically safe water.

1.4.3 Effects of turbidity[a]

Effective disinfection depends upon contact between the disinfecting agent and the microorganisms to be inactivated for an adequate period of time. Various bacteriological and virological studies have demonstrated a marked difference in the extent to which various types of particulate matter in water shield microorganisms from the effects of disinfectants. In general, inorganic particles, such as clay and water-flocculating agents, appear to have little, if any, protective effect. On the other hand, organic particulate matter, whether cell debris, sewage solids, or living or dead organisms such as nematodes or crustaceans, can provide marked protection to microorganisms associated with them. The degree of protection thus afforded is determined to a large extent by

---

[a] For a fuller discussion of the effects of turbidity on drinking-water quality, see Part V, section 16.

the nature of the particulate matter rather than by the amount present as indicated by turbidity measurements (22).

In all processes in which disinfection is practised, the turbidity must always be low, preferably below 1 nephelometric turbidity unit (NTU) and always less than 5 NTU, otherwise the particulate matter will interfere with the efficiency of disinfection, partly by exerting a disinfectant demand and partly by shielding microorganisms, even in the presence of a residual disinfectant otherwise sufficient to ensure a kill. Excessive water turbidity may also interfere with bacteriological examination, especially by membrane filtration. Low turbidity, particularly in treatment works relying on coagulation, can only be achieved by careful control and operation to ensure that the coagulant dose and pH are optimum, that floc-blankets are stable, and that filter runs are optimized by careful monitoring of headloss and turbidity. Filter backwashing is important to ensure that breakthrough of turbidity does not occur. Where coagulation, sedimentation and filtration are essential to ensure the removal of particulate matter, disinfection must invariably be used to ensure microbiologically safe water. Any organic particulate matter present in potable water during distribution exerts a chlorine demand which reduces the available free chlorine residual, especially in dead-end sections of the system. Regular flushing of mains is desirable to avoid such accumulations. Organic turbidity can also serve as a source of nutrients, which may contribute to bacterial growth within the distribution network, especially in slow-flowing parts. Bacterial growth may enhance the accumulation of iron by bioflocculation. This results in the formation of a matrix of slime, calcium carbonate, and other debris attached to the pipe walls, which may result in deterioration in water quality. By-products of bacterial metabolism or decomposition within the slime may lead to taste and odour problems. Surges in water pressure may also dislodge accumulation of slime and sediments.

1.5. Collection, storage, and transport of water samples for bacteriological examination

Care must be taken to ensure that samples are representative of the water to be examined and that no accidental contamination occurs during sampling. Sample collectors should, therefore, be trained and made aware of the responsible nature of their work. Samples should be clearly labelled with the site, date, time, nature of the water and other relevant information and sent to the laboratory for analysis without delay (23–26).

1.5.1 Sample bottles

The size of the sample and bottle depends on the analyses to be undertaken, but 200 ml will usually be sufficient for routine examination for coliform organisms by either the multiple-tube or membrane-filtration method. In special surveys, larger volumes may be necessary. Clean, sterile glass bottles should be used. They should be fitted with ground-glass stoppers or screw caps and the neck of the bottle should be protected from contamination by a suitable cover of paper or aluminium foil. Polypropylene bottles that will withstand autoclaving may also be used (23).

1.5.2 Neutralization of disinfectants

If the water to be examined is likely to contain chlorine, chloramine, chlorine dioxide, or ozone, then 0.1 ml of an 18 g/litre solution of sodium thiosulfate per 100 ml of bottle capacity should be added to neutralize any residual disinfectant. This should neutralize at least 5 mg of available chlorine per litre and will be suitable for routine sampling. In special situations where the residual may be greater, additional thiosulfate is required. This concentration of thiosulfate has no significant effect on the coliform organisms, including *E. coli*, either in chlorinated or in unchlorinated water samples during storage (24).

When samples of disinfected water are taken, the concentration of residual disinfectant at the sampling point should be determined at the same time.

1.5.3 Sampling procedures

When a number of samples are to be taken for various purposes from the same location, the sample for bacteriological examination should be collected first to avoid the danger of contamination of the sampling point.

The sample bottle should be kept unopened until it is required for filling. The stopper should be removed with one hand and, during sampling, neither it nor the neck of the bottle should be allowed to touch anything. The bottle should be held in the other hand near the base. It should not be rinsed out and should be filled so as to leave a small air space when the stopper and cover are replaced.

If it is not possible to collect the sample directly in a sample bottle, a sterile stainless steel jug should be used. This can be sterilized by igniting methylated spirits inside it. After filling, the water should be transferred to the sample bottle.

1.5.3.1 Sampling directly from a source

When collecting samples directly from a river, stream, lake, reservoir, spring, or shallow well, the aim must be to obtain a sample that is representative of the water to be examined (23). It is therefore, undesirable to take samples too near the bank, or too far from the point of draw-off, or from the sediment near the bottom. Areas of stagnation should be avoided.

The sample should be taken by plunging the bottle, preferably neck downwards, into the water to a depth of 15–30 cm so as to avoid floating debris. The bottle should then be turned neck upwards with the mouth facing the direction of the current. Where no current exists, the bottle should be pushed horizontally through the water.

When sampling from a well or from the depths of a lake, reservoir, or cistern, a specially weighted sterile sample bottle or jug should be used.

When sampling raw water in areas where schistosomiasis is endemic, waterproof gloves should be worn to avoid direct contact with the water.

1.5.3.2 *Sampling from wells*

When taking a sample from a well fitted with a hand or mechanical pump, the pump should first be operated to flush out stagnant water from the pipework before the sample is taken. If possible, the mouth of the pump should be flamed, preferably by means of a blow lamp or gas (propane) torch, and a further quantity of water pumped to waste before sampling, which should be done by allowing the water from the pump to flow directly into the bottle.

1.5.3.3 *Sampling in treatment works*

As frequent samples must be taken in treatment works, sampling taps should be provided throughout so that all stages of processing can be monitored. The pipework serving these taps should be short. If long lengths cannot be avoided, then the water should preferably be run continuously. Pipes carrying raw or partially treated water should be cleaned regularly to remove slime and deposits.

If taps are not available, the samples may be taken by lowering the bottle into a tank or channel.

1.5.3.4 *Sampling from storage tanks*

Whenever possible, sampling taps should be provided in all tanks, cisterns, and service reservoirs used for storing water. When these are not available, samples should be taken by lowering a weighted jug or bottle into the tank. Great care should be taken not to contaminate the water during sampling and to avoid taking the sample from the surface layer or the bottom where scum or sediment may be present.

1.5.3.5 *Sampling from taps*

The majority of samples will be taken from taps—in treatment works, storage tanks, in the premises of consumers, or from public standpipes. Public health and water supply authorities should select sampling sites according to an agreed programme. When sampling from the distribution system, taps must be selected with great care. The taps chosen must be clean and should be supplied with water direct from the public main. Additional samples may be needed from tanks supplying high-rise and multiple-occupancy buildings. Taps that leak between the spindle and gland should be avoided as water will run down the outside of the tap and thus contaminate the sample. External fittings, such as rubber or plastic nozzles and filters, should be removed, and the water run to waste for several minutes to ensure that the water in the pipework is flushed out before the sample is taken. Flaming the tap before the sample is taken should be regarded as an optional procedure. To avoid the problems caused by sampling from domestic taps, water supply authorities should consider the installation of protected sampling taps at strategic points in distribution systems.

1.5.3.6 *Sampling from hydrants*

Samples from public supplies should be taken from suitable taps supplying water direct from the mains. If this is not possible, they can be taken from street hydrants. Special care is necessary, however, including flushing and hydrant disinfection.

1.5.4 Transport and storage of samples

The changes that may occur in the bacterial content of water on storage can be reduced to a minimum by ensuring that samples are not exposed to light and are kept cool, preferably between 4 °C and 10 °C— but not frozen. They must not be contaminated by the refrigerant. Examination should begin as soon as possible after sampling, preferably within 24 hours. Any delays before examination must be taken into account when interpreting the results, and must be stated in the report.

If delays are unavoidable, consideration should be given to the possibility of filtering samples on site and transferring them to transport media in airtight containers (*23, 24*). Membranes may be kept satisfactorily on such media for up to 3 days before transfer to conventional media for final examination in the normal way.

1.6 Recommended methods for the detection and enumeration of organisms indicative of pollution

Two basic procedures are available for the detection and enumeration of indicator bacteria in water: the multiple-tube method, in which measured volumes of water are added to replicate tubes of a suitable liquid medium; and the membrane-filtration technique, in which measured volumes of water are passed through a membrane filter that retains bacteria on the surface. It must be appreciated, however, that the results obtained with one method or medium may not be strictly comparable with those given by another for the same group of organisms. The particular method chosen should, therefore, always be applied consistently so as to ensure comparability of results.

1.6.1 Laboratory facilities and safety

Satisfactory facilities and equipment for the examination of drinking-water are important for ensuring reliable and reproducible results, especially for the detection of small numbers of faecal indicator organisms. Clean, comfortable working conditions are essential, not only to prevent cross-contamination of samples, but also for the health and safety of laboratory staff. Attention should be given to good laboratory practice, including the training of staff, the correct operation of incubators and water-baths, and careful preparation of media, as well as the use of quality-control procedures (26).

1.6.2 Detection of coliform organisms

Since coliform bacteria are present in large numbers in excrement and can be detected in concentrations as low as 1 per 100 ml, they are a sensitive indicator of the presence of faecal pollution.

1.6.2.1 *Definition of the coliform organisms*

The term coliform organisms (total coliforms) refers to Gram-negative, rod-shaped bacteria capable of growth in the presence of bile-salts or other surface-active agents with similar growth-inhibiting properties and able to ferment lactose at either 35 °C or 37 °C with the production of acid, gas, and aldehyde within 24–48 hours. They are also oxidase-negative and non-spore-forming. Those that have the same properties at a temperature of 44 °C or 44.5 °C are described as faecal (thermotolerant) coliform organisms. Faecal coliform organisms that ferment both lactose and other suitable substrates, such as mannitol, at 44 °C or 44.5 °C with the production of acid and gas and that also form indole from tryptophan, are regarded as presumptive *E. coli*. Confirmation of *E. coli* may be made by demonstration of a positive result in the methyl red test, by failure to produce acetyl methyl carbinol, and by failure to utilize citrate as the sole source of carbon. These are not taxonomic distinctions but practical working definitions used in water examination and they thus include members of several genera. Some organisms that would be identified taxonomically as "coliform" will, therefore, not be recognized as such in water examination. Examples include both anaerogenic and non-lactose fermenting strains of coliform organisms, as well as occasional strains of *E. coli* that are not thermotolerant. However, such strains are atypical and are greatly outnumbered in the environment by those that give the classical reactions, so that in practice the interpretation of results should not be affected. The choice of tests for the detection and confirmation of the coliform group of organisms should be regarded as part of a professional judgement based partly on the type of water, partly on the objective of the examination, and partly on the laboratory capability.

1.6.2.2 *Techniques for the detection of coliform organisms*

Two basic methods are used for the detection and enumeration of coliform organisms in water: the multiple-tube method and the membrane-filtration method (*1, 23–28*). The two methods do not give strictly comparable results, one reason being that counts on membrane filters give no indication of gas production from lactose, but for practical purposes they do yield comparable information.

For raw water, the detection of coliform bacteria and faecal coliform organisms may be an adequate guide to the microbial quality of the water. In the control of water-treatment processes, coliform organisms (total coliforms) should not be detectable in finished water. Should any coliform organisms be present in a water supply, it is then important that confirmation and differentiation be taken as far as possible in order to determine if the contamination is faecal in origin and to aid in tracing the source.

(*a*) *Multiple-tube method*

The initial test is presumptive because the acid and gas reaction observed may sometimes be caused by some other organism or combination of organisms. The presumption that coliform organisms have given the observed reaction has to be confirmed by additional tests with further confirmatory and differential media. The proportion of false-positive reactions depends on both the bacterial flora of the water under examination and the medium used.

An estimate of the number of presumptive coliform organisms present in a given volume of water can be obtained by inoculation of appropriate volumes into a number of tubes of medium. On incubation, it is assumed that each tube that receives one or more viable organisms in the inoculum will show growth and the positive reactions appropriate to the medium used. Provided that some negative results occur, the most probable number (MPN) of organisms in the original sample may be estimated from the number of tubes giving a positive reaction. Statistical tables of probability are used for this purpose and these, together with 95 % confidence limits, are given in Annex 2 to volume 1.

The MPN procedure is applicable to water of all types and especially to that in which the turbidity is high. The equipment required is relatively cheap and simple, and positive reactions are easy to read. The technique, however, provides only an estimate of the number of bacteria in any sample, and this is subject to considerable inherent error (*24*). Subcultures should be made to confirmatory media and also to solid media to ensure purity before further differentiation can be undertaken.

Volume of water to be examined. The volumes of samples used in tests with liquid media depend on the suspected bacterial content of the water to be examined. With water of good quality, one volume of 50 ml and five of 10 ml should be suitable. If the water is of doubtful or unknown quality, one 50-ml, five 10-ml and five 1-ml quantities should be used. The 50-ml and 10-ml volumes are added to the same quantity of double-strength medium, and the 1-ml volumes to 5 ml of single-strength medium. Heavily polluted raw water should be diluted in order to obtain some negative reactions, and thus a definite result for the MPN.

Choice of media. Four different media are available for use in the presumptive coliform test: minerals-modified glutamate medium (MMGM), lauryl tryptose broth (LTB), MacConkey broth and lactose broth (*1, 23–28*). MMGM is a chemically defined medium with limited nutrients that can be utilized by coliform organisms. The selectivity of MacConkey broth and LTB depends respectively on the presence of bile-salts and the surface active agent, lauryl sulfate; lactose broth is a non-selective medium. After inoculation, tubes of the selected medium are incubated at 35–37 °C and inspected at 24 ± 2 hours and 48 ± 3 hours for the appropriate positive reaction. The gas and acid produced by the fermentation of lactose is detected as follows: in all media, an inverted inner (Durham) tube is used to trap gas, and in MMGM, MacConkey broth, and lactose broth a pH indicator in the medium is also used to demonstrate acidity. Tubes should be tapped to release dissolved gas before they are discarded as negative. The presence of coliform organisms should be confirmed in all tubes giving positive reactions within 24 and 48 hours. This arbitrary time limit excludes occasional members of the coliform group that form gas slowly, but such organisms are generally of limited sanitary significance. The presence of spore-bearing organisms can cause false positive reactions, but they are usually eliminated during subsequent confirmatory tests.

Confirmatory tests. The presence of coliform organisms in positive presumptive reactions should be confirmed, and followed by further differential tests for faecal coliform organisms and *E. coli.*

Tubes exhibiting positive presumptive reactions at 24 hours and 48 hours are subcultured directly to tubes of selective media containing lactose for incubation for 48 ± 3 hours at the same temperatures used in the presumptive test. Since the original reaction may contain a mixture of organisms, a selective medium is necessary and brilliant-green (lactose) bile (BGB) broth may also be used (*1, 23, 24*). Production of gas confirms the presence of coliform bacteria in the original presumptive test. Tubes with negative reactions should be tapped gently to release dissolved gas. By reference to MPN tables, an estimate of the confirmed coliform content in 100 ml of the original sample may be obtained.

To confirm the presence of faecal coliform organisms, subcultures should be made from tubes with positive presumptive reactions to tubes of media, such as BGB or EC[a] broth (*23*) which contain bile-salts for incubation at 44 ± 0.25 °C for 24 ± 2 hours.[b] The production of gas

---

[a] *Escherichia coli.*
[b] If the procedures given in *Standard methods for the examination of water and wastewater (23)* are followed, then 44.5 ± 0.2 °C should be substituted throughout for 44 ± 0.25 °C.

confirms the presence of faecal coliform organisms in the original tubes and a confirmed MPN value can again be obtained. In addition, the numbers of presumptive *E. coli* may be obtained by inoculating tubes of tryptone water and testing for indole formation after 24 ±2 hours at 44 or 44.5 °C (*23, 24*). If positive, the presence of *E. coli* should be confirmed by differential biochemical tests. For convenience, a single tube of medium—in which both gas and indole formation can be demonstrated at 44°C—may be used for the confirmation of presumptive *E. coli* (*29*). To avoid the problems associated with strains of *E. coli* deficient in galactoside permease, mannitol has been used as the fermentable substrate. If only one positive reaction (either gas or indole) occurs, the tests should be repeated in separate tubes of media.

(b) Membrane-filtration technique

The number of coliform organisms in water may also be determined by filtration of measured volumes of the sample through membrane filters (*1, 24–28*). These are normally composed of cellulose esters, typically with pores 0.45 μm in diameter, which retain coliform and many other bacteria present in the sample. The membranes are then incubated face upwards on a selective medium. Characteristic acid- or aldehyde-producing colonies develop on the membrane and these are counted as either presumptive coliform organisms or faecal coliform organisms, depending on the temperature of incubation. Since gas production is not detected on membranes, it is presumed that all colonies that produce acid or aldehyde also produce gas. However, the techniques used in subsequent confirmation will demonstrate gas formation. The results are expressed as the number of organisms in 100 ml of the original sample.

In practice the membrane-filtration technique gives results comparable with those of the multiple-tube method. However, if the sample is filtered through two membranes and one incubated at 35 or 37°C and the other at 44 or 44.5°C, the confirmation procedure is somewhat simplified as a direct estimate of the number of faecal coliform organisms present is possible at the higher temperature.

An advantage of the membrane-filtration method is the promptness with which results can be obtained, thus permitting rapid corrective action and return to normal operation. The technique may be used for the examination of most water, except water of high turbidity, when the membrane will block before a sufficient amount can be filtered. Membranes are also unsuitable for water containing few coliform organisms in the presence of many other organisms capable of growing on the media used, since the latter are liable to cover the membrane and interfere with the growth of the coliform organisms. If non-gas-producing, but lactose-fermenting organisms such as *Aeromonas* predominate in the water, all presumptive coliform colonies on membranes should be confirmed because of the high proportion of false-positive results. The oxidase test will help in the rapid elimination of such false-positive results. Aerobic spore-bearing organisms, which may cause false presumptive reactions in liquid media, will not do so on membranes. The membrane technique may be modified to encourage the recovery of attenuated organisms. Preincubation at lower temperatures, or the use of less selective media, may allow stressed organisms to recover and start growing, after which the test is completed in the normal way (23, 24).

In the membrane-filtration technique, a direct count is made of discrete colonies, but this is still subject to statistical error. In addition, colony morphology can be examined and direct subcultures made, thus reducing the possibility of false-positive reactions from mixed cultures. Although only limited quantities of equipment are required, this is initially more expensive than the equipment used in the multiple-tube test. Membrane filters may be reused provided they remain undamaged after adequate washing and sterilization by boiling, and provided they are only used with the same medium. In addition, it should be appreciated that different membranes have different characteristics; it is thus important to ensure that they are suitable, not only for growth of the organisms sought, but also for the water concerned.

The results given by membrane filtration are not necessarily the same as those obtained by the multiple-tube method, although they usually give comparable results in practice. It is essential, therefore, that an adequate series of parallel tests should be carried out by both methods in order to establish that the membrane-filtration technique is suitable for the water concerned.

Filtration apparatus and technique. The apparatus consists of a sintered disc supported in silicone rubber gaskets fitted in a base to which a graduated funnel can be attached. The sintered or perforated disc supports the membrane filter. For use, the filter-holding assembly is mounted on a flask with a side-arm connected to a vacuum system. A series of filter-holders may be mounted in a manifold, thus permitting several samples to be filtered at the same time. After filtration of the water, the membrane is removed and placed face-upwards either on a suitable agar medium or on a pad soaked in liquid medium in a Petri dish for incubation at the appropriate temperature. Full details of the equipment can be found elsewhere (1, 23–28, 30).

After incubation, the membranes should be examined in a good light. The appearance of colonies will depend on the medium used, but all characteristic colonies should be counted irrespective of size. If necessary, individual colonies may be subcultured into liquid media for confirmation, or on to a solid medium to ensure purity before further differential tests are performed.

For the examination of water samples for coliform organisms (total coliforms) and faecal coliform organisms (presumptive *E. coli*), two separate membranes, appropriate media, and different incubation temperatures are required.

Volume of water to be examined. Counts of the total numbers of coliform organisms and of faecal coliform organisms are made with separate volumes of water, normally of 100 ml. Unless the samples are likely to contain more than 100 coliform organisms in 100 ml, the filtration of 100 ml is necessary for each test. For polluted samples, the volumes should be chosen so that the number of colonies on the membranes lies between approximately 10 and 100. If this volume is less than 10 ml, it should be mixed with a sterile diluent, such as quarter-strength Ringer's solution, 1g/litre peptone water, or a buffered dilution water, so that a minimum of 10 ml is filtered through the membrane.

Choice of medium. Various media can be used in the examination for coliform organisms by the membrane-filtration method. Of these, lactose tergitol agar (*31*), lactose TTC tergitol agar (*31*),[a] and lauryl sulfate lactose broth (*23, 29*) may be used for counts of coliform organisms at 35–37°C, and faecal coliform organisms at 44°C. Endo-type media should be used only for coliform counts at 35 or 37°C, and MFC[b] broth at 44°C for faecal coliform counts. Although all these media rely on the fermentation of lactose for the detection of presumptive coliform organisms, the characteristic reaction varies with each medium. The characteristic metallic sheen of colonies on Endo media depends on the formation of aldehyde.

Confirmatory tests. The extent to which individual colonies should be confirmed depends both on the water and on the reasons for the examination. The information required is similar to that for presumptive positive reactions in the multiple-tube method. To confirm the presence of coliform organisms, gas production from lactose must be demonstrated within 48 hours at 35 or 37°C; and for faecal coliform organisms within 24 hours at 44 or 44.5°C. In addition, the ability to produce indole from tryptophan at 44°C confirms the presence of presumptive *E. coli*. The confirmatory procedures differ slightly from those for the multiple-tube test because colonies from membranes sometimes grow poorly in selective media, such as BGB broth, which should preferably not be used for direct confirmation (*1, 23, 24*). Also, because membranes are normally incubated at both 35 and 44°C or 37 and 44.5°C, it is possible to confirm colonies directly by demonstrating gas production at 44°C without performing the test at a lower temperature (*23, 24*). Gas production at 35 or 37°C may be demonstrated with lactose peptone water (*24*) or lauryl tryptose broth (*23, 24*) and at 44°C with lactose peptone water or EC broth (*23, 24*). Tryptone water should be used for indole production (*23, 24*).

Ideally, all presumptive colonies on membranes should be examined, although this is not always practicable. However, as coliform organisms should not be present in treated waters, all colonies from such samples should be investigated further. With raw water before treatment, the samples should be diluted to yield a manageable number of colonies on membranes. In these circumstances, it is desirable to examine some

---

[a] TTC = 2,3,5-triphenyltetrazolium chloride.
[b] Medium for faecal coliform organisms.

suspicious colonies, preferably at least 10, to determine the nature and extent of the pollution.

1.6.2.3. *Differentiation of coliform organisms*

Because of the importance of confirming the identity of any presumptive *E. coli* in drinking-water, further differential tests are required, including if necessary the use of commercially available identification kits. In temperate climates, such tests may be expected to confirm that organisms that produce acid, aldehyde, gas, or indole at 44 °C are in fact *E. coli*. In hot climates other coliform organisms, such as *Enterobacter* spp. (*32–34*) of lesser hygienic significance, may give the presumptive reaction of *E. coli*, but can be differentiated by these tests. Differential tests must be performed on pure cultures isolated from confirmatory media by subculture on to plates of a non-selective medium. Typical colonies are then used for the indole, methyl red, Voges-Proskauer, citrate, and oxidase tests, and if necessary other biochemical reactions (*27, 28*). The results, together with those from the confirmatory media, are used to identify *E. coli* as defined in section 1.6.2.1. The same tests may also be used for the differentiation of other coliform organisms.

1.6.3 Detection of faecal streptococci

Both the multiple-tube and the membrane-filtration methods may be used for the presumptive detection of this group of organisms, but the application of these techniques is subject to the same limitations noted earlier for coliform organisms. The results of these tests may need confirmation (*1, 23, 24, 26–28*).

1.6.3.1 *Definition of faecal streptococci*

Faecal streptococci belong to Lancefield's serological groups D and Q, which include *S. faecalis* and its varieties, *S. faecium*, *S. durans*, *S. bovis*, and strains with properties intermediate between them. They also include *S. equinus* and *S. avium*. They are capable of growth at 45 °C in the presence of 40 % bile and in concentrations of sodium azide that are inhibitory to coliform organisms and most other Gram-negative bacteria. They are catalase negative. Some species resist heating at 60 °C for 30 minutes, and will grow at pH 9.6 and in media containing 65 g of sodium chloride per litre.

1.6.3.2 *Multiple-tube method*

Appropriate volumes of water are added to tubes of single- or double-strength glucose (dextrose)-azide broth (*1, 23, 24*) and incubated at 35 or 37 °C ± 0.5 °C for 48 ± 3 hours, the incubation being continued if necessary for 72 hours (*24*). Tubes that show acid production and turbidity, often with a sediment, are regarded as containing presumptive faecal streptococci. The production of gas is not sought.

The presence of faecal streptococci in tubes that give positive reactions should be confirmed by inoculation into ethyl violet azide broth for incubation at $35 \pm 0.5\,°C$ for 48 ± 2 hours.

Heavy inoculation of confirmatory media is normally needed (23, 24). Tubes with positive confirmed reaction for faecal streptococci show a blue-purple sediment with turbidity. An alternative confirmatory procedure involves subculture from presumptive tubes on to plates of Pfizer selective enterococcus (PSE) agar. The growth of brown/black colonies with brown haloes in 24 hours at $35 \pm 0.5\,°C$ confirms the presence of faecal streptococci (23).

The number of presumptive and confirmed faecal streptococci in 100 ml of original sample are calculated as described previously for coliform organisms using the probability tables in volume 1, Annex 2.

1.6.3.3 Membrane-filtration technique

Two media, KF agar and m-Enterococcus agar (Slanetz & Bartley), which both contain azide but different carbohydrates, are in general use for the enumeration of faecal streptococci by the membrane-filtration technique. With m-Enterococcus agar, selectivity is linked to the temperature of incubation and when incubated at $37 \pm 0.5\,°C$ for 4 hours, followed by 44 ± 3 hours at $44 \pm 0.25\,°C$ (23, 24), the method is especially selective for *S. faecalis* and *S. faecium* (35). This may be an operational advantage as it may supplement the coliform test in confirming the presence of faecal pollution, especially human (24). Overall recovery of faecal streptococci on this medium is sometimes poor (10, 35). For general use in the detection of all faecal streptococci from water, KF medium, which gives good recovery and selectivity (36, 37), is most often used. Selectivity of KF agar may be enhanced if the medium is sterilized by boiling rather than autoclaving. Membranes are incubated on KF agar at 35 or $37 \pm 0.5\,°C$ for 48 ± 3 hours (23). As this medium is both stable and selective for faecal streptococci, membranes can be transported on it for up to 3 days before completion of the test in the laboratory. Media used in the membrane-filtration method for faecal streptococci are generally very selective and the red colonies formed on both KF and m-Enterococcus agar are usually assumed to be faecal streptococci without further confirmation. If considered necessary, confirmatory tests should demonstrate growth in the presence of 40% bile at $44\,°C$ and a negative catalase reaction (23).

The density of faecal streptococci should be expressed in numbers of colonies per 100 ml of the original sample.

1.6.3.4 Differentiation of faecal streptococci

When the test for faecal streptococci is used to assess the source of pollution, it may be necessary to identify the species present. Following initial examination, all isolates that are catalase-negative may be further identified by tests for growth at 45 °C, at pH 9.6, and in 65 g/litre sodium chloride, for the hydrolysis of aesculin and starch, as well as for lactose fermentation and reduction of 1 g/litre methylene blue milk.

1.6.4 Detection of sulfite-reducing clostridia

Sulfite-reducing clostridia, especially *C. perfringens* (*C. welchii*), may also be used as indicators of faecal pollution (*38*). The organisms in this group are characterized by their ability to form spores and to reduce sulfite to sulfide. This feature is utilized in media for the presumptive detection of clostridia by the formation of a black precipitate of iron sulfide. For the detection of spores, all vegetative cells are inactivated by heating the samples of water at 75–80 °C for 10 minutes (*24*).

1.6.4.1 *Definition of sulfite-reducing clostridia*

This group of organisms consists of anaerobic sulfite-reducing, spore-forming, Gram-positive, catalase-negative, rod-shaped bacteria. In addition, *C. perfringens* ferments lactose, sucrose, and inositol with the production of gas, produces a typical "stormy-clot" reaction in litmus milk, hydrolyses gelatin, and produces lecithinase and acid phosphatase. This organism is non-motile.

Both the multiple-tube and membrane-filtration methods can be used for the detection of sulfite-reducing clostridia.

1.6.4.2 *Multiple-tube method*

The procedure is similar to that described for coliform organisms except that gas production is not sought. In order to maintain anaerobic conditions throughout the period of incubation, screw-capped bottles must be used and filled with broth containing glucose and sulfite, such as differential reinforced clostridia medium (DRCM) (*24*). Appropriate volumes of sample are inoculated into bottles of single- and double-strength medium and incubated at 35 or 37 ± 0.5 °C for 48 ± 3 hours. Tubes that show blackening are presumed to contain sulfite-reducing bacteria and, if the sample was heated before testing, the tubes are presumed to contain sulfite-reducing spore-forming clostridia.

The presence of *C. perfringens* can be confirmed by subculture of growth from bottles with presumptive positive reactions to tubes of litmus milk (*22*) for incubation at 35 or 37 ± 0.5 °C for 48 ± 3 hours. A typical "stormy clot" reaction, together with acidity, confirms the presence of *C. perfringens*. If further confirmation is required, tests for motility and nitrate-reduction may be done (*39*).

1.6.4.3 *Membrane-filtration technique*

No single medium has found general acceptance for the enumeration of sulfite-reducing clostridia in water, but iron sulfite agar (ISA) (*38*), sulfite-polymyxin-sulfadiazine agar (SPS) (*39*), and medium for *Clostridium perfringens* (mCP) (*40*) have been recommended. The incubation time for each medium is 24 hours. Incubation temperatures vary depending on the medium, being 37 °C for SPS, 45 °C for mCP, and 48 °C for ISA. Anaerobic conditions are necessary for growth, and these can be achieved by the use of agar overlays with ISA and SPS, but not with mCP where the need to expose the plates subsequently to ammonia vapour to demonstrate phosphatase reduction precludes this technique. Incubation of mCP must, therefore, be in an anaerobic atmosphere.

If samples are heated at 75–80 °C for 10 minutes before filtering, these techniques will be highly selective for sulfite-reducing clostridia. Confirmation of the presence of *C. perfringens* is necessary only if iron sulfite agar is used, as SPC and mCP, which contain appropriate antibiotics, are very selective for this organism. Nevertheless, media that contain sulfite have stood the test of time and are recommended as standard media; in addition, they have the advantage of making the membrane-filtration technique similar in use and mode of operation to the MPN method.

1.6.5 *Pseudomonas aeruginosa*

P. aeruginosa can be defined and readily distinguished from other fluorescent *Pseudomonas* species by the following differential features: production of pigment, growth at 42 °C, hydrolysis of casein, and the use of unusual sources of organic carbon.

1.6.5.1 *Definition of* P. aeruginosa

P. aeruginosa is a Gram-negative, non-sporing, rod-shaped bacterium that grows at 42 °C, may produce pyocyanin and fluorescent pigments, is oxidase- and catalase-positive, reduces nitrate beyond nitrite, liquefies gelatin, hydrolyses casein but not starch, oxidizes glucose, and reduces acetamide to ammonia.

1.6.5.2 *Multiple-tube method*

Media containing asparagine enhance pigment formation, and this forms the basis of the modification of Drake's broth in current use for the multiple-tube method (*23*). The addition of ethanol (20 ml/l) to the medium prevents the growth of other Gram-negative bacteria during incubation at 35 or 37 °C.

Appropriate volumes of sample are added to tubes of single- and double-strength broth for incubation at 35 or 37 °C for 4 days. All tubes showing growth should be examined daily for pigment production and fluorescence in ultraviolet light. The presence of *P. aeruginosa* in these presumptive reactions should be confirmed by subculture to cetrimide milk agar or acetamide broth (*23*). Casein hydrolysis, together with the production of a blue-green fluorescent pigment after incubation of milk agar at 41.5°C for 24 ±2 hours, and the production of alkaline conditions, as indicated by a purple coloration, within 36 hours at 35°C or 37°C with acetamide broth, confirms the presence of *P. aeruginosa*.

1.6.5.3 *Membrane-filtration technique*

King's A medium modified by the addition of ethanol (*23*) is usually used. After filtration, membranes are placed on modified King's A medium and incubated for 48 ±3 hours at 35 or 37°C. Fluorescent, green-pigmented colonies are considered to be *P. aeruginosa*. Colonies should be confirmed on milk agar at 41.5°C (*23*) for casein hydrolysis and also the production of green pigment and fluorescence within 24 hours.

1.6.6 Colony counts

The usual procedures in current use for estimating the bacterial content of water are the pour-plate and surface-spread methods (*1*, *23–28*, *31*). A membrane-filter technique has also been developed recently (*41*).

For colony counts, the nutrient media used will support the growth of only a proportion of microorganisms present in any water sample, and this will vary with the medium. Anaerobic organisms will not usually grow, even in pour plates. In addition, because microorganisms also occur in clumps and chains, counting those that actually form colonies will underestimate considerably the actual number of viable microorganisms present in the sample. As different microorganisms have different optimum temperatures for growth, it is normal to incubate two plates prepared from the same sample at 35–37°C for 1–2 days and another set at 20–22°C for 3 days. The period of incubation influences the colony count, and it is, therefore, important to adhere strictly to the same practice so that the results are always comparable. For bottled water, it is usual to incubate plates at 35 or 37°C for 3 days (*23*). The count is expressed as the number of colony-forming units in 1 ml of the original sample, stating the medium, the temperature and duration of incubation, and the method used. Briefly, the technique may be outlined as follows: a series of tenfold dilutions of the sample is made, the number depending on the nature and history of the water. From each dilution, 1 ml is added to each of two sterile Petri dishes; molten, nutrient agar at 44–46°C is added (15 ml) to each plate and the sample and medium mixed by rotation. After the agar has set, the plates are inverted and incubated at the desired temperature. The colonies on each pair of plates are then counted, and the number of microorganisms in the original sample obtained by multiplying the arithmetic mean of the counts by the reciprocal of the dilution. Where there are no plates with counts between 30 and 300 colonies, the colonies should still be counted but the result should be recorded as an estimate.

1.6.7 Examination for pathogenic organisms

Although direct search for specific pathogenic bacteria has no place in the routine bacteriological examination of water, there are occasions when examination for intestinal pathogens may be necessary as, for example, during an epidemic or in the evaluation of a new source. The chances of success will then be greater if large samples of water are examined, and if media selective for certain intestinal pathogens are used. Examination will include some, if not all, of the following stages: concentration of the organisms in the sample, inoculation into enrichment broth, subculture on to selective agar media, and bio-chemical and serological examination of suspect colonies. Rather than rely on a single method, it is better to use as many methods as possible so that no opportunity to detect a pathogen is missed (23, 24, 26–28). This is especially so for the detection of *Salmonella* since no single method is suitable for all serotypes.

1.6.7.1 *Concentration of samples*

The technique used will depend largely on the amount of particulate matter in the water. In waters of low turbidity, the sample may be passed through membrane filters. As turbidity increases in raw waters, filtration through diatomaceous earth (23, 24) or cartridge filters (42) may be used to enhance filtration and permit processing of larger sample volumes. Alternatively, use may be made of the gauze-pad technique (23), especially when the numbers of pathogens are small or their presence is not continuous.

1.6.7.2 Salmonella

Sample concentrates may require pre-enrichment in buffered peptone water, followed by enrichment in broth containing either tetrathionate, selenite, magnesium chloride, or malachite green. These may be subcultured on to media such as brilliant green, bismuth sulfite, xylose-lysine desoxycholate (XLD) agar, desoxycholate citrate, or MacConkey agar, and suspect colonies examined biochemically and serologically. Biochemical screening tests should include triple sugar iron agar, indole production, decarboxylase, and β-galactosidase activity. Serological testing should include agglutination with polyvalent anti-O, anti-H, and anti-Vi sera. Prior elimination of auto-agglutinable strains is essential. When *S. typhi* is sought, selenite F medium is preferred. The multiple-tube procedure is used for estimation of the number of *Salmonella* present.

1.6.7.3 Shigella

Since coliform bacteria and most strains of *Proteus vulgaris* are antagonistic to *Shigella*, it is advisable to choose selective enrichment media that minimize accumulations of volatile compounds and by-products derived from these antagonists (23). Nutrient broth adjusted to pH 8.0 (less favourable pH for coliform growth) may be used. Successful enrichment of Shigella may also be achieved with an autocytotoxic medium based on trypticase soy broth containing 1 mmol/litre 4-chloro-2-cyclopentylphenyl β-D-galactopyranoside, 2.5 g/litre lactose, and citrate buffer at pH 6.2 (43). Incubate 6–18 hours at 35 °C. Streak cultures at 6 and 18 hours to XLD agar. Submit suspect colonies to biochemical screening tests and confirm suspect colonies with Shigella antisera (polyvalent and type-specific sera).

1.6.7.4 Cholera and non-cholera vibrios

Alkaline peptone water or taurocholate tellurite peptone water are used for primary enrichment, with subculture to thiosulfate citrate bile salt sucrose agar or taurocholate tellurite gelatin agar as selective media (44). Suspect cultures are inoculated into Kligler iron agar. After 18 hours' incubation, V. cholerae produces a distinctive yellow colour, without any gas production. These cultures are then further screened for urease and oxidase activity; those strains that are urea-negative and oxidase-positive should be submitted to a reference laboratory for further biochemical tests and serological grouping.

1.6.7.5 Enteropathogenic E. coli

The techniques for the detection of faecal coliform organisms in water are used. The colonies are confirmed as E. coli and if the epidemiological evidence warrants it, subcultures may be submitted to a reference laboratory for serological grouping and, if necessary, tests for enterotoxigenicity.

1.6.7.6 Yersinia enterocolitica

M-Endo agar has been found to be the most suitable all-purpose medium owing to the distinct morphological characteristics of colonies at both 25 and 35 °C incubations. Colonies are dark red and well defined after 72 hours' incubation. Growth on MacConkey's agar is equally good provided 25 °C incubation is used (45). All suspect isolates should be screened biochemically at 25 °C and 35 °C on rhamnose, raffinose, and melibiose. If epidemiological evidence warrants it, subcultures should be submitted to a reference laboratory for serological grouping and antibiotic susceptibility tests.

1.6.7.7 Campylobacter fetus

A membrane-filter technique used successfully for isolation of this pathogen involves blood agar containing vancomycin, polymyxin, and trimethoprim (46). Cultures are incubated at 42–43 °C under reduced oxygen tension in an anaerobic jar for 3 days and inspected daily for nonhaemolytic grey mucoid colonies (1–2-mm diameter). Suspect colonies are Gram-stained for typical curved, S-shaped forms and tested for a positive oxidase and catalase reaction, motility, and inability to grow aerobically at 36 °C. Subcultures should be submitted to a reference laboratory for further biochemical tests. Serotyping is not practical on isolates from sporadic outbreaks because of the heterogeneity of the organism and the use of a common antigen or pool of differential serotypes is not yet practical.

REFERENCES

1. *International standards for drinking-water*, 3rd ed. Geneva, World Health Organization, 1971.
2. KABLER, P. W. & CLARK, H. F. Coliform group and fecal coliform organisms as indicators of pollution in drinking water. *Journal of the American Water Works Association*, **52**: 1577 (1960).
3. AKIN, E. W. ET AL. *A virus-in-water study of finished water from six communities.* Cincinnati, US Environmental Protection Agency, 1975. (Environmental protection technology series, EPA-600/1-75-003).
4. GELDREICH, E. E. ET AL. The occurrence of coliforms, fecal coliforms and streptococci on vegetation and insects. *Applied microbiology*, **12**: 63 (1964).
5. PAPAVASSILIOU, J. ET AL. Coli-aerogenes bacteria on plants. *Journal of applied bacteriology*, **30**: 219 (1967).
6. GELDREICH, E. E. ET AL. The faecal coli-aerogenes flora of soils from various geographical areas. *Journal of applied bacteriology*, **25**: 87 (1962).
7. DEANER, D. G. & KERRI, K. D. Regrowth of fecal coliforms. *Journal of the American Water Works Association*, **61**: 465 (1969).
8. GELDREICH, E. E. & KENNER, B. A. Concepts of fecal streptococci in stream pollution. *Journal of the Water Pollution Control Federation*, **41**: R336 (1969).
9. KENNER, B. A. Fecal streptococcal indicators. In: Berg, G., ed., *Indicators of viruses in water and food*, Ann Arbor, Ann Arbor Science, 1978.
10. GELDREICH, E. E. Fecal coliform and fecal streptococcus density relationships in waste discharges and receiving waters. *CRC critical reviews in environmental control*, **6**: 349-369 (1976).
11. KOOL, H. J. Treatment processes applied in public water supply for the removal of micro-organisms. In: *Proceedings of a Symposium on Biological Indicators of Water Quality, Newcastle, 1-15 October 1978*, vol. 2. University of Newcastle, 1978, pp. 17-1 -17-31.
12. FEACHEM, R. ET AL. *Sanitation and disease: Health aspects of excreta and wastewater management.* Baltimore, Johns Hopkins University Press, 1981 (World Bank Studies in Water Supply and Sanitation, No. 3).
13. GELDREICH, E. E. Current status of microbiological water quality criteria. *American Society for Microbiology news*, **47**: 23-27 (1981).
14. HOADLEY, A. W. The significance of fluorescent pseudomonads in water. In: Hoadley, A. W. & Dutka, B. J., ed., *Bacterial indicators of potential health hazards associated with water.* Philadelphia, American Society for Testing and Materials, 1977; pp. 80-114.
15. REITLER, R. & SELIGMAN, R. *Pseudomonas aeruginosa* in drinking water. *Journal of applied bacteriology*, **20**: 145-150 (1957).
16. NEMEDI, L. & LANYI, B. Incidence and hygienic importance of *Pseudomonas aeruginosa* in water. *Acta microbiologica Academiae Scientiarium Hungaricae*, **18**: 319-326 (1971).
17. BURMAN, N. P. & COLBOURNE, J. S. Effects of non-metallic materials on water quality. *Journal of the Institute of Water Engineers and Scientists*, **33**: 11-18 (1979).

18. MOSSEL, D. A. A. ET AL. Microbiological quality assurance for weaning formulae. In: *The microbiological safety of food*, London, Academic Press, 1973, pp. 77–78.
19. MULLER, G. Bacterial indicators and standards for water quality in the Federal Republic of Germany. In: Hoadley, A. W. & Dutka, B. J., ed., *Bacterial indicators, of potential health hazards associated with water*, Philadelphia, American Society for Testing and Materials, 1977, pp. 159–167.
20. TAYLOR, E. W. The pollution of surface and underground waters. *British Water Works Association journal*, **42**: 582–603 (1960).
21. SNEAD, M. C. ET AL. Biological evaluation of benefits of maintaining a chlorine residual in water supply systems. *Water research*, **14**: 403–408 (1980).
22. HOFF, J. C. & GELDREICH, E. E. Effects of turbidity and other factors on the inactivation of viruses by chlorine. In: *Proceedings of the 1978 Annual American Water Works Association Conference and Exposition, Atlantic City, NJ*, Denver, CO, AWWA, 1978 (Paper No. 35–1C).
23. AMERICAN PUBLIC HEALTH ASSOCIATION. *Standard methods for the examination of water and wastewater*, 15th ed., Washington, DC, APHA, 1980, 1134 pp.
24. DEPARTMENT OF HEALTH AND SOCIAL SECURITY. *The bacteriological examination of water supplies*. London, HM Stationery Office, 1969 (Reports on Public Health and Medical Subjects, No. 71).
25. UNION OF SOVIET SOCIALIST REPUBLICS. [*All Union State Standard. Drinking-water methods of sanitary bacteriological analysis*]. Moscow, GOST 18963–73, 1973.
26. GELDREICH, E. E. *Handbook for evaluating water bacteriological laboratories*. Cincinnati, US Environmental Protection Agency, 1975 (EPA-670/9–75–006).
27. [*Methods for the unification of the sanitary microbiological examination of water.*] Bad-Elster, Council for Mutual Economic Assistance, 1979.
28. COUNCIL FOR MUTUAL ECONOMIC ASSISTANCE. [*Standard methods for water quality examination. Part I. Methods of chemical examination of water*], 3rd ed., Moscow, CMEA, 1977. (A summary translation of the 1st edition has been published under the title *Standard methods for the water quality examination for the member countries of the Council for Mutual Economic Assistance*. Prague, The Ministry of Forestry and Water Management in cooperation with the Hydraulic Research Institute, 1968.)
29. PUBLIC HEALTH LABORATORY SERVICE AND STANDING COMMITTEE OF ANALYSTS. Single-tube confirmatory tests for *Escherichia coli*. *Journal of hygiene (Cambridge)*, **85**: 51–57 (1980).
30. CAIRNCROSS, S. & FEACHEM, R. *Small water supplies*. London, The Ross Institute, 1978.
31. VIAL, J. *Bacteriological analysis of drinking water*. Luxembourg, Commission of the European Communities, 1977.
32. RAGHAVACHARI, T. N. S. & IYER, P. V. S. The occurrence of aerogenes group of coliform organisms in faeces and its significance in water analysis. *Indian journal of medical research*, **28**: 55–60 (1940).
33. BOIZOT, G. E. An examination of the modified Eijkman method applied to pure coliform cultures obtained from waters in Singapore. *Journal of hygiene (Cambridge)*, **41**: 566–569 (1941).
34. EVISON, L. M. & JAMES, A. A comparison of the distribution of intestinal bacteria in British and East African water sources. *Journal of applied bacteriology*, **36**: 109–118 (1973).
35. STANFIELD, G. ET AL. *Isolation of faecal streptococci from sewage*. Stevenage, Water Research Centre, 1978.
36. CLAUSEN, E. M. ET AL. Fecal streptococci: indicators of pollution. In: Hoadley, A. W. & Dutka. B. J., ed., *Bacterial indicators of potential health hazards associated with water*, Philadelphia, American Society for Testing and Materials, 1977, pp. 247–264.
37. *Summary report of a Working Group on Bacteriological Examination of Water*. Copenhagen, WHO Regional Office for Europe, 1975.
38. BONDE, G. J. Bacterial indication on water pollution. In: Droop, M. R. & Jannasch, H. W., ed., *Advances in aquatic microbiology*, vol. 1, London, Academic Press, 1977, 381 pp.

39. ANGELOTTI, R. ET AL. Quantitation of *Clostridium perfringens* in foods. *Applied microbiology*, **10**: 193–199 (1962).

40. CABELLI, V. J. *Clostridium perfringens* as a water quality indicator. In: Hoadley, A. W. & Dutka, B. J., ed., *Bacterial indicators of potential health hazards associated with water*, Philadelphia, American Society for Testing and Materials, 1977, pp. 65–79.

41. TAYLOR, R. H. & GELDREICH, E. E. A new membrane filter procedure for bacterial counts in potable water and swimming pool samples. *Journal of the American Water Works Association*, **71**: 402–405 (1979).

42. LEVIN, M. A. ET AL. Quantitative large-volume sampling technique. *Applied microbiology*, **28**: 515–521 (1974).

43. PARK, C. E. ET AL. Improved procedure of selective enrichment of *Shigella* in the presence of *Escherichia coli* by use of 4-chloro-2-cyclopentylphenyl beta-D-galactopyranoside, *Canadian journal of microbiology*, **23**: 563–566 (1977).

44. *Guidelines for the laboratory diagnosis of cholera*. Geneva, World Health Organization, 1974.

45. HIGHSMITH, A. K. ET AL. Isolation of *Yersinia enterocolitica* from well water and growth in distilled water. *Applied and environmental microbiology*, **34**: 745–750 (1977).

46. SKIRROW, M. B. *Campylobacter* enteritis: A "new" disease. *British medical journal*, **2**: 9–11 (1977).

2. THE VIROLOGICAL QUALITY OF DRINKING-WATER

2.1 General description

Viruses of major concern in relation to waterborne transmission of infectious disease are essentially those that multiply in the intestine and are excreted in large numbers in the faeces of infected individuals (1). Concentrations as high as 10^8 viral units per g of faeces have been reported. Even though replication does not occur outside living hosts, enteric viruses have considerable ability to survive in the aquatic environment and may remain viable for days or months (2). Viruses enter the water environment primarily by way of sewage discharges. With the methods at present available, wide fluctuations in the number of viruses in sewage, up to a maximum of 10^6 units/litre, have been found (3). On any given day, many of the 100 or so known enteric viruses can be isolated from municipal sewage, the specific types being those prevalent in the community at that time. Procedures for the isolation of every virus type that may be present in sewage are not yet available. Sewage treatment may reduce the concentration of viruses 10- to 1000-fold, the actual extent depending mainly on the nature and degree of treatment given. Even tertiary treatment does not yield an effluent consistently free from viruses (4). As sewage mixes with receiving water, viruses are carried downstream, remaining viable for varying periods of time depending upon temperature and a number of other less well-defined factors. Consequently, viruses are likely to be present in sewage-polluted water. At the intake to water-treatment plants, counts of up to 49 viral units/litre have been recorded (5).

2.2 Routes of exposure

It is generally believed that the primary route of exposure to enteric viruses is by direct contact with infected persons or by contact with faecally contaminated objects. However, because of the ability of viruses to survive and because of the low infective dose, exposure and consequent infections may occur by less obvious means, including ingestion of contaminated drinking-water.

Explosive outbreaks of viral hepatitis and gastroenteritis resulting from sewage contamination of water supplies have been well documented epidemiologically (6). In contrast, the transmission of low levels of virus through drinking-water of potable quality, although suspected of contributing to the maintenance of endemic enteric viral disease within communities (7), has not yet been demonstrated.

In some developing areas, water sources may be heavily polluted and the water-treatment processes may be less sophisticated and reliable. Because of these factors, as well as the large number of persons at risk, drinking-water must be regarded as having a very significant potential as a vehicle for the environmental transmission of enteric viruses. As with other microbial infections, enteric viruses may also be transmitted by contaminated food and aerosols in addition to the usual mode of direct contact.

2.3 Health effects

Enteric viruses are capable of producing a wide variety of syndromes, including rashes, fever, gastroenteritis, myocarditis, meningitis, respiratory disease, and hepatitis. In general, asymptomatic infections are common and the more serious manifestations rare. However, when drinking-water is contaminated with sewage, two diseases may occur in epidemic proportions – gastroenteritis and infectious hepatitis. Apart from these infections, there is little, if any, epidemiological evidence to show that adequately treated drinking-water is concerned in the transmission of virus infections.

Gastroenteritis of viral origin may be associated with a variety of agents. Many of these have been identified only recently (8), occurring as small particles with a diameter of 27–35 nm in stools of infected individuals with diarrhoea. Most of them have not been characterized chemically or cultured in the laboratory. Viral gastroenteritis, usually of 24–72 hours' duration with nausea, vomiting and diarrhoea, occurs in suspectible individuals of all ages. It is most serious in the very young or very old where dehydration and electrolyte imbalance can occur rapidly and threaten life if not corrected without delay.

Hepatitis, if mild, may require only rest and restricted activities for a week or two, but when severe it may cause death from liver failure, or may result in chronic disease of the liver. Severe hepatitis is tolerated less well with increasing age and the fatality rate increases sharply beyond middle age. The mortality rate is higher among those with pre-existing malignancy and cirrhosis (9).

2.4 Rationale for recommendation

Theoretically, one viral particle is capable of initiating the infectious process; indeed studies with human volunteers have shown that, under somewhat artificial conditions, one unit of vaccine poliovirus detectable by tissue culture could initiate infection (10). However, the ingestion of viral particles—as in drinking-water—will not necessarily ensure that the essential initial contact with susceptible cells will take place. In practice, for most people, the minimum infective dose is likely to be greater than one viral unit and it can best be expressed by the statistical probability of infecting a certain proportion of an exposed population with a given viral dose. In respect of good public health practice, however, it is difficult to disagree with the conclusion of the Safe Drinking Water Committee of the National Academy of Sciences, USA, that "The presence of infective virus in drinking-water is a potential hazard to the public health, and there is no valid basis on which a no-effect concentration of viral contamination in finished water might be established." (*11*). None the less, realistically and economically, it is questionable how far this conclusion can be applied and much further work is clearly required. In the absence of information relating the degree of viral contamination of water to disease, there is no scientific basis on which to set the level of a practical virological standard (*12*). Therefore, the most reasonable approach for controlling the transmission of viruses through drinking-water is to recommend consistently meeting the treatment criteria that have been found through years of experience to be effective in preventing obvious cases of waterborne viral disease. These treatment practices should accord with the criteria that have been shown experimentally to be effective in removing or inactivating seeded viruses. At the present time, the frequent examination of potable water for the occurrence of faecal indicator bacteria remains the only practical and economical approach for routinely monitoring the microbiological safety of drinking-water.

2.5 Methods of examination for viruses

Methods for concentrating viruses from water samples have evolved rapidly during the last few years but their reliability, limits of detection, and precision are not yet well established. For raw water sources, a number of methods, as described in the manual *Examination of water for pollution control* (*13*), may be used. For finished water, the virus concentration method as described in *Standard methods for the examination of water and wastewater* (*14*) is recommended for large volumes. The assessment of water samples for viruses also has the limitation that a minimum of two weeks is required before a result can be given, by which time the water tested would have been distributed and consumed. It must be emphasized that examination of drinking-water for viruses should not supplant bacteriological monitoring or other quality control measures, such as sanitary surveys and physico-chemical measurement of the finished product; it should be regarded as augmenting them. It is obviously desirable that examination for coliform organisms should be done at the same time as viral examination, if necessary on larger volumes of water; if coliform organisms are detected, immediate remedial measures should be instituted without awaiting the virological results.

In general, there are many more faecal bacteria in wastewater than there are viruses. This has led to the hope that suitable bacteria indicative of faecal pollution would also serve as indicators for the presence of viruses in all waters, thereby making direct examination for viral pollution unnecessary. However, bacteria used as conventional indicators to evaluate the safety of potable water supplies have been shown to be less resistant than viruses to environmental factors, and to water and wastewater treatment processes. As a result, enteric viruses could be present in water with little or no signs of bacterial pollution.

Schemes for the recycling of wastewater for domestic use are being considered in some cities, while in many others, water for potable supplies is obtained from contaminated surface sources with a significant proportion of wastewater. In both situations, the risk of viruses penetrating the water-treatment processes—including pretreatment storage and disinfection—must be evaluated carefully. The possible use of bacteriophages of enteric bacteria as indicators for the potential presence of enteroviruses should be considered further. The speed and economy of bacteriophage tests compared with those for the detection of enteroviruses make such a proposition attractive (*15*).

2.6 Interpretation and evaluation of positive findings

In virological work with samples free from viruses or containing only a few viruses, the risk of accidental contamination during sampling and examination in the laboratory is a very real one. The utmost care should therefore be taken and examination of samples of drinking-water preferably separated from work with other viral material. Laboratory quality-control systems should be developed and participating laboratories encouraged to standardize their procedures.

As a matter of course, it is suggested that two separate samples for virological examination should always be collected from the same site. One of these should be examined as soon as possible and the second one placed in cold storage. In the event of a positive finding in the first sample, the likelihood of accidental contamination should be assessed and the second sample should be assayed for virus only if considered necessary in the light of this assessment. Confirmed viral isolates should be type-identified as far as possible. The possible significance to health of positive findings should be evaluated in association with the public health authority.

REFERENCES

1. WHO Technical Report Series, No. 639, 1979 (*Human viruses in water, wastewater and soil*: report of a WHO Scientific Group).
2. AKIN, E. W. ET AL. Enteric viruses in ground and surface waters: a review of their occurrence and survival. In: *Proceedings of the 13th Water Quality Conference*, University of Illinois, 1971.
3. BURAS, N. Concentration of enteric viruses in wastewater and effluent: a two-year survey. *Water research*, **10**: 295–298 (1976).

4. MIELE, R. P. *Pomona virus study – final report*. Los Angeles, Sanitation Districts of Los Angeles County, 1977.
5. SLADE, J. S. Enteroviruses in partially purified water. *Journal of the Institute of Water Engineers and Scientists*, **31**: 219–225 (1977).
6. CRAUN, G. F. & McCABE, L. J. Review of the causes of waterborne disease outbreaks. *Journal of the American Water Works Association*, **65**: 74–84 (1973).
7. BERG, G. Virus transmission by the water vehicle. I. Viruses. *Health laboratory science*, **3**: 86–89 (1966).
8. MADELEY, C. R. Viruses in the stools. *Journal of clinical pathology*, **32**: 1–10 (1979).
9. NATIONAL RESEARCH COUNCIL. *Proceedings of a Symposium on Viral Hepatitis*, Washington, DC, National Academy Press, 1975.
10. PLOTKIN, S. A. & KATZ, M. Minimal infective doses of viruses for man by the oral route. In: Berg, G., ed. *Transmission of viruses by the water route*, New York, Interscience Publishers, 1967.
11. NATIONAL RESEARCH COUNCIL. *Drinking water and health*, vol. 1, Washington, DC, National Academy Press, 1977.
12. GAMBLE, D. R. Viruses in drinking-water: Reconsideration of evidence for postulated health hazard and proposals for virological standards of purity. *Lancet*, **1**: 425–8 (1979).
13. SUESS, M. J., ed. *Examination of water for pollution control*, vol. 3, Oxford, Pergamon Press, 1982.
14. *Standard methods for the examination of water and wastewater*, 15th ed., Washington, DC, AMERICAN PUBLIC HEALTH ASSOCIATION, 1980.
15. KOTT, Y. ET AL. Bacteriophages as viral pollution indicators. *Water research*, **8**: 165–171 (1974).

PART II. BIOLOGICAL ASPECTS

1. PROTOZOA

1.1 General description

Of the intestinal protozoa pathogenic for man, three may be transmitted by drinking-water: *Entamoeba histolytica, Giardia* spp., and *Balantidium coli*. These organisms are the etiological agents of amoebiasis (amoebic dysentery), giardiasis, and balantidiasis, respectively, and they have all been associated with drinking-water outbreaks (1–4). Various, usually free-living, amoeba (e.g., *Naegleria, Hartmannella*, and *Acanthamoeba* spp.) can be waterborne agents of frequently fatal disease. However, waterborne infection with these organisms is almost always associated with recreational contact rather than with the drinking of water.

E. histolytica is distributed worldwide and exists in trophozoite and cyst stages. Infection occurs by ingestion of cysts ranging in size from 10 to 20 μm (average, 12 μm). Since *E. histolytica* is primarily a parasite of primates, man is the reservoir of infection. Dysenteric individuals pass only trophozoites, which are environmentally susceptible to drying and to changes in temperature and salt concentration, and most or all of the parasites in this active amoeboid stage are destroyed by gastric juice (5). Consequently, chronic cases and carriers who excrete cysts are more important sources of infection. Various surveys throughout the world have indicated a prevalence of 0.8–50% for *E. histolytica* infections (6). Carrier rates during epidemics have been estimated at up to 63% (3). The average number of cysts passed per carrier per day has been estimated at 1.5×10^7 and the density of cysts in sewage, assuming a carrier rate of 50% during an epidemic, has been calculated to be 5000 cysts/litre (3).

Giardia spp. have worldwide distribution and are flagellates existing in trophozoite and cyst forms. In addition to man, *Giardia* has been found in numerous mammal species (7, 8) and in psittacine birds (9). The organism that infects man has been designated *Giardia lamblia, Lamblia intestinalis*, or *Giardia intestinalis*. With the exception of *G. muris*, which occurs in mice, there are no features, morphological or otherwise, that allow differentiation of the flagellate occurring in different animal species. It had been considered that the organism was highly host-specific but this has been questioned (10) and recent studies (8) indicate that other animals may act as reservoirs of infection for man. As with *E. histolytica*, infection occurs by ingestion of cysts. Trophozoites are found in the faecal material only when there is acute watery diarrhoea and it is believed these would not survive long in the environment outside an animal host. The cysts are ovoid and are 8–12 μm long by 7–10 μm wide. In various surveys that have been conducted throughout the world, the prevalence of *Giardia* infection in man has ranged from 2.4 to 67.5 % (*6*). The infection is more common in children than in adults (*7*). An asymptomatic carrier state is common but the ratio of cases to carriers has not been determined. Mean daily production of cysts by an infected adult has been calculated to range from 2.1×10^8 to 7.1×10^8 cysts. The density of cysts in domestic raw sewage has been estimated at 10 000–24 000 cysts/litre at a prevalence rate in the human population of 10–25 % (*11*). Beaver (*Castor canadensis*) have been implicated as the source of contamination in at least one community drinking-water outbreak in the USA (*12*).

Balantidium coli is a ciliated organism with worldwide distribution and both the trophozoite and cyst stages can be infective for man (*7*). The spherical to ovoid cysts are 40–60 μm in diameter, are yellowish to greenish, and have a two-membrane cyst wall. Human infections usually occur as the result of ingestion of food or water contaminated with faecal material from infected swine (*5*). Other hosts include lesser primates and, rarely, the dog and rat. *B. coli* is very common in swine, with surveys indicating a prevalence of 21–100 % (*7*). It is considerably less prevalent in man, with twelve surveys throughout the world indicating a prevalence of 0.77 %. Asymptomatic carrier infections can occur in man.

Pathogenic *Naegleria* is the most frequently recognized etiologic agent of primary amoebic meningoencephalitis. Amoebae of the genus *Naegleria* exist in trophozoite, flagellate, and cyst forms (*13*). The trophozoites are usually slug-like or pear-shaped and range in size from 8 to 14 μm. Most have a single central nucleus but occasional bi- or multi-nucleated forms occur. Reproduction is by simple binary fission. The trophozoite will transform into a flagellate stage upon dilution of culture media with distilled water. The flagellate organism is actively motile, pear-shaped, and has 2–4 anterior flagellae. The cysts are circular and about 8–12 μm in diameter. Pathogenic species can be differentiated from non-pathogenic *Naegleria* species by intranasal instillation of the cultured amoebae into mice.

1.2 Routes of exposure

The transmission of pathogenic intestinal protozoa to man occurs through any mechanism by which material contaminated with faeces containing viable organisms from infected individuals can reach the mouth.

1.2.1 Drinking-water

Since man is the primary reservoir for infection with *E. histolytica*, the contamination of water supplies with domestic sewage can lead to the transmission of this organism through drinking-water. Outbreaks traced to sewage contamination of drinking-water have been reported (3). The potential for waterborne transmission may be somewhat greater in the tropics where the carrier rate often exceeds 50% as compared to more temperate regions where prevalence in the general population is generally less than 10%. The cysts can survive several months in water at 0 °C, 3 days at 30 °C, 30 minutes at 45 °C, and 5 minutes at 50 °C (5) and they are one of the most chlorine-resistant pathogens known (14). Waterborne outbreaks of giardiasis have been reported primarily from the USA. In only about 50% of the waterborne outbreaks in the USA has an etiological agent been identified. In those outbreaks where the nature of the etiological agent was established, Giardia lamblia was the most commonly identified pathogen during the period 1972–77; 23 waterborne outbreaks of giardiasis have been reported in the USA since 1965 (1). Drinking-water has also been implicated as the vehicle of transmission in outbreaks occurring in travellers to the USSR (15, 16). Most of the outbreaks have been associated with untreated drinking-water or water receiving disinfection only (2). The infective dose has been shown to be small, with 10 cysts administered in gelatin capsules causing infection in man (17, 18). Information on survival of the organisms in the environment and on resistance to disinfection is incomplete at the present time. However, there are indications that a small percentage of the cysts can survive freezing and that cysts can remain viable in drinking-water up to 77 days at 80 °C (19). Preliminary studies indicate that Giardia spp. fall somewhere between E. histolytica and enteric viruses in their resistance to inactivation by chlorine (14, 20). Considering the distribution of Giardia spp. in man and in a wide variety of domestic and wild animals, and the suggestion that the organisms may not be as host-specific as previously believed, Giardia spp. probably have the greatest potential for transmission through drinking-water of the three intestinal protozoa under discussion.

The only reported waterborne outbreak of balantidiasis occurred in the Truk District of Micronesia in 1971 (4). It was concluded that the epidemic probably resulted from contamination of water supplies by pig faeces when a devastating typhoon destroyed pig pens and precarious water-catchment facilities.

Transmission of amoebic meningoencephalitis has been recorded rarely in those bathing in domestic water supplies and a case occurred from nasal ablutions with water.

1.2.2 Food

E. histolytica cysts may be transmitted by food, and raw vegetables may be a source of infection (7). Although the cysts rarely survive on the hands for more than 10 minutes, except under the fingernails, foodhandlers who are carriers may be an important source of transmission of the organism. Indirect evidence for this route of exposure is provided by one study where inspection and treatment of foodhandlers in Venezuela for *E. histolytica* infection decreased the amoebic dysentery rate from 36.84 to 0.61 per 1000 per year over a three-year period (*7*). Another study in China correlated transmission with the eating of cold bread served with the hands and concluded that transmission by foodhandlers is probably more important than other routes of exposure in that country. Contamination of food may also occur via houseflies, which have been shown to be capable of transmitting the organism in their droppings (*5*).

Only one food-borne outbreak of giardiasis has been reported (*21*), and it is believed that an infected individual who prepared the food was the source of the contamination. Although *Giardia* cysts have been detected on strawberries (*22*) and on vegetables (*23*), the significance of food-borne transmission in the epidemiology of giardiasis is unknown.

Levine (*7*) indicates that *Balantidium* cysts remain alive for weeks in pig faeces if they do not dry out and, again, it seems logical that if food becomes contaminated with faecal material from infected pigs or man, food-borne transmission could occur.

1.2.3 Air

Airborne transmission of intestinal protozoan pathogens directly to man does not seem a likely route of exposure because of the susceptibility of the organisms to inactivation by drying. Although the spread of disease to the lungs and pleura is a complication in untreated amoebiasis patients with liver abscess, primary respiratory infection has not been reported and, if infection does occur as a result of inhalation of contaminated air with subsequent ingestion of mucus-trapped organisms, this route would seem to be several orders of magnitude less important than person-to-person, food, or water transmission.

1.2.4 Other routes of exposure

Sexual transmission of *Giardia* and *E. histolytica* has been reported (*24, 25*), especially among homosexuals. Freeman (5) states that although swimming-pools have not been definitely incriminated in amoebiasis, they are a potential source of *E. histolytica* infection. The significance of the recreational water route for transmission of *Giardia* infections has not been determined, but it is probably low compared with other routes of exposure. Apart from the one drinking-water outbreak of balantidiasis, the only other outbreaks have been in mental institutions and they were attributed to person-to-person spread due to poor personal hygiene and the habit of coprophagy among the patients (*4*).

1.2.5 Relative significance of routes of exposure

Of the three intestinal protozoan pathogens, *E. histolytica* is the most prevalent worldwide. In the USA, person-to-person spread appears to be the most common mode of transmission (26). In other parts of the world, contamination of food by infected foodhandlers appears to be the most significant means of transmission (7). A potential for waterborne outbreaks does exist and the maintenance of a high endemic level of infection in developing countries through drinking-water transmission appears possible, though it is by no means clear that this is so.

The transmission of Giardia infection between children and adults seems to be rare where good personal hygiene is practised. However, the transmission of infections among preschool-age children in day-care centres (27) and similar institutions is probably common. Burke (28) has concluded that the prevalence of endemic infection in many countries is inversely related to socioeconomic level and prevalence is highest where substandard sanitary practices are found. In the USA, waterborne transmission is apparently a significant route of exposure, but evidence on the relative significance of different routes is lacking.

The incidence of balantidiasis in humans is low and direct contact with pigs appears to be the main route of transmission of the causative organism. The potential exists for transmission of the organism in food and water contaminated with pig faeces.

Almost all recorded cases of amoebic meningoencephalitis result from recreational rather than domestic use of water, but epidemiological evidence from developing countries is very scarce.

1.3 Health effects

Though most infections with E. histolytica are asymptomatic or cause only minor symptoms, fatalities can occur (26). The usual clinical manifestations are gastroenteritis with symptoms ranging from mild diarrhoea to fulminating bloody dysentery. Liver abscess is the most common metastatic complication. Pathogenicity appears to depend on strain virulence and on host factors, including nutritional status of the individual and associated bacterial flora (7).

Giardiasis symptoms range from mild self-limiting enteritis to chronic debilitating diarrhoea and asymptomatic infections occur (28). Acute, subacute, and chronic stages of infection have been described (29–31). Mortality from Giardia infection has not been reported in man or any other animal, with the possible exception of psittacine birds (9).

Balantidiasis can present as an acute bloody dysentery (5) but an asymptomatic carrier state also occurs in man (7). Man is highly resistant to infection and when the disease does occur it is usually mild and self-limiting (4).

The amoeba Naegleria fowleri is a widely distributed amoebo-flagellate found in soil and fresh water (32). In the last decade, it has been shown in both tropical and temperate countries to give rise to a lethal meningoencephalitis and about 100 cases have been described (33), usually in those who have been swimming in natural waters. The portal of entry is the nasopharynx from which the amoebae enter the brain by penetration of the olfactory mucosa and cribriform plate. Knowledge about these organisms is still incomplete, but it is clear that the majority are free-living, that multiplication can take place during storage of warm water, and that although infection in man is very rare in relation to the people exposed to possible infection, the prognosis for clinical cases is very bad and treatment difficult (*34*).

1.4 Monitoring

A monitoring programme is not recommended because no standard quantitative or qualitative methods are available. Experimental methods for the concentration and detection of *E. histolytica* and *Giardia* cysts are available (*11, 35*) but they are recommended for use only in conjunction with concurrent epidemiological studies on endemic or epidemic occurrences. The methods available at present are inefficient; the concentration techniques are not reproducible; the identification of organisms in concentrated samples is difficult, and at least in the case of *Giardia* spp., the viability and origin of detected cysts (*11*) cannot be determined. In addition, no recommendations can be made regarding the frequency of sampling.

In situations where disease outbreaks occur from drinking water contaminated with pathogenic intestinal protozoa, boiling of water may provide effective control for inactivation of *Giardia* (*19*), *E. histolytica*, and *B. coli*. Attempts should be made to identify and remove sources of contamination. A sanitary survey should be conducted to identify and correct treatment and distribution system deficiencies. Epidemiological data should be gathered for correlation with physical and chemical data on the treatment plant and distribution system. Guidelines for investigating waterborne illness are available (*36*). The collection of this information could be important in preventing or controlling further outbreaks.

1.5 Rationale for recommendation

As a group, the intestinal pathogenic protozoa occur in large numbers in the faeces of infected individuals in man and a wide variety of domestic and wild animals. The infective dose in man for *Giardia lamblia* is as low as 10 cysts ingested orally, and for both *Giardia* and *E. histolytica*, it has been presumed that one viable cyst may establish infection in a susceptible host (*17, 37*). Simple methods for readily detecting, identifying, and counting intestinal protozoa in drinking-water are not available (*11*). In view of these considerations, protection of sources, adequate treatment, and ensuring integrity of water in distribution are the most effective ways of preventing infection.

Coliform organisms do not appear to be a good indicator for *Giardia* or *E. histolytica* in treated water because of the increased resistance of these protozoans to inactivation by disinfection. In non-disinfected water, the presence of indicator bacteria could suggest the presence of pathogenic protozoa. Previous studies have shown that diatomaceous earth filtration is effective for removal of cysts of *E. histolytica* and *Giardia* and that *E. histolytica* can also be effectively removed by granular media filtration (*38*). Diatomaceous earth achieved up to 99.998 % removal of *Giardia* cysts, and a recent study (*39*) indicated that removal of over 99.99 % of *Giardia* cysts is possible using granular media filtration with attention given to coagulant dose, filtration rate, turbidity, and backwash procedures.

The basis for recommending coagulation, sedimentation, and filtration if protozoa are to be removed from surface waters and unprotected groundwater takes into consideration the inadequacy of the coliform indicator group for this group of organisms, the lack of standard methods for detection, the resistance of the organisms to disinfection, and the demonstrated effectiveness of properly applied filtration for removal of pathogenic intestinal protozoa.

2. HELMINTHS

A great variety of helminth eggs and larvae have been detected in drinking-water and it is clear that all those infective to man should be absent if the drinking-water is to be safe. However, the vast majority of such helminths are not primarily waterborne and it is neither feasible nor necessary to monitor water for them on a routine basis. Two groups of helminths are more directly related to water supply: those transmitted wholly by the ingestion of infected copepod intermediate hosts (Group I) and those whose cercariae are directly infective to man (Group II). Most of the remaining species are grouped into a third category (Group III). This cuts across the formal taxonomy of the organisms involved, which belong to two animal phyla, the Nemathelminthes or roundworms and the Platyhelminthes or flatworms. Within the latter phylum are the Trematoda (flukes) and Cestoda (tapeworms). Helminths that may potentially be transmitted by drinking-water are listed in the table on p. 47.

2.1 Group I (*Dracunculus, Spirometra*)

2.1.1 General description

Group I comprises helminths developing in aquatic copepods and acquired by man drinking water containing the intermediate host crustacea. The most important member of the group is *Dracunculus medinensis*, the guinea-worm, a filarial parasite of man. Female worms mature in the deep tissues and then migrate to lie subcutaneously in a limb. Numerous larvae develop within the body of the female, inducing a blister in the skin of the host, which breaks down. The female exposes her prolapsed uterus there and the larvae are discharged whenever water is sensed by the worm. The life cycle continues if these rhabditiform larvae reach water containing copepods of the genera *Cyclops, Eucyclops, Mesocyclops,* and *Macrocyclops*, which ingest the larvae. Development to the third-stage larvae takes place in the body of the copepod and these are infective to man if ingested (*40*).

It follows that *Dracunculus* is particularly transmitted at unprotected wells and pools rather than through piped water supplies. Infection occurs in a patchy distribution in Africa, especially the West African Sahel, west Asia, and as far east as Afghanistan, India, and south USSR, as well as Indonesia. Man is the predominant definitive host. Tapeworms of the genus *Spirometra*, though much rarer in man, also

46

Helminths potentially transmitted by drinking-water

Main category	Name	Group[a]
Tremoda (flukes)	*Fasciola gigantica*	III
	Fasciola hepatica	III
	Fasciolopsis buski	III
	Schistosoma haematobium	II
	Schistosoma intercalatum	II
	Schistosoma japonicum	II
	Schistosoma mansoni	II
	Schistosoma mekongi	II
Crestoda (tapeworms)	*Echinococcus granulosus*	III
	Echinococcus multilocularis	III
	Hymenolepsis nana	III
	Multiceps sp.	III
	Spirometra mansoni	I
	Spirometra mansonoides	I
	Spirometra proliferum	I
	Spirometra theileri	I
	Taenia solium	III
Nematoda (roundworms)	*Ancylostoma brasiliense*	II
	Ancylostoma duodenale	II
	Ascaris lumbricoides	III
	Dracunculus medinensis	I
	Enterobius vermicularis	III
	Necator americanus	II
	Strongyloides stercoralis	III
	Toxocara spp.	III
	Trichuris trichiura	III
	Uncinaria stenocephala	III

[a] Note that, except for *Dracunculus*, very few indeed of these helminths are chiefly transmitted through drinking-water.

have a stage in aquatic copepods. Adult worms are found in the small intestine of cats. Eggs of *Spirometra* spp. pass out in the faeces and hatch in water to produce coracidium larvae which are ingested by copepods, in which they develop to the procercoid stage (*41, 42*). If man ingests the copepod, the larvae undergo further development in the tissues to the plerocercoid or sparganum stage. This stage may also infect man from the flesh of an alternative intermediate host used as a poultice on lesions, as is the practice in parts of Asia. Species of *Spirometra* are found patchily distributed in the Americas between the USA and Uruguay, and also in east Asia (*44, 45*), Kenya, and the United Republic of Tanzania (*43*).

2.1.2 Routes of exposure

Drinking-water containing infected copepods is the only source of infection with *Dracunculus* and this is the only animal parasite that can be eradicated solely by provision of safe drinking-water. *Spirometra* is an occasional parasite of man and may reach him by other routes than drinking-water.

2.1.3 Health effects

Guinea-worm infection, or dracontiasis (dracunculosis), is a major disabling disease especially in agricultural communities in the Sahel and the Indian subcontinent. The pain of the infection and arthritis of the nearest joint to the worm immobilizes the infected person for several weeks. Since maturation of the worm closely coincides with the rainy planting season, the disability has a disproportionate effect on productivity. Profound allergic reactions may accompany migration of the female, including erythema, urticaria, and intense pruritus, while systemic symptoms may include vomiting and diarrhoea. Sepsis of the blister may occur and an abscess may develop following rupture of the worm during attempted extraction. Fibrosis is sometimes a sequel and ectopic worms may also lead to abscess formation in other viscera. Dracontiasis as a public health threat is a disease of major importance locally, and the local prevalence may exceed 30 %.

Sparganosis is a much less common disease in which the lesions result from subcutaneous larvae which can cause oedema and inflammation, especially upon death of the parasite. Serious consequences follow if the eye is involved, a complication that occurs particularly in South-East Asia.

2.1.4 Monitoring

Dracontiasis is a problem of small unpiped water supplies (e.g., step wells or reservoirs) where regular monitoring is often impracticable. Investigation of such a situation involves the study of the prevalence of infection in man and the collection of copepods using a plankton net or other container, followed by microscopic examination to detect the parasitic filarial larvae. Since prevention is achieved by protecting the water source (see below), the preferred method of monitoring is to determine that sources have been protected. Monitoring for *Spirometra* is not feasible nor indicated.

2.1.5 Rationale for recommendation

A single infected copepod containing a single larva is capable of infecting man with sparganum or with *Dracunculus*, though the worm burden will depend on the number of infective larvae ingested and their sex. Since one fertilized adult female guinea-worm can cause severe disease, infective stages should be absent from drinking-water. As this is the sole route of transmission to man for *Dracunculus*, it is a matter of importance. In view of the way that rhabditiform larvae reach the copepods by being washed into wells from the limbs of those drawing water, it is clear that source protection is the best approach to prevention. The use of well surrounds that rise above ground level and drain away from the well usually suffices, though capping the well and fixing a pump is preferable. In emergency situations the infected copepods may be killed by addition of temephos (Abate) granules to wells in the doses required for insect larva control (46).

2.2 Group II (*Schistosoma, Ancylostoma, Necator*)

2.2.1 General description

Group II comprises a miscellaneous group of flukes and roundworms whose infective larvae are able to penetrate the human skin and mucous membranes. They may therefore be transmitted through drinking-water, but are more of a hazard when water is used for washing or bathing. The main genus concerned is *Schistosoma*.

The schistosomes infecting man belong to three main species: *S. haematobium*, which infects the veins of the vesical plexus and occurs chiefly in Africa and west Asia; *S. mansoni*, which is found in Africa, parts of South and Central America, and certain Caribbean islands; and *S. japonicum*, which is found in China, Indonesia, the Philippines, and other parts of east Asia. Both *S. mansoni* and *S. japonicum* are found in the portal venous system. *S. intercalatum* occurs in west central Africa and *S. mekongi*, a schistosome resembling *S. japonicum*, is found in the Mekong River Basin in South-East Asia. Indeed, it is now being recognized that within each of the major schistosome species there are strain differences related to geographic distribution and host variations.

The adult worms are long-lived and the sexes are separate. The numerous eggs pass through the blood vessel walls and tissues to escape in the urine (*S. haematobium*) or faeces (other species). The eggs hatch on reaching freshwater liberating miracidia which penetrate an appropriate aquatic snail host and undergo development and multiplication for a month or more, following which the cercariae, or larvae infective to man, are shed into the water, and swim about using their bifurcated tails. They are just visible to the naked eye. They rapidly penetrate human skin in contact with the infected water and migrate and mature in the body to complete the cycle.

Cercariae of non-human schistosomes and other related flukes may attempt to penetrate the human skin and die there, producing a rash and intense irritation known as schistosome dermatitis. *S. japonicum* infects a variety of domestic and wild animals as well as man.

The human hookworms *Ancylostoma duodenale* and *Necator americanus*, both with a wide tropical and subtropical distribution, have eggs that hatch and develop in the soil to the third-stage larvae, which reinfect man by penetrating the skin. Hookworms of domestic animals may also invade man and their larvae wander in the skin producing local symptoms.

2.2.2 Routes of exposure

Schistosome infections are acquired when infected water is used for domestic activities, bathing, or washing. Ingested cercariae can penetrate the buccal mucous membranes, but this is a minor route of entry (*47*). The relevance of drinking-water is that it is used for washing if readily available and the benefits of safe water are gained only if there is reduced contact with the infected sources previously used.

While the possibility of piped untreated surface water transmitting schistosomiasis is real, most transmission is from unpiped sources such as pools, wells, and also cisterns used for religious ablutions.

Schistosome dermatitis is a hazard of recreational and occupational water use rather than drinking-water. *Ancylostoma* larvae have been shown to be infective in drinking-water and this may be an appreciable but not major route of transmission (*48, 49*).

2.2.3 Health effects

The human schistosomes are a cause of severe morbidity and sometimes death in a number of the 200 million people infected worldwide. Pathology is due mainly to the host's reaction to eggs that failed to escape. Primary lesions are mainly in the liver, intestine, and around the bladder, but the most severe consequences are due to secondary damage of the upper urinary tract, to bladder cancer, and to liver fibrosis and its haemodynamic consequences.

Hookworms chiefly lead to iron-deficiency anaemia, while the other helminths in this group produce skin lesions.

2.2.4 Monitoring

Detection of schistosome cercariae in water is a research procedure unsuited to routine monitoring. The cercariae may be concentrated from water by filtration or detected by immersing small suitable rodents, allowing time for development, and dissecting them. Cercariae on paper, glass fibre, or acetate filters tend to lose their shape and species identification is not reliable, though various differential staining methods have been devised. It is more feasible to look for appropriate vertebrate hosts and determine infection rates in them.

2.2.5 Rationale for recommendation

Since a single cercaria is infective, cercariae should be absent from drinking-water and there is no safe level. In the absence of routine monitoring assays, reliance has to be placed on preventive measures if a significant risk from drinking-water is suspected in an area. The cercariae have a free-living life of under 48 hours and storage for this period renders water safe (*50*). It is likely that storage for 24 hours will greatly reduce infectivity. Slow sand filters, provided they are properly operated, will remove the majority of cercariae, and disinfection at a residual level of 0.5 mg of free chlorine per litre for 1 hour will kill cercariae of the human schistosomes (*51*). A sounder approach is to use a source that does not contain the host snails and that is not subject to excretal contamination.

2.3 Group III (*Ascaris, Trichuris, Strongyloides, Enterobius,* Fasciolids, *Hymenolepis, Echinococcus*)

2.3.1 General description

A large number of helminths have resistant eggs or cysts infective to man. If these gain access to drinking-water and are ingested, man becomes infected. In the case of all the species listed in this group in the table on p. 47, other methods of transmission, such as food and direct faecal-oral routes are far more important than the drinking-water route and for practical purposes safe disposal of excreta is of major importance, while attention to drinking-water alone will not significantly reduce transmission in most parts of the world.

The most widespread intestinal helminths, *Ascaris lumbricoides* and *Trichuris trichiura*, produce resistant eggs of characteristic appearance within which the embryo must undergo development in the external environment before it becomes infective. When eggs are ingested by man, the larvae emerge in the intestine and undergo complex migrations before returning there to mature and lay eggs, at the rate of some 200 000 daily. The eggs are excreted in the faeces and, being heavy, settle relatively rapidly in water.

Less ubiquitous are *Strongyloides stercoralis*, whose larvae are the infective form, *Enterobius*, the pinworm, whose sticky, less resilient eggs are more suited to direct faecal-oral transmission, and some animal nematodes that undergo limited development in man.

The liver flukes of the genera *Fasciola* and *Fasciolopsis* that can infect man and other mammals develop in snails and the emerging cercariae then encyst on water plants; man is infected by eating the plants or peeling them with his teeth. The cysts may gain access to drinking-water (*45*). The human tapeworms of the genera *Hymenolepis*, with a direct life cycle, and *Echinococcus*, where man is infected by ingesting eggs usually acquired from dogs, have the potential for spread in drinking-water.

2.3.2 Routes of exposure

These helminths all have a faecal-oral transmission and the infective stages tend to be of fairly high relative density. In no case is drinking-water the predominant vehicle of transmission, although the eggs do get into water from time to time, particularly those of the common *Ascaris* and *Trichuris*.

2.3.3 Health effects

The intestinal helminths produce a great range of symptoms. Many infections are subclinical, a few are fatal. It is likely that the majority produce low-grade chronic effects that are hard to quantify in the individual but are collectively important. Diversion of nutrients from the host to feed the worms is considerable. The intestinal helminths make up by their very high prevalence (52) for the limited pathology they produce in many individuals infected.

2.3.4 Monitoring

While in outbreak or research situations helminth eggs and larvae may be extracted from drinking-water by filtration and most eggs can be identified microscopically, these procedures are unsuitable for routine use and the frequency of water contamination does not justify them.

2.3.5 Rationale for recommendation

A single fertilized egg, mature larva, or encysted cercaria can cause infection. Therefore these should be absent from drinking-water and this is best achieved by protecting the source from faecal contamination. Should they gain access to raw water, most will be removed by filtration procedures, especially using slow sand filters, while all are relatively resistant to chlorination, especially *Ascaris* (53).

3. FREE-LIVING ORGANISMS

3.1 General description

The free-living organisms generally considered significant in water supplies include plankton and macroinvertebrates and these are the groups discussed in detail in this section. Plankton consists of microscopic or small organisms that live primarily in suspension in the water column. Phytoplankton consists of the free-living bacteria, fungi, and algae. The algae are chlorophyll-bearing organisms, having different colours. They are autotrophic, unicellular or colonial, and motile or non-motile. The fungi and bacteria are largely heterotrophic. Zooplankton consists of free-living protozoa, rotifers, cladocera, copepods, worms and, in their early stages of development or during brief excursions, the larvae of some aquatic insects and fish. The macroinvertebrates include the larger benthic animals, such as aquatic insect larvae, crustacea, and gastropods.

Plankton organisms are important in water supplies because they interfere with water-treatment processes, produce substances toxic to humans, harbour human pathogens, and contribute organic matter, which may be converted to halogenated organic compounds during chlorination. The macroinvertebrates may affect the efficiency of water distribution systems and the acceptance of water by consumers.

The algae may present problems because of their occurrence in raw surface water, and uncovered storage reservoirs containing treated (finished) water. In addition to these sites, zooplankton may occur in open wells. The occurrence of macroinvertebrates related to water supply problems is limited to distribution systems.

The abundance and species composition of algae in surface water supplies are governed by a combination of natural environmental conditions (54). Maximum algal counts in surface water range from only a few to many millions of organisms per ml. The algae in surface waters are most frequently dominated by the diatoms, followed by the coccoid green and blue-green algae. Winter populations are generally dominated by the pennate diatoms, whereas in summer, the centric diatoms, coccoid greens, and blue-greens are most abundant. In late summer, or where temperatures remain above 25 °C throughout the year, the blue-green algae often comprise the bulk of the algal biomass (55–58). Extensive studies of algae in water supplies have been published by Palmer (59).

A wide variety of free-living organisms has been found in water supply distribution systems, including fungi, algae, protozoa, rotifers, worms, water fleas, shrimps and other crustacea, spring-tails, midge larvae, mussels, and snails. Initial reports on these occurrences were based on investigations following complaints of the appearance of animals at consumers' taps. More recently, systematic studies have revealed the presence of organisms in numerous pipe networks even where there was no obvious reason to suspect their presence. It is now believed that most systems will contain some animals and a list of organisms found to infect certain water mains has been compiled by Collingwood (60).

3.2 Health effects

There is mounting evidence that some toxic substances produced by some algae in water supplies may have a significant adverse effect on public health. Poisonous species occur in two major groups of algae: yellow-greens (Xanthophyta) and blue-greens (Cyanophyta). The toxic blue-green algae are the most important in fresh water supplies (61–64).

Toxic substances released by algae may penetrate the treatment works. In laboratory experiments, aluminium coagulation, filtration, and chlorination were not effective in the removal of algal toxins. Even treatment by activated carbon in quantities similar to those used in water treatment was unsuccessful (65).

The occurrence of public health problems related to algae is quite limited. A relationship between high concentrations of blue-green algae and outbreaks of gastroenteritis in humans has been reported in India (69), the Philippines (68), and the USA (66, 67). Substances similar to the endotoxins of Gram-negative bacteria have been isolated from several blue-green algae (70).

One of the most severe recent outbreaks of waterborne gastroenteritis, which affected approximately 5000 persons in Sewickley, Pennsylvania (71, 72), was linked to a bloom of the ubiquitous filamentous blue-green alga, Schizothrix calcicola, in open finished-water reservoirs.

Algae and their extracellular products can represent a large part of the organic matter in surface water supplies and may be important sources of precursors in the formation of trihalomethanes during chlorination (73–76).

The free-living animals that have been reported to occur in water mains in temperate countries are not known to cause disease directly, but some have been shown to ingest and protect pathogenic organisms under laboratory conditions (77). Where domestic water is stored on site it is essential to cover it and prevent breeding of mosquito vectors of disease in the storage containers.

3.3 Other effects—taste and odour, colour and turbidity, interference with treatment, and infestation of distribution systems

Algae in water supplies may be aesthetically unpleasant, and may interfere with water treatment by increasing the chlorine demand, causing taste and odour problems, and clogging filters (59, 78).

The growth of either aquatic plants or animals may constitute a natural source of odour and taste. In surface waters, algae constitute the main problem, whereas animal forms may proliferate in underground waters, reservoirs, or pipelines. Many algae secrete oils that are liberated either during metabolic activity or when dead cells disintegrate. These oils impart typical odours and tastes to the water (79, 80).

Coloration and turbidity may be a problem (81) in unpiped and treated waters when the reservoirs are open or when the treatment is insufficient. In addition, the growth of algae and other organisms can interfere with the maintenance and operation of water-treatment systems by clogging filters (82, 83).

3.4 Monitoring

Adequate methods are available for the collection and analysis of free-living organisms in water supplies (35, 84–89).

3.4.1 Sampling

Algae and other microorganisms may undergo rapid temporal changes, varying significantly in abundance and species composition from day to day (90). The microbial biomass and species composition should therefore be determined frequently if the data are to be used effectively to modify water-treatment processes, control tastes and odours, and detect harmful concentrations of microorganisms in finished water supplies.

3.4.2 Biomass

The size, shape, and volume of different algae differ greatly, and counts alone do not provide an accurate estimate of the amount of organic matter (biomass) contributed by each taxon. The biomass is commonly estimated by measuring cell surface area, cell volume, or chlorophyll content of the algae (35, 84, 85).

3.4.3 Endotoxins

The *Limulus* amoebocyte lysate test for endotoxins is relatively simple and is being used with increasing frequency (70, 72, 91). However, this test does not differentiate between endotoxins of bacterial and algal origin; a quantitative relationship between endotoxin concentrations and health effects has not been established; and the significance and interpretation of test results remain to be determined.

3.5. Rationale for recommendation

Organisms present in water supplies may cause adverse effects on health, aesthetic problems, objectionable odours and taste, and can also interfere with water treatment. Although the organisms found to infest distribution systems in temperate climates are not associated with known adverse health effects, it is desirable for aesthetic reasons that their appearance at consumer taps be minimized. The present state of the art does not permit the establishment of maximum allowable limits, but it is recommended that, wherever possible, free-living organisms be removed from drinking-water. This may be achieved by protection of the source, implementing good practices of treatment, periodic and systematic swabbing and flushing of pipelines, and monitoring of water quality.

REFERENCES

1. CRAUN, G. F. Waterborne giardiasis in the United States: a review. *American journal of public health*, **69**: 817–819 (1979).
2. CRAUN, G. F. Waterborne outbreaks of giardiasis. In: Jakubowski, W. & Hoff, J. C., ed. *Waterborne transmission of giardiasis*, Cincinnati, US Environmental Protection Agency, 1979 (EPA-600/9–79–001).
3. CHANG, S. L. & KABLER, P. W. Detection of cysts on *Entamoeba histolytica* in tap water by the use of membrane filter. *American journal of hygiene*, **64**: 170–180 (1956).
4. CENTER FOR DISEASE CONTROL. Balantidiasis – Truk District Micronesia. *Morbidity and mortality weekly report*, **21**: 59 (1972).
5. FREEMAN, B. A. *Burrows textbook of microbiology*, 21st ed. Philadelphia, Saunders, 1979.
6. BELDING, D. L. *Textbook of parasitology*, 3rd ed. New York, Meredith Publishing Co., 1965.
7. LEVINE, N. D. *Protozoan parasites of domestic animals and of man*, 2nd ed. Minneapolis, Burgess Publishing Co., 1973.
8. DAVIES, R. B. & HIBLER, C. P. Animal reservoirs and cross-species transmission. In: Jakubowski, W. & Hoff, J. C., ed. *Waterborne transmission of giardiasis*, Cincinnati, US Environmental Protection Agency, 1979 (EPA-600/9-79-001).
9. PANIGRAHY, B. ET AL. Zoonotic diseases in psittacine birds: Apparent increased occurrence of chlamydiosis (psittacosis), salmonellosis and giardiasis. *Journal of the American Veterinary Medical Association*, **175**: 359–361 (1979).
10. FILICE, F. P. Studies on the cytology and life history of a *Giardia* from the laboratory rat. *University of California publications in zoology*, **57**: 53–145 (1952).
11. JAKUBOWSKI, W. & ERICKSEN, T. H. Methods for detecting *Giardia* cysts in water supplies. In: Jakubowski, W. & Hoff, J. C., ed. *Waterborne transmission of giardiasis*, Cincinnati, US Environmental Protection Agency, 1979 (EPA-600/9-79-001).
12. KIRNER, J. C. ET AL. A waterborne outbreak of giardiasis in Camas, Washington. *Journal of the American Water Works Association*, **70**: 35–40 (1978).
13. LOCKEY, M. W. Primary amebic meningoencephalitis. *Laryngoscope*, **88**: 484–503 (1978).
14. HOFF, J. C. Disinfection resistance of *Giardia* cysts: Origins of current concepts and research in progress. In: Jakubowski, W. & Hoff, J. C., ed. *Waterborne transmission of giardiasis*, Cincinnati, US Environmental Protection Agency, 1979 (EPA-600/9-79-001).

15. JOKIPII, L. & JOKIPII, A. M. M. Giardiasis in travelers: A prospective study. *Journal of infectious diseases*, **130**: 295–299 (1974).
16. CENTER FOR DISEASE CONTROL. Giardiasis – in residents of Rome, N.Y., and in travelers to the Soviet Union. *Morbidity and mortality weekly report*, **24**: 371 (1975).
17. RENDTORFF, R. C. The experimental transmission of human intestinal protozoan parasites. II. *Giardia lamblia* cysts given in capsules. *American journal of hygiene*, **59**: 209–220 (1954).
18. RENDTORFF, R. C. The experimental transmission of *Giardia lamblia* among volunteer subjects. In: Jakubowski, W. & Hoff, J. C., ed. *Waterborne transmission of giardiasis*, Cincinnati, US Environmental Protection Agency, 1979 (EPA-600/9-79-001).
19. BINGHAM, A. K. ET AL. *Giardia* spp.: Physical factors of excystation *in vitro*, and excystation vs. eosin exclusion as determinants of viability. *Experimental parasitology*, **47**: 284–291 (1979).
20. HOFF, J. C. ET AL. Inactivation of *Giardia muris* cysts by chlorine. In: *Abstracts from the 79th Annual Meeting of the American Society for Microbiology, Los Angeles*, ASM, 1979.
21. OSTERHOLM, M. T. ET AL. An outbreak of foodborne giardiasis. *New England journal of medicine*, **304**: 24–28 (1981).
22. JACKSON, G. Comment made in open discussion of paper by G. R. Healy on the presence and absence of *Giardia lamblia* in studies on parasite prevalence in the USA. In: Jakubowski, W. & Hoff, J. C., ed. *Waterborne transmission of giardiasis*, Cincinnati, US Environmental Protection Agency, 1979 (EPA-600/9-79-001).
23. TAY, J. ET AL. Search for cysts and eggs of human intestinal parasites in vegetables and fruits. In: *Conference on Cooperative Research Needs for the Renovation and Reuse of Municipal Wastewater in Agriculture, 15–19 December 1980, Cocoyoc, Morales, Mexico* (Sponsored by the Institute of Water Research, East Lansing).
24. MILDVAN, D. ET AL. Venereal transmission of enteric pathogens in male homosexuals. *Journal of the American Medical Association*, **238**: 1387–1389 (1977).
25. HURWITZ, A. L. & OWEN, R. L. Venereal transmission of intestinal parasites (medical information). *Western journal of medicine*, **128**: 89–91 (1978).
26. KROGSTAD, D. J. ET AL. Current concepts in parasitology—amebiasis. *New England journal of medicine*, **298**: 262–265 (1978).
27. BLACK, R. E. ET AL. Giardiasis in day-care centers: Evidence of person-to-person transmission. *Pediatrics*, **60**: 486–491 (1977).
28. BURKE, J. A. The clinical and laboratory diagnosis of giardiasis. *CRC critical reviews in clinical laboratory sciences*, **7**: 373–391 (1977).
29. WOLFE, M. S. Giardiasis. *Journal of the American Medical Association*, **233**: 1362–1365 (1975).
30. WOLFE, M. S. Current concepts in parasitology—giardiasis. *New England journal of medicine*, **298**: 319–321 (1978).
31. WOLFE, M. S. Managing the patient with giardiasis: Clinical, diagnostic and therapeutic aspects. In: Jakubowski, W. & Hoff, J. C., ed. *Waterborne transmission of giardiasis*, Cincinnati, US Environmental Protection Agency, 1979 (EPA-600/9-79-001).
32. LAWANDE, R. V. The seasonal incidence of primary amoebic meningoencephalitis in northern Nigeria. *Transactions of the Royal Society of Tropical Medicine and Hygiene*, **74**: 141 (1980).
33. THONG, Y. G. Primary amoebic meningoencephalitis: fifteen years later. *Medical journal of Australia*, **1**: 352–354 (1980).
34. KASPRZAK, W. ET AL. Studies on some pathogenic strains of free-living amoebae isolated from lakes in Poland. *Annales de la Société belge de Médecine tropicale*, **54**: 351–357 (1974).
35. AMERICAN PUBLIC HEALTH ASSOCIATION. *Standard methods for the examination of water and wastewater*, 15th ed., Washington, DC, 1980.
36. INTERNATIONAL ASSOCIATION OF MILK, FOOD, AND ENVIRONMENTAL SANITARIANS, INC. *Procedures to investigate waterborne illness*, 1st ed. Ames, IA, IAMFES, 1979.
37. BEAVER, P. C. ET AL. Experimental *Entamoeba histolytica* infections in man. *American journal of tropical medicine and hygiene*, **5**: 1000–1009 (1956).

38. LOGSDON, G. S. ET AL. Water filtration techniques for removal of *Giardia* cysts and cyst models. In: Jakubowski, W. & Hoff, J. C., ed. *Waterborne transmission of giardiasis*, Cincinnati, US Environmental Protection Agency, 1979 (EPA-600/9-79-001).

39. LOGSDON, G. S. ET AL. Alternative filtration methods for removal of *Giardia* cysts and cyst models. *Journal of the American Water Works Association*, **73**: 111–118 (1981).

40. MULLER, R. *Dracunculus* and dracunculiasis. *Advances in parasitology*, **9**: 73–151 (1971).

41. MUELLER, J. F. The laboratory propagation of *Spirometra mansonoides* (Mueller, 1935) as an experimental tool. II. Culture and infection of the copepod host, and harvesting the procercoid. *Transactions of the American Microscopical Society*, **78**: 245–255 (1959).

42. MUELLER, J. F. Host-parasite relationships as illustrated by the cestode *Spirometra mansonoides*. In: McCauly, J. E., ed. *Host-parasite relationships*, Corvallis, Oregon State University Press, 1966 (Proceedings of the 26th Annual Biology Colloquium, Corvallis, Oregon, 23–24 April 1965).

43. OPUNI, E. K. & MULLER, R. L. Studies on *Spirometra theileri* (Baer, 1925) n. comb. 1. Identification and biology in the laboratory. *Journal of helminthology*, **48**: 15–23 (1974).

44. FAUST, E. C. ET AL. *Animal agents and vectors of human disease*, 4th ed. Philadelphia, Lea & Febiger, 1975.

45. FAUST, E. C. ET AL. *Craig and Faust's clinical parasitology*, 8th ed. Philadelphia, Lea & Febiger, 1970.

46. MULLER, R. Laboratory experiments on the control of cyclops transmitting guinea worm. *Bulletin of the World Health Organization*, **42**: 563–576 (1970).

47. MALDONADO, J. F. & PERKINS, K. W. *Schistosomiasis in America*. Barcelona, Editorial Científico-médica, 1967.

48. OKAMOTO, K. An experimental study of the migration route and development of *Ancylostoma duodenale* in pups after oral infection. *Journal of the Kyoto Prefectural Medical University*, **70**: 135–152 (1961).

49. HIGO, A. An experimental study on the migration route and the development of *Ancylostoma duodenale* in pups after cutaneous infection. *Journal of the Kyoto Prefectural Medical University*, **70**: 851–874 (1962).

50. JONES, M. F. & BRADY, F. J. Survival of *Schistosoma japonicum* cercariae at various temperatures in several types of water. *National Institute of Health bulletin*, **189**: 131–136 (1974).

51. COLES, G. C. & MANN, H. Schistosomiasis and water works practice in Uganda. *East African medical journal*, **48**: 40–43 (1971).

52. STOLL, N. R. This wormy world. *Journal of parasitology*, **33**: 1–18 (1947).

53. KELLER, P. Sterilization of sewage sludges. I. A review of the literature pertaining to the occurrence and viability of parasitic ova in sewage with special reference to *Ascaris lumbricoides*. *Journal and proceedings. Institute of Sewage Purification*, Part 1, 92–99 (1951).

54. LUND, J. W. G. The ecology of the freshwater phytoplankton. *Biological reviews of the Cambridge Philosophical Society*, **40**: 231–293 (1965).

55. US PUBLIC HEALTH SERVICE. Annual Compilation of Data. National Water Quality Network, Basic Data Branch, Division of Water Supply and Pollution Control, Department of Health, Education and Welfare, Cincinnati, OH (1957–1963).

56. WEBER, C. I. ET AL. Phytoplankton. In: Flynn, K. C. & Mason, W. T., Jr., ed. *The freshwater Potomac, aquatic communities and environmental stresses*, Interstate Commission on the Potomac River Basin, Rockville, MD, 1978.

57. WILLIAMS, L. G. & SCOTT, C. Principal diatoms of major waterways of the United States. *Limnology and oceanography*, **7**: 365–379 (1962).

58. WILLIAMS, L. B. Possible relationships between plankton-diatom species numbers and water quality estimates. *Ecology*, **45**: 809–823 (1964).

59. PALMER, C. M. *Algae and water pollution*. Cincinnati, US Environmental Protection Agency, 1977.

60. COLLINGWOOD, R. W. *Occurrence, significance and control of organisms in distribution systems.* Barcelona, International Water Supply Association Congress, 1966, 13 pp.
61. GORHAM, P. R. Toxic algae. In: Jackson, D. F., ed. *Algae and man*, New York, Plenum Press, 1964, pp. 307–336.
62. SCHWIMMER, M. & SCHWIMMER, D. Algae and medicine. In: Jackson, D. F., ed. *Algae and man*, New York, Plenum Press, 1964, pp. 368–412.
63. SCHWIMMER, M. & SCHWIMMER, D. Medical aspects of phycology. In: Jackson, D. F., ed. *Algae, man and the environment*, Syracuse, Syracuse University Press, 1968, pp. 279–358.
64. MOORE, R. E. Toxins from blue-green algae. *BioScience*, **27**: 797–802 (1977).
65. INGRAM, W. M. & PRESCOTT, B. W. Toxic fresh water algae. *The American midland naturalist*, **52**: 75–87 (1954).
66. TISDALE, E. S. Epidemic of intestinal disorders in Charleston, West Virginia, occurring simultaneously with unprecedented water supply conditions. *American journal of public health*, **21**: 198–200 (1931).
67. VELDEE, M. V. Epidemiological study of suspected waterborne gastroenteritis. *American journal of public health*, **21**: 1227–1235 (1931).
68. DEAN, A. G. & JONES, T. C. Seasonal gastroenteritis and malabsorption at an American military base in the Philippines. *American journal of epidemiology*, **95**: 111–127 (1972).
69. GUPTA, R. S. & DASHORA, M. S. Algal pollutants and potable water. In: Pajasek, R. B., ed. *Drinking water quality enhancement through source protection*, Ann Arbor, Ann Arbor Science, 1977, pp. 431–459.
70. KELETI, G. ET AL. Composition and biological properties of lipopolysaccharides isolated from *Schizothrix calcicola* (ag.) Gomont (Cyanobacteria). *Applied and environmental microbiology*, **38**: 471–477 (1979).
71. LIPPY, E. C. & ERB, J. Gastrointestinal illness at Sewickley, Pennsylvania. *Journal of the American Water Works Association*, **68**: 606–610 (1976).
72. KAY, G. P. ET AL. Algal concentration as a quality parameter of finished drinking waters in and around Pittsburg, Pennsylvania. *Journal of the American Water Works Association*, **72**: 170–176 (1980).
73. MILLER, S. Drinking water and its treatment (An interview with Professor Sontheimer, University of Karlsruhe). *Environmental science and technology*, **14**: 510–514 (1980).
74. HOEHN, R. C. ET AL. Algae as sources of trihalomethane precursors. *Water and sewage works*, **126**: 66–67 (1979).
75. BRILEY, K. F. ET AL. Trihalomethane production from algal precursors. *Paper presented at the Third Conference on Water Chlorination: Environmental Impact and Health Effects, Colorado Springs, 31 October–2 November 1979*, Ann Arbor, Ann Arbor Science, 1980, Vol. 3.
76. TARDIFF, R. G. Health effects of organics: Risk and hazard assessment of ingested chloroform. *Journal of the American Water Works Association*, **69**: 658–661 (1977).
77. CHANG, S. L. ET AL. Survival and protection against chlorination of human enteric pathogens in free-living nematodes isolated from water supplies. *American journal of tropical medicine and hygiene*, **9**: 136–142 (1960).
78. PALMER, C. M. *Algae in water supplies.* Cincinnati, US Department of Health, Education and Welfare, 1959.
79. COX, C. R. *Operation and control of water treatment processes.* Geneva, World Health Organization, 1964.
80. COLLINGWOOD, R. W. The effect of algal growth on the quality of treated water. In: *Proceedings of a Symposium on Biological Indicators of Water Quality, Newcastle, October 1–15, 1978*, University of Newcastle, 1978.
81. FOGG, G. E. Extracellular products of algae in freshwater. *Archiv für Hydrobiologie, Beiheft: Ergebnisse der Limnologie*, **5**: 1–25 (1971).
82. PALMER, C. M. Algae and other interference organisms in New England water supplies. *Journal of the New England Water Works Association*, **72**: 27–46 (1958).
83. DEGRÉMONT. *Mémento technique de l'eau*, 8th ed. Rueil-Malmaison, Degrémont, 1978.
84. WEBER, C. I. *Biological field and laboratory methods for measuring the quality of*

surface waters and wastes. Cincinnati, US Environmental Protection Agency, 1973.

85. WEBER, C. I. Recent developments in the measurement of the response of plankton and periphyton to changes in their environment. In: Glass, G., ed. *Bioassay techniques and environmental chemistry*, Ann Arbor, Ann Arbor Publishers, 1973, pp. 119–138.

86. WELCH, P. S. *Limnological methods.* New York, McGraw-Hill, 1948.

87. *Determinacôes biológicas.* São Paulo, Companhia de Tecnologia de Saneamento Ambiental, 1978.

88. BRANCO, S. M. *Hidrobiologia aplicada a engenharia sanitaria.* 2nd ed., São Paulo, Companhia de Tecnologia de Saneamento Ambiental, 1978.

89. *International standards for drinking-water*, 3rd ed. Geneva, World Health Organization, 1971.

90. WEBER, C. I. & MOORE, D. R. Phytoplankton, seston and dissolved organic carbon in the Little Miami River at Cincinnati, Ohio. *Limnology and oceanography*, **12**: 311–318 (1967).

91. JORGENSEN, J. H. ET AL. Comparison of *Limulus* assay, standard plate count, and total coliform count for microbiological assessment of renovated wastewater. *Applied and environmental microbiology*, **37**: 928–931 (1979).

PART III. HEALTH-RELATED INORGANIC CONSTITUENTS

1. ARSENIC

1.1 General description

1.1.1 Sources

Arsenic occurs naturally in all environmental media and is usually present in the form of compounds with sulfur and with many metals (copper, cobalt, lead, zinc, etc.) (*1*). The average concentration in the earth's crust is about 2 mg/kg (*2*). Although arsenic exists in various valency states and in both organic and inorganic forms, the levels of environmental arsenic are normally reported in terms of total arsenic (*3*). In some localized geographic areas, commercial use and production of arsenic compounds have resulted in significant elevation in the amounts of environmental arsenic above natural background levels.

1.1.2 Occurrence in water

Many arsenic compounds are water-soluble and, thus, contamination of water can occur. The chemical form of arsenic in water has not been fully elucidated, but both tri- and pentavalent forms have been identified; some forms of organic arsenic have been found in water (*1*). Geothermal discharges in New Zealand have been found to contain significant quantities of arsenic. In spite of its ubiquitousness in nature, most of the arsenic found in water derives from industrial discharges; the highest concentrations other than those occurring naturally in spring-waters are usually in areas of high industrial activity (*4*).

1.2 Routes of exposure

1.2.1 Drinking-water

A large number of water supplies contain very low levels of arsenic, i.e., well below 10 μg/litre (*1–3*). In special situations, gross contamination of well supplies has occurred and several thousand micrograms of arsenic per litre of water have resulted (*5*). There is little information concerning the form or species of arsenic in water supplies.

1.2.2 Food

Arsenic is present in most foodstuffs at concentrations generally below 1 mg/kg of dry weight (*1*). Marine fish can contain higher levels than this, and shellfish may contain levels well over 50 mg/kg (*2,5*); the arsenic in seafood is mainly in an organic form (*1*). The average dietary sources of arsenic seem to give rise to an intake of about 30 µg/kg of body weight per day (*2, 6, 7*). Somewhat higher estimates have been quoted for certain countries (*3, 8*).

1.2.3 Air

The air in non-urban and non-industrial areas contains very low levels of arsenic, i.e., generally less than 0.01 µg/m³ (*1*). In urban areas and particularly industrial towns, levels of arsenic have exceeded 1 µg/m³ (*1*). A typical airborne exposure might be about 0.2 µg/m³ of air, mainly as inorganic arsenic.

1.2.4 Other routes of exposure

1.2.4.1 *Industrial exposure*

Industrial exposure to fumes containing arsenic compounds does arise, especially in smelting operations; levels of arsenic exceeding 1 mg/m³ of air have been recorded (*1*). Levels of arsenic as high as 380 mg/kg have been reported in soil near smelters (*9*).

1.2.4.2 *Tobacco*

Tobacco contains some arsenic, although, as a result of the decreasing use of arsenical sprays, the levels nowadays are low (*1,8*).

1.2.4.3 *Miscellaneous routes*

Certain pharmaceutical products contain arsenic and, for some individuals, the exposure to arsenic via these sources can be high. For the general population, the contribution from such routes of exposure to arsenic is negligible in comparison with those from food, water, and air.

1.2.5 Relative significance of different routes of exposure

Based simply on intake of total arsenic, the contribution from the normally low levels in water in relation to general overall exposure is relatively small; however, at an arsenic level of 0.05 mg/litre in drinking-water, this may equal or even exceed the total dietary intake in a non-occupationally exposed individual.

It is not possible to make estimates of the amounts of arsenic taken up by the body via various routes because very little is known about the particular forms of arsenic in food and water; this governs the proportions absorbed (*2, 6*).

1.3 Metabolism

The form of arsenic affects its absorption. Elemental arsenic is very poorly absorbed. Some trivalent and pentavalent inorganic arsenic compounds are readily absorbed (2); organic arsenic, too, is generally well absorbed (6). Precise figures for the fractions absorbed of the different forms of arsenic do not appear to have been reported. Following exposure to arsenic, it enters the blood (8) and subsequently it is found mainly in the liver, muscles, kidneys, spleen, and skin (1); smaller quantities are also found in the brain, heart, uterus, thyroid, and pancreas, as well as in the hair and nails (10). Arsenic is transferred across the placenta (1). The biological half-life of arsenic appears to be between about ten hours and a few days (1). There are no data indicating that arsenic accumulates with age. Human subjects are able to transform inorganic arsenic into monomethyl and dimethyl compounds (1); the overall mechanism of biotransformation is little understood, however (5). The excretion of arsenic compounds is mainly via urine (1, 2, 5). Trivalent arsenic may inhibit the activity of many enzymes by reacting with sulfydryl groups; reactions involving such groups are considered to be responsible for the toxic action of arsenic compounds (5).

1.4 Health effects

There is no firm evidence that arsenic in any form is essential to man, although it is known that some organic arsenic compounds are beneficial as a growth stimulant for animals (2).

The toxicity of arsenic compounds depends on the chemical and physical form of the compound, the route by which it enters the body, the dose and duration of exposure, dietary levels of interacting elements, and the age and sex of the exposed individual (2). Inorganic arsenic is more toxic than organic arsenic; trivalent inorganic arsenic is more hazardous than the pentavalent form. It is recommended that, when water is found to contain arsenic at levels of 0.05 mg/litre, an attempt should be made to ascertain the valency and chemical forms of the element.

Acute poisoning by arsenic involves the central nervous system, leading to coma and, for doses of 70–180 mg, to death (11). The gastrointestinal tract, nervous system, the respiratory tract, and the skin can be severely affected (2). Chronic poisoning is manifested by general muscular weakness, loss of appetite and nausea, leading to inflammation of the mucous membranes in the eye, nose, and larynx; skin lesions may also occur. Neurological manifestations and even malignant tumours in vital organs may also be observed (2). Poisoning may appear with doses as low as 3–6 mg/day over extended periods (12). One fatality was reported where for two and a half years well-water containing up to 7.6 mg/litre was drunk (13). Even as little as 0.6 mg of arsenic per litre of water might have been responsible for some infant deaths in Chile

(*14–16*), although there is some uncertainty about this incident (*2*). Although there is no evidence of adverse health effects with a concentration of 0.05 mg/litre, absorbed amounts of arsenic can be detected in the hair of people drinking water at this concentration (*2*). In China (Province of Taiwan) (*17, 18*), skin cancer in some villages was found to be associated with drinking well-water containing average arsenic concentrations of about 0.5 mg/litre; there is some doubt about the actual levels of arsenic, however (*2*). Evaluations of the risk of getting skin cancer have been made by a WHO Task Group and the US Environmental Protection Agency (*19*).[a] Applying a linear non-threshold model, a WHO Task Group (*19*) estimated that a lifetime exposure to arsenic in drinking-water at a concentration of 0.2 mg/litre gave a 5% risk of getting cancer of the skin (this model assumes that the metabolism of arsenic is the same at low as at high exposures).

Various manifestations, such as hyperpigmentation, keratoses, and lung cancer, have been observed where high occupational exposure to arsenic has occured (*9*).

REFERENCES

1. COMMISSION OF THE EUROPEAN COMMUNITIES. *Trace metals: exposure and health effects.* Oxford, Pergamon Press, 1979.
2. *Guidelines for Canadian drinking water quality, 1978.* Quebec, Ministry of Supply and Services, 1979 (Supporting documentation).
3. *Quality criteria for water.* Washington, DC, US Environmental Protection Agency, 1976.
4. *Some metals and metallic compounds.* Lyon, International Agency for Research on Cancer, 1980 (IARC Monographs on the evaluation of the carcinogenic risk of chemicals to humans, vol. 23).
5. NATIONAL RESEARCH COUNCIL. *Drinking water and health.* National Academy of Sciences, Washington, DC, 1977.
6. UNDERWOOD, E. J. *Trace elements in human and animal nutrition.* New York, Academic Press, 1977.
7. WHO Technical Report Series, No. 532, 1973 (*Trace elements in human nutrition*), pp. 49–50.
8. *Toxicology of metals, vol. II.* Washington, DC, US Environmental Protection Agency, 1977 (Environmental Health Effects Research Series).
9. NATIONAL RESEARCH COUNCIL. *Arsenic. Medical and biological effects of environmental pollutants.* Washington, DC, National Academy Press, 1977.
10. MEALEY, J. ET AL. Radioarsenic in plasma, urine, normal tissues, and intracranial neoplasms. *Archives of neurology and psychiatry,* **81**: 310 (1959).
11. VALLEE, B. L. ET AL. Arsenic toxicology and biochemistry. *A.M.A. archives of industrial health,* **21**: 132 (1960).
12. LISELLA, F. S. ET AL. Health aspects of arsenicals in the environment. *Journal of environmental health,* **30**: 157 (1972).
13. WYLLIE, J. Investigation of source of arsenic in well water. *Canadian journal of public health,* **28**: 128 (1937).
14. ROSENBERG, H. G. Systemic arterial disease with myocardial infarction. Report on two infants. *Circulation,* **47**: 270 (1973).

---

[a] GREATHOUSE, D. G. *Maximum acceptable limit for arsenic in drinking-water.* A criteria document prepared for the World Health Organization, 1980, unpublished.

15. ROSENBERG, H. G. Systemic arterial disease and chronic arsenicism in infants. *Archives of pathology*, **97**: 360 (1974).
16. ZALDIVAR, R. Arsenic contamination of drinking water and foodstuffs causing endemic chronic poisoning. *Beiträge zur Pathologie*, **151**: 384 (1974).
17. TSENG, W. P. ET AL. Prevalence of skin cancer in an endemic area of chronic arsenicism in Taiwan. *Journal of the National Cancer Institute*, **40**: 453 (1968).
18. YEH, S. Skin cancer in chronic arsenicism. *Human pathology*, **4**: 469 (1973).
19. *Arsenic*. Geneva, World Health Organization, 1981 (Environmental Health Criteria 18).

2. ASBESTOS

2.1 General description

2.1.1 Sources

Asbestos is a general term for fibrous silicate minerals of the serpentine or amphibole mineral groups. Six minerals have been characterized as asbestos: chrysotile, crocidolite, anthophyllite, tremolite, actinolite, and amosite. Chrysotile is a member of the serpentine group; the others belong to the amphibole group. These various forms of asbestos are composed of 40–60% silica, as well as oxides of iron, magnesium, and other metals.

Asbestos is introduced into natural waters by the dissolution of asbestos-containing minerals and ores and from industrial effluents. There is some indication that atmospheric pollution may also contribute to the asbestos content of natural waters (*1*). Sedimentation, resuspension, migration, and chemical reactions affect the movement, abundance, and fate of asbestos fibres in water. The length of time from introduction of asbestos fibres into water until their disappearance is unknown (*2*).

The use of asbestos-cement (which contains 170 g of asbestos per kg—80% chrysotile and 20% crocidolite) for pipes in distribution systems could contribute to the asbestos content of drinking-water.

2.1.2 Occurrence in water

Asbestos is commonly found in domestic water supplies. Typical background levels in rivers and lakes are considered to be about 1 million fibres per litre (*3*), although reported values range from less than 1 million to 10 million fibres per litre. Levels vary considerably, depending upon proximity to industrial sources. The asbestos content of untreated water from the Ottawa River has been reported to be 9.5 million fibres per litre (*1*).

Generally, ordinary sand filtration removes about 90% of the individual asbestos fibres from water supplies (*2*). The most effective method for removal involves chemical coagulation with iron salts and polyelectrolytes followed by filtration. Based on surveys of asbestos concentrations in the drinking-water supplies in Canada, it was found that about 5% of the Canadian public consume water with fibre concentrations exceeding 10 million fibres per litre and about 0.6%

consume water with concentrations exceeding 100 million fibres per litre. Levels ranged up to 2000 million fibres per litre in some asbestos-mining communities (4).

2.2 Routes of exposure

2.2.1 Drinking-water

Since reported levels for the asbestos content of drinking-water supplies vary tremendously and since wide margins of error are associated with fibre-to-mass conversions, it is difficult to make an accurate estimate of the average daily intake from this source. Based on the results of a Canadian national survey and the assumption that the daily water consumption is 2 litres, the intake of asbestos from drinking-water for 95 % of the population is probably less than 0.0001 mg/day.

2.2.2 Food

The extent of asbestos contamination of solid foodstuffs has not been fully researched, owing to the lack of a practical, reliable analytical method. Foods that contain soil particles, dust, or dirt almost certainly contain asbestos fibres. Foodstuffs can also derive asbestos from water or impure talc used in their preparation (talc may be used as a dusting powder in chewing gum, on coated rice, and as an antisticking agent for moulded foods) (5). Asbestos may also be introduced into foods from impure mineral silicates, such as talc, soapstone, or pyrophyllite, which are used as carriers for spray pesticides (6). Asbestos fibres make excellent filter materials, and at one time asbestos was widely used in the food industry for clarification of beverages and other liquids. In some cases, the use of asbestos filters increases the concentration of asbestos in the finished product. Concentrations of 0.151 million fibres per litre have been found in some English beers (7). For Canadian beers, concentrations of 4.3–6.6 million fibres per litre have been recorded; in soft drinks, levels were between 1.7 million and 12.2 million fibres per litre (1). However, use of asbestos filters in the food industry has declined.

2.2.3 Air

Asbestos is present in the air as a result of natural processes, such as weathering of rocks, or from industrial emissions. In the USA, typical levels of asbestos in air range from 1×10^{-5} to 10×10^{-5} mg/m^3 (8, 9). It has been estimated that 1×10^{-6} mg of chrysotile asbestos could contain up to 1 million fibrils (8).

If the concentration of asbestos in air is assumed to be less than 3×10^{-5} mg/m^3 (concentrations recorded in urban areas in the USA are usually below this level (10)) and the daily respiratory volume to be

$20 \, m^3$, then the daily intake of asbestos from air would be less than 0.0006 mg.

2.2.4 Relative significance of different routes of exposure

It is apparent then that exposure to asbestos varies considerably, depending on proximity to industrial and natural sources. In general, however, intake from air is greater than intake from water. It should be noted, however, that fibre size distribution in air may be quite different from that in water. There are few quantitative data available on levels of asbestos present in food.

2.3 Absorption and distribution

2.3.1 Ingestion

Asbestos fibres are ingested with food, beverages, and drinking-water, and a significant proportion of inhaled asbestos cleared from the airways by mucociliary action is subsequently swallowed (*11*). The fate of ingested asbestos fibres is the subject of some controversy; a number of authors have concluded that asbestos fibres cross the walls of the gastrointestinal tract and penetrate into other tissues (*12–15*), but others claim that there is no clear evidence that this occurs (*11, 16, 17*). When a suspension of chrysotile fibres was injected directly into the stomachs of anaesthetized rats, it appeared that asbestos fibres were capable of penetrating the gut wall and migrating to the blood, spleen, omentum, brain, and other tissues (*13*). There was no evidence of lesions or transmigration of fibres across the gut wall in a well-controlled study in which rats consumed a diet containing 50 g of chrysotile per kg of feed for 21 months (*11*). Further work is required to determine the amount of asbestos fibres that cross the walls of the human gastrointestinal tract under normal conditions and if the number of absorbed fibres is sufficient to cause adverse local or systemic effects (*12*).

2.3.2 Inhalation

Evidence from several sources, including animal experiments, pathological observations, and physical studies, indicates that asbestos can be deposited in airways of the respiratory tract by sedimentation or by interception (*18*). Deposition by sedimentation is determined mainly by the diameter of the fibre; deposition by interception is determined by fibre length.

After inhalation of small amounts of asbestos, short fibres are ingested by macrophages and then removed by mucociliary mechanisms; longer fibres are usually coated and subsequently fragment, disappearing within 18 months (*16*). Others are coated and remain as asbestos or ferruginous bodies. The clearance of asbestos from the lungs depends therefore on the fibres involved (shape, size, distribution, and chemical type), the presence of any factor that might affect the activity of the alveolar macrophages, and the level of exposure (16).

Animal experiments have shown that inhaled asbestos fibres can move from the lung or trachea to other tissues (16). Fibres have been found in lymph nodes of guinea pigs exposed to crocidolite. The fibres presumably pass through the lymphatic system. Asbestos bodies have also been observed in the thyroid of guinea-pigs that had inhaled anthophyllite. In tissues of human mesothelioma patients, asbestos has been found in the lung, lymph nodes, and peritoneum; asbestos bodies have also been identified in the spleen and small bowel (19).

Few data on levels of asbestos in tissues are available, and there is a need to determine the number of fibres that must accumulate in various tissues before disease develops (20). In one study, there were few asbestos fibres found in the lungs of normal patients; an intermediate number were present in asbestos-exposed patients with mesothelioma, and many thousands of such fibres were identified in the lung tissue of severely asbestotic patients (21).

2.4 Health effects

2.4.1 Ingestion

2.4.1.1. *Animal studies*

The presence of asbestos fibres in many sites in the colonic epithelium and lamina propria has been noted in rats fed chrysotile asbestos (12).

Groups of 32 Wistar SPF rats were fed Italian talc or Canadian chrysotile at the rate of 100 mg/day in malted milk powder on 5 days a week for 100 days over a six-month period; 16 controls were fed only malted milk. One gastric leiomyosarcoma was observed in each of the two groups fed chrysotile and talc, but none in the controls (22). In a study in which rats consumed filter paper containing chrysotile (approximately 25 mg/kg of body weight) daily for life, malignancies appeared in 8–14 months (23).

2.4.1.2 *Epidemiological studies*

Although there have been a number of epidemiological studies conducted to date, the results of only one have shown a marginally significant association between asbestos levels in drinking-water and cancers of the digestive tract (24). Cancer incidence data for 721 census tracts of 5 Bay Area counties in California were examined in relation to chrysotile fibre concentrations in drinking-water, which were reported to range from not detectable to 36 million fibres per litre. The proportion of stomach cancers in males attributed to asbestos exposure was 10%, based on the comparison of incidence data in low-exposure areas to incidence data in high-exposure areas, with due allowance for socioeconomic factors. These findings need to be confirmed.

A study of 22 municipalities in Quebec did not reveal excess cancer mortality that could be related to the presence of asbestos in drinking-water (25). The areas studied included Thetford Mines and Asbestos, where asbestos mining operations have been established for almost a century. Epidemiologists in the USA are currently investigating gastrointestinal cancer mortality data for Duluth, Minnesota, where asbestos concentrations of up to 100 million fibres per litre have been present in the public water supply since 1955; no evidence to suggest an increase in deaths due to gastrointestinal cancers has yet been found (26–28). The latency period for development of cancer from occupational inhalation of asbestos is 20–40 years, and therefore more complete data are required before any firm conclusions can be drawn from the Duluth studies. No association has been found in a study conducted in Pensacola, Florida, where levels up to 38 million fibres per litre have been recorded (24). Similarly, preliminary results of a study conducted in the Puget Sound region of Washington State have demonstrated no association between asbestos concentrations in drinking-water and cancer of the colon, stomach, kidney, or total alimentary tract. A more precise case-control study in this area is still under way (29).

As mentioned previously, the use of asbestos-cement pipes in distribution systems is a potential source of asbestos contamination of drinking-water. Generally, it has been concluded that the concentrations of asbestos in drinking-water resulting from the use of asbestos-cement pipes do not present a hazard to human health (30–32). A study of cancer incidence in Connecticut over a 35-year period failed to show a relationship between the use of asbestos-cement pipes and the incidence of gastrointestinal cancer (33). A further study on asbestos-cement pipes and drinking-water in relation to cancer incidence was published by Meigs et al. (34), also with apparently negative results. However, the hypothesis that ingested asbestos fibres cause cancer cannot be ruled out at the present time.

2.4.2 Inhalation

Occupational exposure to airborne asbestos has resulted in pulmonary fibrosis (asbestosis), pleural calcification, bronchogenic carcinoma of the lung, malignant mesothelioma of the pleura and peritoneum, and cancer of the gastrointestinal tract. Although data are inconclusive, it has also been suggested that there is a relationship between occupational inhalation of asbestos and malignancies of the ovary and larynx (35–38). Asbestosis has been described as a "chronic inflammatory reaction in the terminal bronchioles and alveoli of the lung, with considerable fibrosis, leading to distortion and eventual obliteration of the alveoli" (39). The latency period ranges from 7 to 20 years. Generally, it is believed that all types of asbestos can cause asbestosis (*40, 41*) and that fibre-type is of less importance than in the etiology of asbestos-induced cancer (*41, 42*).

A significant relationship between occupational exposure to airborne asbestos and incidence of bronchial carcinoma has been repeatedly noted (*43–47*). From 14% to 50% of patients with asbestosis die from bronchial carcinoma (*40, 41*). Evidence strongly suggests a synergistic relationship between cigarette smoking and asbestos in the causation of bronchial carcinoma (*40, 46, 48, 49*).

An increasing incidence of asbestos-induced malignant mesotheliomas of the pleura and peritoneum has been noted in the Federal Republic of Germany, Great Britain, South Africa, the USA, and elsewhere (*40*). Few mesotheliomas have been identified in Canada (*43*). The latency period between first asbestos exposure and appearance of the tumour is from 20 to 40 years (*50*).

There appears to be a relationship between prolonged inhalation of asbestos and the incidence of gastrointestinal cancer (*47, 51*). It has also been reported that gastrointestinal cancers account for one third of all malignant neoplasms in workers in Quebec chrysotile mines and mills and that the proportion is higher in those exposed to more than 400 million particles per cubic foot (14000 million particles/m^3) (*41, 48*). In a review of the subject (*52*), only one paper did not support the conclusion that occupational exposure to inhaled asbestos leads to an increased incidence of cancer of the digestive system.

REFERENCES

1. CUNNINGHAM, H. M. & PONTEFRACT, R. D. Asbestos fibres in beverages and drinking water. *Nature*, **232**: 332 (1971).
2. *Asbestos in the Great Lakes with emphasis on Lake Superior*. A report to the International Joint Commission, Great Lakes Research Advisory Board, 1974.
3. KRAMER, J. R. & MURDOCH, O. Asbestos research at McMaster University. *Canadian research and development*, Nov.–Dec.: 31 (1974).
4. *A national survey for asbestos fibres in Canadian drinking water supplies*. Ottawa, Department of National Health and Welfare, 1979 (Environmental Health Directorate Publication 79-EMD-34).
5. EISENBERG, W. V. Inorganic particle content of food and drugs. *Environmental health perspectives*, **9**: 183 (1974).
6. KAY, K. Inorganic particles of agricultural origin. *Environmental health perspectives*, **9**: 193 (1974).
7. BILES, B. & EMERSON, T. R. Examination of fibres in beer. *Nature*, **219**: 93 (1968).
8. SELIKOFF, I. J. ET AL. Asbestos air pollution. *Archives of environmental health*, **25**: 1 (1972).
9. NICHOLSON, W. J. & PUNDSACK, F. L. Asbestos in the environment. In: *Biological effects of asbestos*, Lyon, International Agency for Research on Cancer, 1972 (IARC Scientific Publications, No. 8).
10. BRUCKMAN, L. & RUBINO, R. A. Asbestos: rationale behind a proposed air quality standard. *Journal of the Air Pollution Control Association*, **25**: 1207 (1975).
11. GROSS, P. ET AL. Ingested mineral fibers: do they penetrate tissue or cause cancer? *Archives of environmental health*, **29**: 341 (1974).

12. LEE, D. H. K. Biological effects of ingested asbestos: report and commentary. *Environmental health perspectives*, **9**: 113 (1974).
13. CUNNINGHAM, H. M. & PONTEFRACT, R. D. Penetration of asbestos through the digestive tract of rats. *Nature*, **243**: 352 (1973).
14. WESTLAKE, G. E. Asbestos fibers in the colonic wall. *Environmental health perspectives*, **9**: 227 (1974).
15. AMACHER, D. E. ET AL. Effects of ingested chrysotile on DNA synthesis in the gastrointestinal tract and liver in the rat. *Environmental health perspectives*, **9**: 319 (1974).
16. HOLT, P. F. Small animals in the study of the pathological effects of asbestos. *Environmental health perspectives*, **9**: 205 (1974).
17. BOLTON, R. E. & DAVID, J. M. G. The short-term effects of chronic asbestos ingestion in rats. *Annals of occupational hygiene*, **19**: 121 (1976).
18. TIMBRELL, V. The inhalation of fibrous dusts. *Annals of the New York Academy of Sciences*, **132**: 255 (1965).
19. GODWIN, M. C. & JAGATIC, J. Asbestos and mesotheliomas. *Environmental research*, **3**: 391 (1970).
20. BROWN, A. Lymphohematogenous spread of asbestos. *Environmental health perspectives*, **9**: 203 (1974).
21. FONDIMARE, A. & DESBORDES, J. Asbestos bodies and fibers in lung tissues. *Environmental health perspectives*, **9**: 147 (1974).
22. WAGNER, J. C. ET AL. Animal experiments with talc. In: *Inhaled particles and vapors, IV*, New York, Pergamon Press, 1977.
23. GIBEL, W. ET AL. Animal experimental investigations of the carcinogenic activity of asbestos filter material following oral administration. *Archiv für Geschwulstforschung*, **46**: 437 (1976).
24. MCCABE, L. J. & MILLETTE, J. R. Health effects and prevalence of asbestos fibers in drinking water. *Proceedings of the American Water Works Association Annual Conference, San Francisco, 24–29 June 1979*, Denver, CO, AWWA.
25. WIGLE, D. T. Cancer mortality in relation to asbestos in municipal water supplies. *Archives of environmental health*, **32**: 185 (1977).
26. KAY, G. H. Asbestos in drinking water. *Journal of the American Water Works Association*, **66**: 513 (1974).
27. LEVY, B. S. ET AL. Investigating possible effects of asbestos in city water: surveillance of gastrointestinal cancer in Duluth, Minnesota. *American journal of epidemiology*, **103**: 362 (1976).
28. MASSON, T. J. ET AL. Asbestos-like fibers in Duluth water supply: relation to cancer mortality. *Journal of the American Medical Association*, **228**: 1019 (1974).
29. SEVERSON, R. K. *A study of the effects of asbestos in drinking water on cancer incidence in the Puget Sound region.* Seattle, University of Washington, 1979 (M.Sc. thesis).
30. The American Water Works Research Foundation. A study of the problem of asbestos in water. *Journal of the American Water Works Association*, **66**: 1 (1974).
31. ELZENGA, C. H. J. ET AL. *A preliminary investigation into the appearance of asbestos in Dutch drinking water.* Netherlands, 1972 (report of Drinking Water Group of the TNO Health Organization Support Committee).
32. OLSON, H. L. Asbestos in potable water supplies, *Journal of the American Water Works Association*, **66**: 515 (1974).
33. HARRINGTON, J. M. ET AL. An investigation of the use of asbestos cement pipe for public water supply and the incidence of gastrointestinal cancer in Connecticut, 1935–1973. *American journal of epidemiology*, **107**: 96 (1978).
34. MEIGS, J. W. ET AL. Asbestos cement pipe and cancer in Connecticut 1955–1974. *Journal of environmental health*, **42**: 187–191 (1980).
35. STEEL, P. M. & MCGILL, T. Asbestos and laryngeal carcinoma. *Lancet*, **2**: 416 (1973).
36. SHETTIGARA, P. T. & MORGAN, R. W. Asbestos, smoking and laryngeal carcinoma. *Archives of environmental health*, **30**: 517 (1977).
37. MORGAN, R. W. & SHETTIGARA, P. T. Occupational asbestos exposure, smoking and laryngeal carcinoma. *Annals of the New York Academy of Sciences*, **271**: 308 (1976).

38. GRAHAM, H. & GRAHAM, R. Ovarian cancer and asbestos. *Environmental research*, 1: 115 (1967).
39. DAVIES, P. ET AL. Asbestos induced selective release of lysosomal enzymes from mononuclear phagocytes. *Nature*, 251: 423 (1974).
40. PARKES, W. R. *Occupational lung disorders*. London, Butterworths, 1974.
41. HARINGTON, J. S. The biological effects of mineral fibres, especially asbestos, as seen from *in vitro* and *in vivo* experiments. *Annales d'anatomie pathologique*, 21: 155 (1976).
42. ANDERSON, H. A. ET AL. Asbestos disease resulting from household exposure to occupational dusts. *Chest*, 66: 318 (1974).
43. MCDONALD, J. C. Cancer in chrysotile mines and mills. In: *Biological effects of asbestos*, Lyon, International Agency for Research on Cancer, 1972 (IARC Scientific Publications, No. 8).
44. WEBSTER, I. Malignancy in relation to crocidolite and amosite. In: *Biological effects of asbestos*, Lyon, International Agency for Research on Cancer, 1972 (IARC Scientific Publications, No. 8).
45. MEURMAN, L. D. ET AL. Mortality and morbidity of employees of anthophyllite asbestos mines of Finland. In: *Biological effects of asbestos*, Lyon, International Agency for Research on Cancer, 1972 (IARC Scientific Publications, No. 8).
46. NEWHOUSE, M. L. Cancer among workers in the asbestos textile industry. In: *Biological effects of asbestos*, Lyon, International Agency for Research on Cancer, 1972 (IARC Scientific Publications, No. 8).
47. SELIKOFF, I. J. ET AL. Cancer risk of insulation workers in the United States. In: *Biological effects of asbestos*, Lyon, International Agency for Research on Cancer, 1972 (IARC Scientific Publications, No. 8).
48. SELIKOFF, I. J. ET AL. Asbestos exposure, smoking and neoplasia. *Journal of the American Medical Association*, 204: 104 (1968).
49. SELIKOFF, I. J. Air pollution and asbestos carcinogenesis: investigation of possible synergism. In: Mohr, L. L. et al., ed., *Air pollution and cancer in man*, Lyon, International Agency for Research on Cancer, 1977 (IARC Scientific Publications, No. 16, p. 247).
50. *Asbestos*. Lyon, International Agency for Research on Cancer, 1977 (IARC Monographs on the evaluation of the carcinogenic risk of chemicals to humans, vol. 14) pp. 62–65.
51. SELIKOFF, I. J. ET AL. Asbestos exposure and neoplasia. *Journal of the American Medical Association*, 188: 22 (1964).
52. SCHNEIDERMAN, M. A. Digestive system cancer among persons subjected to occupational inhalation of asbestos particles: A literature review with emphasis on dose response. *Environmental health perspectives*, 9: 307 (1974).

3. BARIUM

3.1 General description

3.1.1 Sources

Barium is present in the earth's crust in a concentration of 0.5 g/kg and the mineral barytes, barium sulfate, is the commonest source. Barium carbonate (witherite) is another form, although less common. Traces of barium are present in most soils. Barium compounds are used in oil drilling, production of paints, the processing of diesel fuels, the manufacture of paper, rubber, linoleum and similar products, the manufacture of ceramic glazes and enamels, and in medical diagnostics.

3.1.2 Occurrence in water

Most waters contain some barium, but the concentration is generally well below 0.1 mg/litre (1, 2), although some underground sources may contain levels as high as 10 mg/litre in geothermal brines (3). The source is normally natural mineral matter; although barium sulfate, the predominant form, is only slightly soluble in water, in the presence of certain common anions the solubility of barium can be markedly enhanced. The chemical form of barium in tap-water is not definitely known (4, 5).

3.2 Routes of exposure

3.2.1 Drinking-water

Relatively little is known about levels of barium in tap-water throughout the world. However, a survey of water quality in 100 cities in the USA found levels of barium in the range 0.002–0.38 mg/litre, with a median level of 0.043 mg/litre (6) and, in a survey of 2595 water samples, less than 0.1 % contained barium at a level exceeding 1 mg/litre (4). In the USSR, the maximum allowable concentration for barium in drinking-water has been set at 0.1 mg/litre (7).

3.2.2 Food

Barium is present in traces in many foodstuffs (1); brazil nuts are an especially rich source and contain up to several thousand μg/g (1). There is relatively little published information on dietary intake, but some values ranging between less than 0.1 to nearly 2 mg/day have been published (8). Dietary intakes are estimated to be about 0.06–1.2 mg/day in the USA (9, 10) and 0.4–1.2 mg/day in Canada (5).

3.2.3 Air

Little published information is available, but an average concentration of 0.005 ng/m³ air has been reported for cities in the USA (1).

3.2.4 Other routes of exposure

3.2.4.1 Industrial exposure

Barium can be detected in the working atmosphere of industrial environments, and a threshold limit (based on health effects) has been set at 0.5 mg/m³ of air for soluble barium compounds in industrial situations in the USA (4).

3.2.4.2 Smoking

Traces of barium can be detected in tobacco, but because little barium is inhaled during smoking, the exposure from this source is unimportant.

3.2.5 Relative significance of different routes of exposure

Two simplified examples of exposure (where the levels in water are elevated) have been calculated to illustrate the relative intake.

Assuming a daily water consumption of 2 litres with a barium concentration of 0.05 mg/l, the daily intake of barium from water would be 0.1 mg. With a daily air intake of 20 m³ and ambient air levels of 0.005 ng of barium per m³, the daily exposure to barium via inhalation is 0.1 ng/day. This amount is negligible compared with barium intake via food or water. Assuming a typical daily food intake of 0.5–1.2 mg of barium, the total daily intake from all sources is 0.6–1.3 mg. No reliable information is available for the exposure of children to barium.

As the levels of barium in drinking-water are unlikely to exceed 0.05 mg/litre, the intake in relation to normal total intake is unlikely to be more than about 15 %. However, where the barium levels in water are high, i.e., over 1 mg/litre, more than three-quarters of the total barium intake could result from drinking-water.

The uptake of barium is difficult to quantify accurately. Barium in food is poorly absorbed (8), but since a high proportion could be absorbed if ingested as a soluble salt in water, then the contribution made by drinking-water to the absorption of the element could be especially large.

3.3 Metabolism

There is no evidence that barium is essential for human nutrition. Insoluble forms of barium, such as barium sulfate (used medicinally), are very poorly absorbed, and have a very low toxicity. Soluble barium salts are readily and rapidly absorbed (*4*); in this form, 50 % or more of ingested barium is absorbed (*1*). Only a small fraction of barium in normal foodstuffs appears to be absorbed (*8*). Most absorbed barium is present in bone (*8*), but it has also been detected in a number of other tissues, such as kidney, liver, and heart (*8*).

The metabolic pathways for barium are similar to those for calcium (*4*). Excretion of barium occurs more readily than with calcium; about a quarter of any ingested barium is excreted within 24 hours (*4, 5*).

3.4 Health effects

Animal experiments using drinking-water suggest that 5 mg of barium per litre, as barium acetate, would cause no toxic effects even after prolonged exposure (*11*). In rats, a total dose of 250 mg of barium over a period of 4–13 weeks also caused no adverse effects (*12*).

Barium is acutely toxic when soluble salts are ingested in excess; if taken as the chloride, the fatal dose of barium for an adult is about 550–600 mg (*4*). In high doses, it induces a strong prolonged stimulant action on all muscles, including those of the heart and gastrointestinal tract (*4*). The acute toxic dose has been found by one research group to be 200–500 mg (*1*). In another study, 125 mg as a single dose has been suggested to be the threshold for acute toxic effects (*13*).

In certain industries, it is claimed that a benign form of pneumoconiosis is associated with the inhalation of dusts containing barium (*4*). A recent epidemiological study indicated that, where levels of barium in water were up to 10 mg/litre, a statistical association was found with cardiovascular death rates, but because there were population changes in the communities studied, the finding must be treated with a certain degree of caution (*3*). Epidemics of barium poisoning have been reported in China (*14*). "Pa Ping" disease was attributed to the prolonged ingestion of table salt containing up to 250 g of barium chloride per kg (*14*).

REFERENCES

1. *Toxicology of metals, vol. II*. Washington, DC, US Environmental Protection Agency, 1977 (Environmental Health Effects Research Series).
2. *Quality criteria for water*. Washington, DC, US Environmental Protection Agency, 1976.
3. BRENNIMAN, G. R. ET AL. Cardiovascular disease death rates in communities with elevated levels of barium in drinking water. *Environmental research*, **20**: 318 (1979).
4. NATIONAL RESEARCH COUNCIL. *Drinking water and health*. Washington, DC, National Academy of Sciences, 1977.

5. *Guidelines for Canadian drinking water quality, 1978.* Quebec, Ministry of Supply and Services, 1979 (Supporting documentation).
6. DURFOR, C. N. & BECKER, E. Selected data on public supplies of the 100 largest cities in the United States, 1962. *Journal of the American Water Works Association,* **56**: 237 (1964).
7. KRASOVSKY, G. N. ET AL. [A proposed maximum allowable concentration for barium in water.] *Gigiena i sanitarija,* (6): 86 (1980) (in Russian).
8. UNDERWOOD, E. J. *Trace elements in human and animal nutrition.* New York, Academic Press, 1977.
9. GORMICA, A. Inorganic elements in food used in hospital menus. *Journal of the American Dietetic Association,* **56**: 397 (1970).
10. SCHROEDER, H. A. ET AL. Trace metals in man: strontium and barium. *Journal of chronic diseases,* **25**: 491 (1972).
11. SCHROEDER, H. A. & MITCHENER, M. Lifetime studies in rats: effects of aluminum, barium, beryllium and tungsten. *Journal of nutrition,* **105**: 420 (1975).
12. *Guidance for the issuance of variances and exemptions.* Washington, DC, US Environmental Protection Agency, Office of Drinking Water, 1979.
13. BROWNING, E. Barium. In: *Toxicity of industrial metals,* London, Butterworths, 1969.
14. POLSON, C. J. & TATTERSALL, R. N. Barium. In: *Clinical toxicology,* London, Pitman Medical, 1969.

4. BERYLLIUM

4.1 General description

4.1.1 Sources

Beryllium is commonly found as part of feldspar mineral structures and may exist as the mineral beryl in small localized deposits (*1*). The amounts mined have been very small. The primary source of beryllium in the environment is the burning of fossil fuels, although contamination is normally slight (*2*). Because of its light weight and high tensile strength (*3*), the metal is valuable as a constituent of special alloys for applications such as space vehicles, X-ray windows, and certain electrical components.

4.1.2 Occurrence in water

Beryllium can enter waterways through the weathering of rocks, atmospheric fallout, and industrial and municipal discharges. However, levels in fresh water appear to be very low, generally below 1 μg/litre (*4*).

4.2 Routes of exposure

4.2.1 Drinking-water

Few surveys have been carried out to determine levels of beryllium in drinking-water. Concentrations ranging from 0.01 to 1.2 μg/litre, with a mean concentration of 0.2 μg/litre, were found in a survey in the USA (*5*).

4.2.2 Food

Relatively little information is available concerning the beryllium content of food. Data from the USA indicate that about 100 μg/day could be a typical dietary intake (*6*). A study in the United Kingdom estimated that the average dietary intake could be less than 15 μg/day (*7*). The beryllium content of various foodstuffs collected in New South Wales, Australia, ranged from 0.01–0.12 mg/kg (*8*).

4.2.3 Air

Data from the USA indicate that atmospheric beryllium is detected infrequently and usually in small amounts (*4*); levels generally are within the range 0.3–3 ng/m³ of air (*6*).

4.2.4 Other routes of exposure

4.2.4.1 *Industrial exposure*

Industrial processes in which exposure to beryllium can occur include the mining and extraction of beryllium, aerospace equipment, alloy machining, electroplating, and atomic energy industries. Exposure to beryllium in industrial settings is via inhalation and skin contact. Without ventilation, concentrations of beryllium in air as high as 23 $\mu g/m^3$ have been reported (*3*). In the USA, the current occupational standard for exposure to beryllium is 2 $\mu g/m^3$ of air (*9*).

4.2.5 Relative significance of different routes of exposure

The contribution of beryllium in air to the total body burden is negligible compared with the contribution from food and water combined. The calculations below, therefore, consider the contributions made by food and water only.

(*a*) *Daily diet: 10 µg of beryllium per day* (*adults*)

Beryllium concentration in water	Weekly beryllium intake (μg)			Total water contribution (%)
	Water only	Food only	Total	
1 µg/litre	14	70	84	17
2 µg/litre	28	70	98	28

(*b*) *Daily diet: 100 µg of beryllium per day* (*adults*)

Beryllium concentration in water	Weekly beryllium intake (μg)			Total water contribution (%)
	Water only	Food only	Total	
1 µg/litre	14	700	714	2
2 µg/litre	28	700	728	4

4.3 Metabolism

Dermal absorption of beryllium is negligible, since beryllium becomes bound to certain epidermal constituents, but there have been reports of beryllium dermatitis and granulomatous ulcerations of the skin at sites where insoluble beryllium compounds were embedded; conjunctivitis has also been described (*10*). Compounds of beryllium are not readily absorbed when ingested, since they tend to form insoluble compounds at physiological pH levels (*3*). Less than 1% of ingested beryllium is absorbed (3, 11). Inhaled soluble compounds of beryllium are absorbed from the lungs into the blood and are transported mainly as orthophosphate colloids, being ultimately deposited in the bone as insoluble hydroxides. Insoluble beryllium compounds tend to remain in the lungs indefinitely after inhalation (4, 6). Absorbed beryllium is either excreted in the urine or deposited in the kidneys or bone (3). Excretion of beryllium is fairly rapid (11, 12).

4.4 Health effects

The inhalation of beryllium has been demonstrated to be harmful to human beings (11). Acute respiratory exposure can result in severe health effects, including rhinitis, pharyngitis, pneumonitis, and pulmonary oedema (11).

Beryllium is very poorly absorbed from the gastrointestinal tract, and its toxicity via this route of entry is low; no reports of oral toxicity in humans have been found in the available literature.

Inhalation, intratracheal instillation, and intravenous injection have been shown to induce tumours in experimental animals. Although firm evidence of human cancer induction by exposure to beryllium is lacking, the evidence is suggestive (4). Some studies of tumour induction in experimental animals as a result of ingestion of beryllium compounds at levels up to 500 mg/kg of diet have been inconclusive. In two studies, rats and mice ingesting beryllium in drinking-water at a concentration of 5 mg/litre for life did not show a statistically significant increase in tumours as compared with controls (13, 14). In another study, rats were fed beryllium in the diet at levels of 5, 50, and 500 mg/kg of feed. The authors concluded that there was no evidence of a carcinogenic response related to beryllium ingestion (15).

Epidemiological studies conducted in the USA have not shown a significant correlation between ingestion of beryllium and human cancer. However, the International Agency for Research on Cancer (16) concluded: "There is sufficient evidence that beryllium metal and several beryllium compounds are carcinogenic to three experimental animal species . . . Taken together, the experimental and human data indicate that beryllium should be considered suspect of being carcinogenic to humans." This conclusion is related essentially to inhalation of beryllium.

The only standard limits for beryllium in water are those issued by the USSR, where a maximum allowable concentration of 0.2 µg/litre has been set (1).

REFERENCES

1. SASHINA, L. A. [Experimental data to substantiate the maximum permissible concentration of beryllium in water bodies.] *Gigiena i sanitarija*, (2): 10 (1965) (in Russian).

2. TEPPER, L. B. Beryllium. In: Lee, D. H. K., ed. *Metallic contaminants and human health*, New York, Academic Press, 1972.

3. OAK RIDGE NATIONAL LABORATORY. *Reviews of the environmental effects of pollutants VI. Beryllium*. Cincinnati, Health Effects Research Laboratory, 1978 (EPA-600/1-78-028).

4. *Ambient water quality criteria for beryllium*. Washington, DC, US Environmental Protection Agency, 1980 (EPA-440/5-80-024).

5. KOPP, J. F. & KRONER, R. C. *Trace metals in waters of the United States*. Cincinnati, US Department of the Interior, 1967.

6. *Toxicology of metals, vol. II*. Washington, DC, US Environmental Protection Agency, 1977 (Environmental Health Effects Research Series).

7. HAMILTON, E. I. & MINSKY, M. J. Abundance of the chemical elements in man's diet and possible relations with environmental factors. *Science of the total environment*, 1: 375 (1973).

8. MEEHAN, W. R. & SMYTHE, L. E. Occurrence of beryllium as a trace element in environmental materials. *Environmental science and technology*, 1: 839 (1967).

9. *TLVs-Threshold limit values for chemical substances and physical agents in the workroom environment with intended changes for 1979*. Cincinnati, American Conference of Governmental Industrial Hygienists, 1979.

10. NICHIMURA, M. Clinical and experimental studies on acute beryllium disease. *Nagoya journal of medical science*, 28: 17 (1966).

11. NATIONAL RESEARCH COUNCIL. *Drinking water and health*. Washington, DC, National Academy of Sciences, 1977.

12. STOCKINGER, H. E. *The toxicology of beryllium*. Washington, DC, US Department of Health, Education and Welfare, 1972 (Publication 2173).

13. SCHROEDER, H. A. & MITCHELL, M. Lifetime studies in rats: effects of aluminum, barium, beryllium and tungsten. *Journal of nutrition*, 105: 420 (1975).

14. SCHROEDER, H. A. & MITCHELL, M. Lifetime effects of mercury, methylmercury and nIne other trace metals on mice. *Journal of nutrition*, 105: 452 (1975).

15. MORGAREIDGE, K. ET AL. *Chronic feeding studies with beryllium sulphate in rats*. Pittsburgh, Food and Drug Research Laboratories, Inc., 1975 (Final report to the Aluminum Company of America, 15219).

16. *Some metals and metallic compounds*. Lyon, International Agency for Research on Cancer, 1980 (IARC Monographs on the evaluation of the carcinogenic risk of chemicals to humans, vol. 23), p. 190.

5. CADMIUM

5.1 General description

5.1.1 Sources

Cadmium-containing minerals are found in specific parts of the world, although the metal is uniformly distributed in trace amounts in the earth's crust. Practically all zinc ores contain small amounts of cadmium. Although rare, the predominant ore of cadmium is greenockite (cadmium blende), i.e., cadmium sulfide; this form is often associated with deposits of sphalerite (zinc sulfide). Cadmium production began slowly at the end of the nineteenth century. It is produced normally as a by-product of zinc extraction. The use of the element has been increasing steadily during this century, but it is only in the last 20 years that it has been of major interest. Cadmium has begun to contaminate the environment, and it has been found in air, food, soil, plants, and water. The principal uses of cadmium are in the fabrication of alloys and solders, metal plating, as pigments, as stabilizers in plastic materials, and in batteries.

5.1.2 Occurrence in water

The solubility of cadmium in water is influenced by the nature of the source of the cadmium and the acidity of the water. Surface-waters that contain more than a few micrograms of cadmium per litre have probably been contaminated by discharges of industrial wastes or by leaching from areas of landfill, or from soils to which sewage sludge has been added (1). Unpolluted waters generally contain less than 1 μg/litre (2-4).

The levels of cadmium in public water supplies are normally very low, since generally only tiny amounts exist in raw water and even where levels are somewhat elevated, many conventional water-treatment processes will remove much of the cadmium (5).

Higher levels of cadmium in tap-water are associated with plated plumbing fittings, silver-base solders, and galvanized iron piping materials (1).

5.2 Routes of exposure

5.2.1 Drinking-water

Drinking-water normally contains very low concentrations of cadmium, of the order of 1 μg/litre or less (*1, 6–8*); occasionally, levels up to 5 μg/litre have been reported (*1*) and on rare occasions levels up to 10 μg/litre have been detected (*9*). In some areas, well-water may contain elevated concentrations of cadmium (*1*). It is probable that the levels of cadmium could be higher in areas supplied with soft water of low pH, since this would tend to be more corrosive to any plumbing systems containing cadmium. The level of cadmium in a sample of water is likely to be a function of how long the water has been in contact with the plumbing, and as a consequence there is likely to be a variation in concentration when water is drawn at different times of day from the same tap. It would be difficult to define the average exposure unless a large number of samples were collected.

Estimated daily exposure to cadmium via water, based on a water consumption of 2 litres per day, ranges from substantially less than 1μ g to over 10 μg per day. Such estimates are, of course, based on the presumption that all the cadmium from the water is ingested; it is likely, however, that in the preparation of certain beverages, e.g., tea, not all the cadmium will be consumed since some could become attached to the surfaces of the utensils, etc.

5.2.2 Food

Most foodstuffs contain traces of cadmium; crops grown in polluted soil (from industrial contamination and from use of sewage sludge as fertilizer) or irrigated with polluted water may contain increased concentrations, as may meat from animals grazing on contaminated pastures. The kidneys and livers of animals concentrate cadmium, and people who eat these food items will tend to ingest more cadmium than those who do not (*1*). Shellfish also tend to accumulate the metal (*1*). An additional source of cadmium is phosphate fertilizer.

A comprehensive review of the cadmium content of foodstuffs worldwide has been published (*1*). Most foodstuffs contain less than 0.1 mg/kg (wet weight). Typical dietary intakes range from 15 to 60 μg of cadmium per day (*1, 10*). The higher value is derived from a typical Japanese diet (*9*). Japan has the highest natural levels as well as the highest levels in polluted areas. Virtually nothing is known of the chemical form of cadmium in foodstuffs; such information would be valuable to ascertain the precise absorption of the metal when food is ingested.

5.2.3 Air

The levels of cadmium in ambient air are generally low (*11*). Long-term average concentrations may vary from less than 0.001 to 0.5 μg/m^3

depending on the degree of industrialization and presence of cadmium-emitting industries (10). A comprehensive review of the cadmium levels in the air worldwide has been published (1). It has been estimated that members of the general population will generally inhale less than 0.05 μg/day (10). For unusually polluted areas, maximum values as high as 3.5 μg/day have been estimated (1), and even in cities where the levels are 30 times as high as rural background values, this would still be small relative to the food source.

The fraction of the particles containing cadmium that are deposited and retained in the lung varies with the particle size, but an average deposition of 25% has been estimated (1) for the range of particles that exists in ambient air.

5.2.4 Smoking

Cadmium is present in tobacco, one cigarette normally containing 1–2 μg. Because cadmium is volatile at elevated temperatures, some of the metal will be inhaled during smoking. Estimates have been made of the contribution made by various types of tobacco in different countries (1). Typically, 2–4 μg of cadmium per 20 cigarettes is inhaled; probably 50% will be deposited in the lungs (1).

5.2.5 Industrial exposure

Workers in industry may inhale concentrations of cadmium ranging from a few micrograms up to several thousand micrograms per cubic metre of air (10); the highest levels, however, are associated mainly with exposures that occurred some years ago.

5.2.6 Relative significance of different routes of exposure

From the information provided in sections 5.2.1–5.2.5, it can be seen that, for different individuals and population groups, there can be a wide range of exposures to cadmium from water, food, air, and occupational exposure, etc. For such exposures, a WHO Expert Committee (8) recommended that the weekly intake of cadmium should not exceed 0.5 mg per person. To give some idea of particular situations, a few simplified examples are given here.

For low concentrations of cadmium in water (up to 1 μg/litre), the typical contribution made by water to overall intake for adults is less than 5 to 10% (based on dietary intakes of 15–60 μg/day and inhalation of 0.05 μg/day from ambient air).

Since some information is available on the absorption of ingested and inhaled cadmium, it is relevant to consider the absorption of cadmium; some estimates for uptake in adults are provided in the table on page 87. No reliable information is available for children.

(a) *Food intake 20 μg of Cd per day (6% absorption); inhalation 0.05 μg/day with 25% retention by the lungs and 64% absorption; drinking-water containing 1 μg of Cd per litre or 5 μg of Cd per litre (6% absorption)*

Cadmium concentration in water	Weekly cadmium uptake (μg)				Total water contribution (%)
	Water only	Air only	Food only	Total	
1 μg/litre	0.8	0.1	8.4	9.3	9
5 μg/litre	4.2	0.1	8.4	12.7	33

Smoking 20 cigarettes per day would reduce the water contributions to 5% and 21% respectively.

(b) *Food intake 50 μg of Cd per day (6% absorption); inhalation 0.05 μg/day with 25% retention by the lungs and 64% absorption; drinking-water containing 1 μg of Cd per litre or 5 μg of Cd per litre (6% absorption)*

Cadmium concentration in water	Weekly cadmium uptake (μg)				Total water contribution (%)
	Water only	Air only	Food only	Total	
1 μg/litre	0.8	0.1	21.0	21.9	4
5 μg/litre	4.2	0.1	21.0	25.3	17

Smoking 20 cigarettes per day would reduce the water contributions to 3% and 13% respectively.

5.3 Metabolism

Cadmium is fairly readily absorbed through ingestion or through the lungs. Alimentary absorption is affected by a number of factors, such as age, calcium, iron, zinc, and protein deficiency (*1, 7, 10*), and the chemical form of the cadmium ingested. As with lead (*12*), the state of the stomach is likely to influence the amount absorbed, with a fasting stomach probably providing the maximum uptake in contrast with a full stomach. In human subjects given labelled cadmium, between 4.7% and 7% (mean 6%) of orally administered cadmium was absorbed (*1, 10*). Dietary factors, such as iron, calcium, and protein deficiency, may increase the gastrointestinal absorption rate (*13*). In iron-deficient women, up to 20% of ingested cadmium was found to be absorbed. Pulmonary absorption is dependent on the size and solubility of the particles containing cadmium and will be affected, too, by the depth and rate of breathing. Approximately 50% of 0.1-μm particles will be retained by the lung in contrast to 20% for particles as large as 2 μm (*10*). From information on the particle size distribution of cadmium in the general air and for tobacco smoke, it has been estimated that typically 25% and 50% of the particles will be retained, respectively (*10*).

Absorbed cadmium will enter the blood and become concentrated in certain parts of the human body (*1, 14*). Both the liver and the kidneys act as stores of cadmium (about 50 % of any accumulated cadmium is found in these organs) (*1*). The cadmium is to a large extent bound to a protein of low relative molecular mass, known as metallothioneine (*11*); this metal-binding protein is believed to be involved in cadmium transport and absorption (*15*). Cadmium has a long biological half-time in the body (13–38 years) and accumulates with age (*1*). It has been found that, because the placenta acts as a fairly efficient barrier to cadmium, the newborn are virtually free of cadmium (approximately 1 μg only (*1, 5*)) in contrast to non-occupationally exposed people of age 50 who may have 10–50 mg (*5*) stored in their bodies. Grossly exposed industrial workers have been discovered to have levels of over 1000 mg (*1*). Blood levels are usually below 20 μg/litre for non-smoking members of the general population (*1, 10*). Cadmium in blood reflects recent exposure rather than body burden. Excretion of cadmium is usually rather slow, mainly via the urine (*1*). On a group basis, urinary cadmium is generally regarded as a good indication of the body burden. Cadmium interacts with other metals, especially zinc, and may influence the relative distribution of zinc in the body.

5.4 Health effects

Acute effects have been seen where food has been contaminated by cadmium from plated vessels; severe gastrointestinal upsets have been reported (*1, 5*). The acute oral lethal dose of cadmium for man has not been established, but it is estimated to be several hundred milligrams (*16*). Health effects have been demonstrated in industrial workers heavily exposed to cadmium oxide fumes and dust (*10*). Bronchitis, emphysema, anaemia, and renal stones have been reported (*5*). The renal cortex is generally accepted to be the critical organ for cadmium accumulation in man (*8, 10, 17*). The classical renal effects of cadmium poisoning are associated with proteinuria, glucosuria, and aminoaciduria (*7*). Where the exposure has been high, as was the case in Japan in the outbreak of Itai-Itai disease (a bone disease), irreversible renal injury occurred in those most severely exposed, especially in elderly women (*1*). There have been no reported effects from the low levels of cadmium that can be found in drinking-water, although contaminated beverages made from dispensers containing cadmium-plated fittings have caused acute effects in children (*1*).

There have been many studies in which animals have been dosed with cadmium. There is evidence of hypertension after long-term low-level oral exposure and teratogenic, mutagenic and carcinogenic effects after injection of high doses (*1*). Short-term administration of cadmium in drinking-water at a level of 10 mg/litre has been found to cause partial inhibition of the gastrointestinal absorption of iron (*1*). Krasovsky's work indicates general toxic and gonadotoxic effects due to cadmium (*18*).

There are suggestions that there is a relationship between ingestion of cadmium and hypertension in man; however, at the present time, this relationship is inconclusive (*19*).

The evidence that cadmium may be carcinogenic to man is rather weak (*7, 20*) although prolonged and heavy industrial exposures may constitute an increased risk of prostate cancer (*21, 22*). In the opinion of a WHO Study Group (*15*) the epidemiological studies on carcinogenicity of cadmium are not conclusive because most of them involved only a small number of workers. For this and other reasons the possible carcinogenicity of cadmium cannot be considered in deriving health-based occupational exposure limits. The relationship between exposure level and cadmium concentration in blood is not yet sufficiently understood to derive a biological limit for blood with satisfactory precision. However, a WHO Study Group (*15*) agreed that the value of 10 µg/litre of whole blood should be accepted as a tentative no-adverse-effect level.

It has been estimated that 3 mg of cadmium is the no-effect level for cadmium administered as a single oral dose to man (*1*). At the sixteenth meeting of the Joint FAO/WHO Expert Committee on Food Additives and Food Contaminants held in April 1972, it was recommended that the provisional tolerable intake of cadmium should not exceed 400–500 µg per week for an adult, i.e., 57–71 µg/day (*9*).

Estimates have been made of a no-effect level for long-term exposure to cadmium (*1*). A threshold-effect dose of 200 µg/day has been suggested on the basis of epidemiological studies; this represents 12 µg/day absorbed (assuming 6 % oral absorption). Taking the critical concentration for the human kidney cortex as between 200 and 250 mg/kg, it has been calculated that a person ingesting 248 µg/day (close to the 200 µg/day threshold) would achieve this critical kidney concentration at age 50 (*1*). With a daily intake within the range 57–71 µg, as recommended in the report mentioned above (*9*), the renal cortex would receive about a quarter of the critical concentration of 200 mg/kg.

REFERENCES

1. COMMISSION OF THE EUROPEAN COMMUNITIES. *Criteria (dose/effects relationships) for cadmium.* Oxford, Pergamon Press, 1978.
2. HIATT, V. & JUFF, J. E. The environmental impact of cadmium: an overview. *International journal of environmental studies*, 7: 277 (1975).
3. FLEISCHER, M. ET AL. Environmental impact of cadmium: a review by the panel on hazardous trace substances. *Environmental health perspectives*, 7: 253 (1974).
4. FRIBERG, L. ET AL. *Cadmium in the environment*, 2nd ed. Cleveland, CRC Press, Inc., 1974.

5. NATIONAL RESEARCH COUNCIL. *Drinking water and health.* Washington, DC, National Academy of Sciences, 1977.

6. *Guidelines for Canadian drinking-water quality, 1978.* Quebec, Supply and Services, 1979 (supporting documentation).

7. *Ambient water quality criteria for cadmium.* Washington, DC, US Environmental Protection Agency, Office of Water Regulations and Standards, Criteria and Standards Division, 1980.

8. WHO Technical Report Series, No. 505, 1972 (*Evaluation of certain food additives and the contaminants: mercury, lead and cadmium*).

9. *The hazards to health of persistent substances in water*: annexes to a report of a WHO Working Group. Copenhagen, WHO Regional Office for Europe, 1972.

10. COMMISSION OF THE EUROPEAN COMMUNITIES. *Trace metals: exposure and health effects.* Oxford, Pergamon Press, 1979.

11. *Environmental health criteria for cadmium.* Geneva, World Health Organization, 1979 (interim report).

12. CHAMBERLAIN, A. C. ET AL. *Report R.9198.* Harwell, UK Atomic Energy Research Establishment, 1978.

13. FLANAGAN, P. R. ET AL. Increased dietary cadmium absorption in mice and human subjects with iron deficiency. *Gastroenterology*, **74**: 841 (1978).

14. UNDERWOOD, E. J. *Trace elements in human and animal nutrition.* New York, Academic Press, 1977.

15. WHO Technical Report Series, No. 647, 1980 (*Recommended health-based limits in occupational exposure to heavy metals*: report of a WHO Study Group).

16. GLEASON, M. *Clinical toxicology of commercial products.* Baltimore, Williams & Williams, 1969.

17. *Toxicology of metals, vol. II.* Washington, DC, US Environmental Protection Agency, 1977 (Environmental Health Effects Research Series).

18. KRASOVSKY, G. N. ET AL. Toxic and gonadotropic effects of cadmium and boron relative to standards for these substances in drinking water. *Environmental health perspectives*, **13**: 69 (1976).

19. NATIONAL INSTITUTE FOR OCCUPATIONAL SAFETY AND HEALTH. *Criteria for a recommended standard. Occupational exposure to cadmium.* Washington, DC, US Department of Health, Education and Welfare, 1977.

20. *Cadmium, nickel, some epoxides, miscellaneous industrial chemicals and general considerations on volatile anaesthetics.* Lyon, International Agency for Research on Cancer, 1976 (IARC Monographs on the evaluation of the carcinogenic risk of chemicals to humans, vol. 11).

21. KIPLING, M. D. & WATERHOUSE, J. A. H. Cadmium and prostatic carcinoma. *Lancet*, **1**: 730 (1967).

22. LEMAN, R. A. ET AL. Cancer mortality among cadmium production workers. *Annals of the New York Academy of Sciences*, **271**: 273 (1976).

6. CHROMIUM

6.1 General description

6.1.1 Sources

Most rocks and soils contain small amounts of chromium. The commonest ore is chromite in which the metal exists in the trivalent form; the ore is present in commercial quantities only in a few countries. Hexavalent chromium also exists naturally but infrequently. Chromium in its naturally occurring state is in a highly insoluble form; however, weathering, oxidation and bacterial action can convert it into a slightly more soluble form. Most of the more soluble forms in soil, especially any hexavalent chromium, are mainly the result of contamination by industrial emissions. Some contamination arises from the use of sewage sludge added to land. Contamination of air, water, and food has occurred as a result of man's use of chromium; traces of some natural chromium are present in food. Trivalent and hexavalent chromium occur in biological media, but only the trivalent form is stable, since hexavalent chromium is readily reduced by a variety of organic species (*1*).

The major uses of chromium are for chrome alloys, chrome plating, oxidizing agents, corrosion inhibitors, manufacture of chromium compounds, such as pigments, and in the textile, ceramic, glass, and photographic industries.

6.1.2 Occurrence in water

Because of the low solubility of chromium generally, the levels found in water are usually low (9.7 μg/litre) (*2*); however, there are examples of contamination of water, in some cases serious, in which effluents containing chromium compounds have been discharged to rivers; the chromium may be in the trivalent or hexavalent form, either as a soluble salt or as insoluble particles and often as a chemical complex. The valency of the chemical form in natural waters is influenced by the acidity of the water (*3*). Trivalent chromium is converted to the insoluble hydroxide at neutral pH (*4*). Total chromium levels in raw water are usually 10 μg/litre or less; rarely do levels exceed 25 μg/litre except in highly contaminated situations (*1–6*). There is a tendency for the naturally occurring higher levels of chromium to be associated with waters of the greatest hardness (*5*).

The levels of chromium in finished water entering the public supply are normally about the same as, or perhaps slightly lower than, those in raw source water.

6.2 Routes of exposure

6.2.1 Drinking-water

Drinking-water normally contains very low concentrations of chromium (i.e., 5 μg/litre or less) (2). It would be very rare to find chromium levels as high as 20 μg/litre in tap-water. The level in tap-water may be a function of the period the water has been standing in contact with any plumbing fittings and as a consequence there could be variations in chromium levels when water is drawn at different times of the day at the same tap.

Based on a water consumption of 2 litres per day, it can be estimated that the daily exposure to chromium in water might vary from substantially less than 10 μg to perhaps 40 μg per day on rare occasions. Since apparently trivalent chromium rarely occurs in drinking-water that is chlorinated, it is assumed that most waterborne chromium is in the hexavalent form (1).

6.2.2 Food

Foodstuffs vary considerably in their chromium content, which ranges from 20 to 590 μg/kg. Because varying and conflicting figures have been quoted it is not possible to make a proper comparison between different foodstuffs; a review of the chromium content of 45 items of food has been published (5). Some seafoods appear to contain elevated levels of 0.02–0.21 mg/kg (8). Wines contain chromium and concentrations up to 60 μg/litre have been recorded (3). The chromium in foodstuffs is in both the trivalent and the hexavalent form. Some contamination of food could result during its preparation where plated or stainless steel utensils are used. There is a dearth of information on total dietary intake of chromium. In the USA it has been estimated to vary from 5 μg to 500 μg per day (2, 4, 9) and this range probably covers the vast majority of diets throughout the world. Average dietary chromium is probably about 100–300 μg/day. Virtually no information on the dietary intake for children exists.

6.2.3 Air

There is only a limited amount of information on the levels of chromium in the air. Reported values (5) would suggest that mean concentrations in air in towns are typically about 0.02 μg/m^3. In heavily industrialized areas however, concentrations of over 20 times this value have been recorded (5).

Most of the chromium in the air will be in the form of fine particles, of which perhaps one half of those inhaled could become deposited in the respiratory tract. Based on a daily inspired volume of 22.8 m^3 and 50 % alveolar retention, the daily quantity deposited in the lungs would be about 0.2 μg.

6.2.4 Smoking

Cigarettes contain traces of chromium; a value of 1.4 μg per cigarette has been quoted (3) and some of this will be inhaled and absorbed. It is extremely difficult to give the precise exposure for smokers, but since only a small fraction of the chromium will be inhaled and only perhaps one half will be deposited in the lung, it can be estimated that the chromium retained in this way by smoking 20 cigarettes per day would not exceed a few micrograms per day.

6.2.5 Occupational exposure

Levels of airborne chromium in a number of industrial situations, particularly in plating plants and where welding occurs, can be very much higher than in the ambient environment. Concentrations as high as hundreds of micrograms per cubic metre of air have been recorded.

6.2.6 Ingestion of dirt, dust, etc.

Little is known about the levels of chromium in dust, but in general it is unlikely to be a very important source of exposure even where young children have the opportunity to ingest dirt and dust.

6.2.7 Relative significance of different routes of exposure

From the discussion in section 6.2.1–6.2.6, it would appear that for different individuals and population groups there is the opportunity for a fairly wide range of exposure to chromium in water, food, air, and the environment. To illustrate the potential relative contributions of chromium in water in relation to overall intake, a few examples are given here.

6.2.7.1 *Estimates of chromium uptake*

The estimates have been based on 10 % absorption of chromium when ingested from food and water and 50 % overall absorption and retention of chromium from respired air; it is further assumed that on a daily basis a person consumes 2 litres of water and breathes 20.0 m^3 of air per day containing chromium at a concentration of 0.02 μg/m^3.

No reliable information is available for assessing the absorption of chromium by children or other sensitive groups.

(a) *Food intake: 100 µg of chromium per day*

Chromium concentration in water	Weekly absorption of chromium (µg)				Total water contribution (%)
	Water only	Air only	Food only	Total	
20 µg/litre	28	2	70	100	28
50 µg/litre	70	2	70	142	49
100 µg/litre	140	2	70	212	66

(b) *Food intake: 300 µg of chromium per day*

Chromium concentration in water	Weekly absorption of chromium (µg)				Total water contribution (%)
	Water only	Air only	Food only	Total	
20 µg/litre	28	2	210	240	12
50 µg/litre	70	2	210	282	25
100 µg/litre	140	2	210	352	40

6.3. Metabolism

Chromium is absorbed through both the gastrointestinal and respiratory tracts. The amount absorbed differs in each system and depends on the form of chromium (1). Trivalent chromium is an essential form of the element for human beings. Hexavalent chromium is toxic.

Discrepancies in values reported for chromium absorption from the digestive tract appear in the literature; exact values are not known. Trivalent chromium is poorly absorbed. From 0.1 % to 1.2 % of trivalent chromium salts are absorbed, whereas 25 % of glucose tolerance factor (GTF), a chromium complex necessary for normal glucose tolerance, is absorbed. Natural chromium complexes in the diet seem to be more available for absorption than simple salts (1). It appears that at least 10 % of the chromium in food is absorbed (10). Tissue chromium levels in rats exposed for one year to hexavalent chromium in drinking-water at a level of 25 mg/litre were approximately 9 times higher than the levels in tissues of rats similarly exposed to trivalent chromium (11). Hence it is assumed that the absorption rate for waterborne hexavalent chromium is at least 9 times that for trivalent chromium, i.e., approximately 10 %, and elevated waterborne chromium is usually hexavalent. No quantitative information was found concerning the rates of absorption through the respiratory tract. Respiratory absorption would be expected to be dependent on particle size and solubility (2, 10). A reasonable estimate for the absorption of inhaled chromium is 50 % of that inhaled.

Chromium is distributed in human tissues in variable, low concentrations. Chromium levels in tissues other than the lungs decline with age (*1*). The largest stores of chromium in man are in skin, muscle, and fat; tissue levels are a function of sex, age, and geographical location (*3*). A homoeostatic mechanism, involving hepatic or intestinal transport mechanisms, prevents accumulation of excess trivalent chromium (*12*).

Chromium is excreted slowly, mainly in the urine but also in the faeces.

6.4 Health effects

Chromium appears to be necessary for glucose and lipid metabolism and for utilization of amino acids in several systems. It also appears to be important in the prevention of mild diabetes and atherosclerosis in humans (*1*). The harmful effects of waterborne chromium in man are associated with hexavalent chromium; trivalent chromium, which is regarded as a form of chromium essential to man, is considered practically non-toxic and no local or systemic effects appear to have been reported. People living in areas of the world where atherosclerosis is mild or virtually absent tend to have higher chromium levels in tissues than people from areas where the disease is endemic (*13*).

Hexavalent chromium at 10 mg/kg of body weight will result in liver necrosis, nephritis, and death in man; lower doses will cause irritation of the gastrointestinal mucosa (*14*).

Toxic effects have been observed in rats and rabbits when their drinking-water contained more than 5 mg of hexavalent chromium per litre (*3*), although in other studies up to 25 mg/litre produced no ill effects. A study of the effects of hexavalent chromium and cholesterol on the development of atherosclerosis in rabbits appears consistent with the hypothesis that chromium inhibits the development of experimentally induced atherosclerosis (*15*). Serum cholesterol levels, too, are higher in rats fed diets low in available chromium (*1*).

Hexavalent chromium in high doses has been implicated as the cause of digestive tract cancers in man (*3, 16*), and there is firm evidence that there is an increased risk of lung cancer for workers who are exposed to high levels of chromium (*3, 5*). Two studies of the chromate (VI) pigment industry suggest a risk of lung cancer similar to that seen in the production industry in which there is a large excess risk of the disease (*7*), the greatest risk occurring in workers involved with processing either dichromate(VI) or chromium(VI) trioxide. Prostate cancer and maxillary sinus cancers have been reported in workers in other chromium-using industries (chromium platers); however, the risk of cancers at sites other than the lung cannot be assessed on the basis of the current data. Exposure to a mixture of chromium(VI) compounds of different solubilities (as found in the chromate production industry) carries the greatest risk for human beings. Epidemiological data do not allow an evaluation of the relative contributions to carcinogenic risk of metallic chromium, chromium(III) and chromium(VI), or of soluble versus insoluble chromium compounds (7). Other health effects related to industrial exposures have been reported, e.g., hexavalent chromium can produce cutaneous and nasal mucous-membrane ulcers and dermatitis (from skin contact) (4). The threshold level of exposure to hexavalent chromium needed to produce health effects is very unclear. According to an IARC monograph: "There is no evidence that at current levels of non-occupational exposure to chromium a health hazard exists" (17).

REFERENCES

1. TOWILL, L. E. ET AL. Reviews of the environmental effects of pollutants, III: Chromium. Cincinnati, US Department of Commerce, National Technical Information Service, 1978 (PB-282-796).
2. NATIONAL RESEARCH COUNCIL. Chromium. Washington, DC, National Academy of Sciences, 1974.
3. Guidelines for Canadian drinking water quality, 1978. Quebec, Ministry of Supply and Services, 1979 (supporting documentation).
4. NATIONAL RESEARCH COUNCIL. Drinking water and health. Washington, DC, National Academy of Sciences, 1977.
5. COMMISSION OF THE EUROPEAN COMMUNITIES. Trace metals: exposure and health effects. Oxford, Pergamon Press, 1979.
6. KOPP, J. F. & KRONER, R. C. Trace metals in waters of the United States. Cincinnati, US Department of the Interior, 1967.
7. Some metals and metallic compounds. Lyon, International Agency for Research on Cancer, 1980 (IARC Monographs on the evaluation of the carcinogenic risk of chemicals to humans, vol. 23), p. 303.
8. TEHERANI, D. K. ET AL. Determination of heavy metals and selenium in fish from Upper Austrian waters. II. Lead, cadmium, scandium, chromium, cobalt, iron, zinc and selenium. Berichte der Oesterreichischen Studiengesellschaft für Atomenergie (1977) (SGAE No. 2797, pp. 1–21 (Chemical abstracts, 88, No. 49150e)).
9. UNDERWOOD, E. J. Trace elements in human and animal nutrition. New York, Academic Press, 1977.
10. FRIBERG, L. ET AL. Chromium. In: Handbook on the toxicology of metals, Amsterdam, Elsevier/North-Holland Biomedical Press, 1979.
11. MACKENZIE, R. D. ET AL. Chronic toxicity studies. II. Hexavalent and trivalent chromium administered in drinking water to rats. A.M.A. archives of industrial health, 18: 232 (1958).
12. SCHROEDER, H. A. ET AL. Abnormal trace metals in man – chromium. Journal of chronic diseases, 15: 941 (1962).
13. SCHROEDER, H. A. The role of chromium in mammalian nutrition. American journal of clinical nutrition, 21: 230 (1968).
14. KAUFMAN, D. B. ET AL. Acute potassium dichromate poisoning in man. American journal of diseases of children, 119: 374 (1970).
15. NOVAKOVA, S. ET AL. [The content of hexavalent chromium in water sources and the effect on the development of experimental arteriosclerosis in warm blooded animals.] Gigiena i sanitarija, 39(5): 78–80 (1974) (in Russian).
16. TELEKY, L. Krebs bei Chromarbeiten. Deutsche medizinische Wochenschrift, 62: 1353 (1936).
17. Some inorganic and organometallic compounds. Lyon, International Agency for Research on Cancer, 1973 (IARC Monographs on the evaluation of the carcinogenic risk of chemicals to humans, vol. 2), p. 100.

7. CYANIDE

7.1 General description

7.1.1 Sources

Cyanide is present wherever life and industry occur. Both inorganic and organic forms of cyanide exist; the latter are normally classified as nitriles. Cyanides form part of life processes, in particular as metabolic intermediates. The commonest forms of cyanide include hydrogen cyanide (hydrocyanic acid in solution), those cyanide salts that are readily soluble in water, and metallocyanide complexes (*1*). The cyanide ion can combine with heavy metal ions to form complexes, some of which are very stable (*2*). Cyanide salts hydrolyse to give hydrocyanic acid (*1*).

Cyanides are used in many industrial processes, e.g., in the production of acrylonitrile, adiponitrile, and methylmethacrylate. Cyanides are also used for extracting gold and silver, in the production of steel, for electroplating, and for the preparation of some intermediates in chemical synthesis (*2*). In some of these processes, contamination of the air and water can arise. The occasional use of cyanide for pest extermination may be a source of water contamination.

7.1.2 Occurrence in water

Hydrocyanic acid dissociates to give the cyanide ion in water (*1, 2*); its dissociation is pH dependent, with the ionic form predominating above pH 8.2 (*2*). Conversion of cyanide to the much less toxic cyanate will occur at pH levels of 8.5 and above (*3*). In general, the levels in raw water appear to be low (i.e., less than 0.1 mg/litre) except in the case of serious contamination, mainly by industrial discharges to river or other sources (*2*). Metal-treating industries, coke and gas manufacture, and a variety of chemical producers can be major sources of cyanide contamination of water (*4*). Chlorination of potable water to a free chlorine residual under neutral or alkaline conditions will reduce the concentration of cyanide in the finished water to very low levels (*3*). Chlorination of water (at pH > 8.5) converts cyanides to innocuous cyanates (*3, 5*), which can ultimately be decomposed to carbon dioxide and nitrogen gas.

7.2 Routes of exposure

7.2.1 Drinking-water

Comprehensive information on levels of cyanide in drinking-water is not available; in general, it appears that the concentrations are well below maximum acceptable levels.

7.2.2 Food

Most foods contain traces of cyanides. Some foods of plant origin contain elevated natural levels of cyanide (e.g., almonds); cyanides can be found in fish living in contaminated waters (2). Cyanides are decomposed on heating and, thus, cooked foods will tend to contain lower levels (6). Typical daily intakes of cyanide from food do not appear to be properly known; in general, dietary intake is regarded to be low. The acceptable daily intake of cyanide residues via ingestion of fumigated foods has been set at 0.05 mg/kg of body weight (7).

7.2.3 Air

Representative values do not appear to have been published, but it is generally considered that levels are extremely low.

7.2.4 Other routes of exposure

It has been reported that exposure to cyanide may be high in certain industrial situations (8). In such cases this may constitute a major exposure pathway.

7.2.5 Relative significance of different routes of exposure

As insufficient information is available to deduce precise exposure from food sources (the main natural source of cyanide other than drinking-water), it is not possible to estimate reliably the relative contributions made by water and food.

7.3 Metabolism

The cyanide ion is readily absorbed in animal species, and its highly poisonous effects are induced rapidly. Cyanide blocks oxidative processes in the cells of the carotid and aortic bodies, allowing anaerobic products, e.g., lactic acid, to accumulate in them, thus stimulating respiration. This reaction is due to the combination of cyanide with the catalytic iron group of cytochrome oxidase, preventing the donation of electrons to molecular oxygen. The absorption of oxygen by the cells is inhibited, i.e., cellular oxidation cannot proceed, and the main supply of energy to the cells ceases. Without oxidation of glucose, neurones convert glucose to lactic acid at a greatly increased rate. In animals the increase in the lactic acid content of the brain, even with small doses of cyanide, may cause coma and convulsions with irreversible cerebral damage, although the rest of the body is unharmed (9). Low exposures to cyanide are not fatal to human beings who have an efficient detoxification system whereby the cyanide is converted to the thiocyanate ion, which is non-toxic at low levels (2).

7.4 Health effects

A single dose of 50–60 mg for a human being is usually fatal (3). Exposures of 2.9–4.7 mg of cyanide per day are regarded as non-injurious to humans, owing to the highly efficient detoxification system in the human body in which the cyanide ion is converted to the relatively non-toxic thiocyanate ion through the rhodanese and thiosulfate enzyme system (1). Higher exposures may be fatal. On the basis of animal experiments, it has been calculated that an acceptable daily intake for man is 8.4 mg of cyanide (8).

REFERENCES

1. *Quality criteria for water*. Washington, DC, US Environmental Protection Agency, 1976 (EPA-440/9-76-023).
2. *Guidelines for Canadian drinking water quality, 1978*. Quebec, Ministry of Supply and Services, 1979 (supporting documentation).
3. *National interim primary drinking water regulations*. Washington, DC, US Environmental Protection Agency, 1976.
4. GOTTS, R. M. ET AL. *Treatment of industrial wastes at municipal water pollution control plants*. Ontario, Ontario Water Resources Commission, 1966 (Proceedings, Ontario Industrial Waste Conference), p. 151.
5. Cyanides. In: Kirk, R. E. & Othmer, D. F., ed. *Encyclopedia of chemical technology*, 2nd ed. New York, John Wiley and Sons, 1965. vol. 6, p. 574.
6. LEDUC, G. ET AL. The use of sodium cyanide as a fish eradicant in some Quebec lakes. *Naturaliste canadien*, **100**: 1 (1973).
7. *International standards for drinking water*, 3rd ed. Geneva, World Health Organization, 1971.
8. US Environmental Protection Agency. Water quality criteria; availability. *Federal register*, **44**: 43 667 (1979).
9. PASSMORE, R. & ROBSON, J. S. *A companion to medical studies*, vol. 2. Oxford, Blackwell Scientific Publications, 1970, pp. 15–25.

8. FLUORIDE

8.1 General description

8.1.1 Sources

Fluorine is a fairly common element, representing about 0.3 g/kg of the earth's crust (*1*). It exists in the form of fluorides in a number of minerals, of which fluorspar, cryolite, and fluorapatite are the commonest; many rocks contain fluoride minerals. Fluorides are used industrially in the production of aluminium and are commonly present in phosphate fertilizers, bricks, tiles, and ceramics; they are also used in metallurgy (*2*). Fluorides are now frequently added to certain pharmaceutical products, including toothpastes and vitamin supplements (*2*). Owing to industrial activity, involving the use of so many fluorine-containing substances, fluoride contamination of the environment is ubiquitous. Thus, plants, foodstuffs, and water all contain traces of fluoride.

8.1.2 Occurrence in water

Traces of fluorides occur in many waters and higher concentrations are often associated with underground sources. In areas that are rich in fluoride-containing minerals, e.g., fluorapatite, well-waters may contain up to about 10 mg of fluoride per litre or even more (*3, 4*). The highest natural level reported is 2800 mg/litre (*1*). Most waters contain below 1 mg of fluoride per litre (*1*). Occasionally, fluorides may enter a river as a result of industrial discharges.

8.2 Routes of exposure

8.2.1 Drinking-water

The levels of fluoride in tap-water are very similar to those found in the source water, except where fluoridation of the supply is practised. In general, unfluoridated supplies contain less than 1 mg of fluoride per litre, but, depending on the type and situation of the source, may very occasionally contain up to 10 mg/litre (*2*). In most parts of the world such sources have by now been identified. Where fluoridation of water supplies is practised, fluoride concentration is normally within the range 0.6–1.7 mg/litre, ambient air temperature usually being the deciding

100

factor (5). With a consumption of 2 litres of water per day, between 1.2 and 3.4 mg of fluoride per day could thus be ingested from drinking-water in those areas where fluoridation is practised. Elsewhere, the daily exposure will range from a fraction of a milligram to perhaps 20 mg in very exceptional circumstances.

8.2.2 Air

Fluorine-containing compounds are present in the air, mainly arising from industrial emissions. Concentrations in air will vary, depending on the type of industrial activity, but it has been estimated that the general exposure, equivalent to less than 1 $\mu g/m^3$ of air (2), is insignificant in comparison with ingested fluorine (6).

8.2.3 Food

Virtually all foodstuffs contain at least traces of the element. All vegetation contains some fluoride, which is absorbed from soil and water. Some foods may contain high levels, particularly fish, some vegetables, and tea (1, 2). For example, the fluorine content of some fish can be as high as 100 mg/kg, and tea may contain more than twice that concentration, in contrast to most other foodstuffs, which rarely exceed 10 mg/kg (1). The use of fluoridated water in food processing plants can often double the level of fluoride in prepared foodstuffs. Estimates have been made in various countries to ascertain the daily dietary intake of fluoride, which for adults ranges from 0.2 to 3.1 mg (1). For children, e.g., the 1–3 years age group in the USA, an intake of about 0.5 mg/day has been estimated (1).

8.2.4 Other routes of exposure

8.2.4.1 Industrial exposure

A number of industrial processes are known to release fluorine-containing compounds into the air of the work place, and there are numerous documented examples of human exposure, in particular in aluminium smelters and glass manufacture (1). Such exposure to fluorine under these conditions (i.e., concentrations of up to several milligrams per cubic metre of air) could provide the major contribution to the total exposure. However, with improvements in working conditions, this route of exposure becomes far less significant.

There is little information on the exposure to fluoride from tobacco smoke; however, it is not considered to be an important source of fluoride in comparison with the other routes of exposure.

8.2.4.2 *General population exposure*

Various products, such as toothpaste, tooth powders, mouth washes, chewing gum, vitamin supplements, and drugs, may contain added quantities of soluble fluoride, mainly in inorganic form. Such compounds are commonly added to dentifrices, typically at concentrations of about 1 g/kg (*1*). Studies have shown that significant quantities of fluoride can be absorbed by this route, and a possible absorption of about 50 µg of fluoride per "brushing" has been demonstrated (*1*). The use of topical applications of fluoride solutions can contribute to an increased absorption (*7*). Mouth washes can provide as much as 2 mg of fluoride (*1*). There is a range of different products that incorporate fluorides, including tablets containing sodium fluoride, used as an anticariogenic agent. Regular use of such tablets can provide up to about 1 mg of fluoride per day (*1*).

8.2.5 Relative significance of different routes of exposure

The following table shows the influence of fluoridated water on the total dietary intake in adults.

Food intake 1 mg of fluoride per day; drinking-water containing 0.5, 1.0, and 1.5 mg of fluoride per litre; water consumption 2 litres per day

Fluoride concentration in water	Weekly intake of fluoride (mg)				Total water contribution (%)
	Water only	Air only[a]	Food only	Total	
0.5 mg/litre	7.0	0.0	7.0	14.0	50
1.0 mg/litre	14.0	0.0	7.0	21.0	67
1.5 mg/litre	21.0	0.0	7.0	28.0	75

[a] Negligible contribution from ambient air (input by other routes of exposure not considered in these calculations).

Estimates of the *uptake* of fluoride from food and water are not provided here, since the efficiency of absorption is generally quite high (*8*) and figures quoted for *intake*, given above, would be fairly close to those relevant to absorption.

8.3 Metabolism

Fluoride ingested with water is almost completely absorbed (*1*); fluoride in the diet is not as fully absorbed as from water, but the absorption is still rather high, although in the case of certain foods (e.g., fish and some meats) only about 25 % of the fluorides may be absorbed (*8*).

Absorbed fluoride is distributed rapidly throughout the body. It is retained mainly in the skeleton, and a small proportion is retained in the teeth (*7*). The amount of fluoride in bone increases up to the age of 55 years (9). At high doses, fluoride can interfere with carbohydrate, lipid, protein, vitamin, enzyme, and mineral metabolism (1). Many of the symptoms of acute fluoride intoxication are the result of its binding effects with calcium (6). Fluoride is excreted primarily in the urine. The excretion is influenced by a number of factors, including the general health of the person and his previous history of fluoride exposure (2); the rate of retention decreases with age, and most adults can be regarded for practical purposes as "in balance" (6). Under this "steady state" condition, the fluoride present in the body is sequestered in calcified tissues; most of the remainder is present in plasma and thus available for excretion. Skeletal sequestration and renal excretion are the two major ways by which the body prevents the accumulation of toxic amounts of the fluoride ion.

8.4 Health effects

Fluorine has been fairly conclusively demonstrated to be an essential element for some animal species (7); in particular, fertility and growth rate are improved as a result of relatively small doses of fluorine (7).

Once fluoride is incorporated into teeth, it reduces the solubility of the enamel under acidic conditions and thereby provides protection against dental caries. There is good evidence to show that the presence of fluoride in water results in a substantial reduction of dental caries in both children and adults (2). The incidence of caries decreases as the concentration of fluoride increases to about 1 mg/litre, although mottling may sometimes occur even to an objectionable degree when the level rises to 1.5–2.0 mg/litre (6). Long-term consumption of water containing 1 mg of fluoride per litre may lead to such mottling in patients with long-standing renal disease or polydipsia (6), but only in persons in whom the teeth are under mineralization, i.e., children 0–7 years of age. Skeletal fluorosis has been observed in persons when water contains more than 3–6 mg of fluoride per litre depending on intake from other sources. Intakes of 20–40 mg of fluoride per day (or more where water contained in excess of 10 mg/litre (1)) over long periods have resulted in crippling skeletal fluorosis (6). It has been accepted that 1 mg/litre is a safe level in relation to the fluoridation of water supplies, and the recommended control limits in water (5, 10) are around this figure, the exact concentrations depending on the air temperature.

In high doses, fluoride is acutely toxic to man. Pathological changes include haemorrhagic gastroenteritis, acute toxic nephritis, and various degrees of injury to the liver and heart muscle (1). The acute lethal dose is about 5 g as sodium fluoride, i.e., about 2 g of fluoride (2). In animals, a variety of severe symptoms have been observed as a result of environmental exposure to fluoride in highly contaminated areas (7). Chronic effects from high exposure in man are primarily related to mottling of teeth and fluorosis, in which bone structure is affected, sometimes to a very alarming degree, producing serious crippling (1).

Chronic effects on kidneys have also been observed, generally in persons with renal disorders (*1*). Other less common problems, including effects on the thyroid, are known as a result of high exposure (*1*). Initial signs and symptoms of intoxication are vomiting, abdominal pain, nausea, diarrhoea, and even convulsions (*6*).

Epidemiological studies in areas where the concentration of fluoride in the water is naturally high have only rarely shown adverse effects (*6*); these have included mottling of teeth and skeletal fluorosis in areas of exceptionally high levels in water (*2*). In areas having optimal levels of fluoride in water, obvious effects (i.e., objectionable dental fluorosis) were observed in two children known to be suffering from diabetes insipidus (*6*).

It has been suggested that mongolism and cancer are associated with elevated levels of fluoride in water. The idea that mongolism could be related to fluoride exposure stems from one limited study where the prevalence of mongolism registered at institutions was recorded in relation to fluoride levels in the water (*11*); the study has been very severely criticized, however, by the Royal College of Physicians in the United Kingdom (*6*). For nearly 30 years, various epidemiological studies have been carried out to assess whether there is a link between cancer and fluoride in water. In a few cases, claims were made that there was a positive association, but these have been criticized (*2*). It is now generally considered that there is no acceptable evidence whatever that fluoride in water is carcinogenic to human beings (*2, 6, 12, 13*).

The question of sensitivity to fluoride has been raised (*6*). In general, claims that there are some people who are sensitive to fluoride have been dismissed. For example, no particular sensitivity has been recorded in the millions of tea drinkers who would ingest considerable quantities of fluoride from the infusion of tea leaf. However, the possibility of sensitivity or some idiosyncratic reaction cannot be completely dismissed (*6*), although from the evidence available the incidence of such cases would be expected to be small. Suggestions that fluoride is mutagenic or teratogenic or that it is related to birth defects have been thoroughly reviewed and have not proved justified (*6*).

REFERENCES

1. *Fluorides and human health*. Geneva, World Health Organization, 1970 (Monograph Series, No. 59).
2. *Guidelines for Canadian drinking water quality, 1978*. Quebec, Ministry of Supply and Services, 1979 (supporting documentation).
3. *Fluorides*. Washington, DC, National Academy of Sciences, 1971.
4. BULUSU, K. R. ET AL. Fluorides in water, defluoridation methods and their limitations. *Journal of the Institution of Engineers (India)*, **60** (1979).
5. *International standards for drinking water*. Geneva, World Health Organization, 1971.
6. NATIONAL RESEARCH COUNCIL. *Drinking water and health*. Washington, DC, National Academy of Sciences, 1977.

7. UNDERWOOD, E. J. *Trace elements in human and animal nutrition.* New York, Academic Press, 1977.
8. NEWBURN, E. & ZIPKIN, I. *Fluoride metabolism. Fluoride and dental caries.* Springfield, IL, Charles Thomas, 1976.
9. JACKSON, D. & WEIDMANN, S. M. Fluorine in human bone related to age and the water supply of different regions. *Journal of pathology and bacteriology,* **76**: 451 (1958).
10. *European standards for drinking water.* Geneva, World Health Organization, 1970.
11. RAPAPORT, I. Nouvelles recherches sur le mongolisme. A propos du rôle pathogénique du fluor. *Bulletin de l'Académie nationale de Médecine,* **143**: 367 (1959).
12. *Some aromatic amines, anthraquinones and nitroso compounds, and inorganic fluorides used in drinking-water and dental preparations.* Lyon, International Agency for Research on Cancer, 1982 (IARC Monographs on the evaluation of the carcinogenic risk of chemicals to humans, vol. 27).
13. CLEMMESEN, J. The alleged association between artificial fluoridation of water supplies and cancer: a review. *Bulletin of the World Health Organization,* **61**: 871–883 (1983).

9. HARDNESS[a]

9.1 General description

Hardness of water is not a specific constituent but is a variable and complex mixture of cations and anions. Hardness is predominantly due to calcium and magnesium, although strontium, barium, and other polyvalent ions contribute. Hardness is commonly expressed as mg of calcium carbonate equivalent per litre, and this is the unit adopted throughout this document. Several other units are used in various countries. Traditionally, hardness is a measure of the capacity of water to react with soap. It is often divided into carbonate (temporary) and non-carbonate (permanent) types of hardness.

9.1.1 Sources

Calcium and magnesium are common elements present in many minerals. Among the commonest sources of calcium and magnesium in water are limestones, including chalk (calcium carbonate). Calcium and magnesium are present in a great number of industrial products and they are common constituents of food.

A minor contribution to the total hardness of water is made by such polyvalent ions as zinc, manganese, aluminium, strontium, barium, and iron, dissolved from minerals such as sphalerite, armangite, bauxite, strontianite, witherite, and phosphosiderite.

9.1.2 Occurrence in water

Although most calcium compounds are not easily soluble in pure water, the presence of carbon dioxide readily increases their solubility, and sources of water containing up to 100 mg of calcium per litre are fairly common (*1–4*); sources containing over 200 mg of calcium per litre are rare (*1–4*). Many salts containing magnesium are easily soluble and water sources containing levels of magnesium at concentrations up to 10 mg/litre are common (*1–4*). Water sources rarely contain more than 100 mg/litre (*1–4*), and calcium hardness usually predominates.

The buffering capacity of water, normally described as alkalinity, is closely associated with water hardness. Thus, such anions as hydroxide,

---

[a] A discussion of some other aspects of the effect of hardness on water quality will be found in Part V, section 5, p. 264.

bicarbonate, and carbonate have a significant influence, and so to a lesser degree do phosphate and silicate; molecular species of weak acids also contribute.

9.2 Routes of exposure

9.2.1 Drinking-water

Although water is sometimes artificially softened at the water-treatment plant, generally the raw water hardness is similar to that found in drinking-water piped to a household. Hardness may range from less than 10 mg/litre to over 500 mg/litre (*4*); water sources may have a hardness of less than 50 mg/litre; values above 500 mg/litre are relatively uncommon in most countries (*2, 3, 5*).

9.2.2 Food

Virtually all foods contain calcium and magnesium. Typical diets provide about 1000 mg of calcium per day (*6*) and 200–400 mg of magnesium (*1, 7*). The predominant source of ingested calcium and magnesium is normally food. Dairy products are a particularly rich source of calcium (*2*); magnesium tends to be associated more with meat and foodstuffs of plant origin (*6*).

9.2.3 Air, occupational exposure, and cigarette smoking

Although all these provide a route of exposure to man, the contribution in relation to food is trivial.

9.2.4 Relative significance of different routes of exposure

Only food and water provide important routes of exposure and therefore only these sources are considered. Some estimates of possible situations are given below. All the estimates relate to adults.

(a) Food intake: 1000 mg of calcium per day

Concentration of calcium in water	Weekly intake of calcium (mg)			Water contribution (%)
	Water only	Food only	Total	
25 mg/litre	350	7000	7350	5
100 mg/litre	1400	7000	8400	17
200 mg/litre	2800	7000	9800	29

Typical water contribution in relation to total input of calcium is about 5–20%.

(b) Food intake: 200 mg of magnesium per day

Concentration of magnesium in water	Weekly intake of magnesium (mg)			Water contribution (%)
	Water only	Food only	Total	
10 mg/litre	140	1400	1540	9
50 mg/litre	700	1400	2100	33
100 mg/litre	1400	1400	2800	50

(c) Food intake: 400 mg of magnesium per day

Concentration of magnesium in water	Weekly intake of magnesium (mg)			Water contribution (%)
	Water only	Food only	Total	
10 mg/litre	140	2800	2940	5
50 mg/litre	700	2800	3500	20
100 mg/litre	1400	2800	4200	33

Typical contribution from water in relation to the total input of magnesium is about 5–20%.

Although some information on dietary intake is available for children, these contributions have not been calculated since it can be estimated that the relative proportions of the different routes of exposure in children are roughly similar to those in adults.

Separate calculations have not been made for the uptake of calcium and magnesium. Although about 30% and 35% respectively of calcium and magnesium appears to be absorbed from the diet, little reliable information is available on the uptake of calcium from tap-water, and therefore only crude estimates of uptake would be possible.

9.3 Health effects

There is some suggestive evidence that drinking extremely hard water might lead to an increased incidence of urolithiasis. This has been suggested to account for urolithiasis in a small human population in the USSR where the local tap-water contained 300–500 mg of calcium per litre (8); it has also been demonstrated in animals drinking extremely hard water (200–400 mg of calcium per litre), when they had been subjected to a high ambient temperature of 30 °C (9). The occurrence of drinking-water containing as much as 500 mg of calcium per litre must, however, be very rare.

Thus, there appears to be no firm evidence that water hardness causes ill effects in man (2). Conversely there have been a number of studies, where the results suggest that water hardness protects against disease.

9.3.1 Water hardness and cardiovascular disease

In 1957 in Japan, it was demonstrated that there was a close association between death rates from strokes and the acidity of river-derived drinking-water (4). Since that time, a number of studies in various parts of the world have demonstrated that there is a highly statistically significant negative association between water hardness and cardiovascular disease (3, 10). In most studies, the calcium concentration has shown the strongest correlation, but the magnesium content of the water has been indicated as the most significant correlating factor in some Canadian studies (4, 7). Some small-scale studies, however, have not confirmed the relationship (11–13). Uncertainty remains about the magnitude of a possible "water effect" and the extent to which confounding factors might account for it, e.g., factors such as air temperature, rainfall, latitude, longitude, socioeconomic factors, town type, and air pollution. Nevertheless, a strong statistical association can be demonstrated when allowance is made for a number of confounding variables (2, 14). A special international scientific colloquium on this particular subject was held in 1975 (3). In a recent retrospective large-scale study (14) of 253 towns in Great Britain, after allowing for climatic conditions and certain social factors, mortality from stroke and ischaemic heart disease was found to be strongly related to water hardness, but only up to about 170 mg/litre as $CaCO_3$. Of many water factors analysed, the correlation with hardness and the calcium content of water was high, although other water parameters, many of which are intercorrelated with hardness, also provided strong statistical associations. Such retrospective studies are limited in what they can achieve and research is continuing; particular hope is placed on prospective studies of cardiovascular risk factors in selected population groups to assess the significance of a wide range of water parameters (14).

Several hypotheses have been proposed in an attempt to account for the relationship, but there is at present no definite evidence that hardness, or its major constituents calcium and magnesium, are involved. The two most quoted hypotheses relate to (a) a constituent (or constituents) in hard water being protective in some way and (b) a substance (or substances) in soft water (e.g., metals leached from piping materials) promoting the disease. In the case of the protective hypothesis it is often considered that the diet provides an adequate supply of calcium and magnesium, although for magnesium there is the possibility of a dietary deficiency in some situations (7). However, the presence of other elements, e.g., lithium, chromium, vanadium, and silicon, could have a protective role (15, 16). Lead and cadmium, which can be leached from plumbing material, have been suggested as possibly promoting the disease, but there is no firm evidence that they are involved in this way (3).

9.3.2 Water hardness and other diseases

The results of several studies have suggested that a variety of other diseases are correlated with the hardness of water. These include certain nervous system defects, anencephaly, perinatal mortality, and various types of cancer (2, 17–20). Although some of these findings have been demonstrated in different countries (2), there is still considerable doubt about their significance. These associations may merely reflect disease patterns that can be explained by social, climatological, and various environmental factors, rather than by the hardness of water.

REFERENCES

1. NATIONAL RESEARCH COUNCIL. *Drinking water and health.* Washington, DC, National Academy of Sciences, 1977.
2. *Guidelines for Canadian drinking water quality.* Quebec, Ministry of Supply and Services, 1979 (supporting documentation).
3. AMURIS, R. ET AL., ed. *Hardness of drinking water and public health.* Oxford, Pergamon Press, 1975 (Scientific colloquium, Luxembourg, 1975).
4. MARIER, J. R. ET AL. *Water hardness, human health, and the importance of magnesium.* Ottawa, Canada, National Research Council, 1979.
5. *Quality criteria for water.* Washington, DC, US Environmental Protection Agency, 1976.
6. WHO Technical Report Series, No. 532, 1973 (*Trace elements in human nutrition*: report of a WHO Expert Committee).
7. NERI, L. C. & JOHANSEN, H. L. Water hardness and cardiovascular mortality. *Annals of the New York Academy of Sciences*, **304**: 203 (1978).
8. BOKINA, A. I. ET AL. [Hygienic assessment of drinking water hardness as a factor favouring the development of urolithiasis.] *Gigiena i sanitarija*, **30**(6): 3 (1965) (in Russian).
9. BOKINA, A. I. & YURIEVA, V. K. [Shifts of certain biochemical indices in persons after long-term use of hard drinking water.] *Gigiena i sanitarija*, **31**(12): 33 (1966) (in Russian).
10. KOBAYASHI, J. On the geographical relationship between the chemical nature of river water and death-rate from apoplexy. *Berichte des Ohara Instituts für Landwirtschaftliche Biologie*, **2**: 12 (1957).
11. ALLWRIGHT, S. P. A. ET AL. Mortality and water hardness in three matched communities in Los Angeles. *Lancet*, **2**: 860 (1974).
12. BIERENBAUM, M. L. ET AL. Possible toxic water factor in coronary heart disease. *Lancet*, **1**: 1008 (1975).
13. MEYERS, D. Ischaemic heart disease and the water factor. A variable relationship. *British journal of preventive and social medicine*, **29**: 98 (1975).
14. POCOCK, S. J. ET AL. British regional heart study: geographic variations in cardiovascular mortality, and the role of water quality. *British medical journal*, **280**: 1243 (1980).
15. VOORS, A. W. Lithium in the drinking water and atherosclerotic heart death: epidemiological argument for a protective effect. *American journal of epidemiology*, **92**: 164 (1970).
16. SCHWARTZ, K. Silicon, fibre, and atherosclerosis. *Lancet*, **1**: 454 (1977).
17. STOCKS, P. Incidence of congenital malformations in the regions of England and Wales. *British journal of preventive medicine*, **24**: 67 (1970).
18. HART, J. T. The distribution of mortality from coronary heart disease in South Wales. *Journal of the Royal College of General Practitioners*, **19**: 258 (1970).
19. FEDRICK, J. Anencephalus and the local water supply. *Nature*, **227**: 177 (1970).
20. LOWE, C. R. ET AL. Malformations of the central nervous system and softness of local water supplies. *British medical journal*, **2**: 357 (1971).

10. LEAD

10.1 General description

10.1.1 Sources

Lead is a natural constituent of the earth's crust at an average concentration of about 16 mg/kg (*1*). It is present in a number of minerals, the principal one being galena (lead sulfide); most countries have lead deposits of one sort or another. Lead has been widely used for many centuries, and in many places some contamination of the environment has occurred as a result of the mining and smelting processes used or from the use of products made from it. Consequently it is present in air, food, water, soil, dust, and snow. Lead in the environment exists almost entirely in the inorganic form, but small amounts of organic lead result from the use of leaded gasoline and from natural alkylation processes that produce methyl lead compounds (*2*). Lead is used widely for a variety of purposes, including the manufacture of acid accumulators, alkyl lead compounds for gasoline, solder, pigments, ammunition, caulking, and cable sheathing. Its use as roofing materials and piping materials, including pipes used for potable water, is currently being discontinued and discouraged.

10.1.2 Occurrence of lead in water

The natural lead content of lake and river water worldwide has been estimated to be 1–10 μg/litre (*1, 3*). Although higher values have been recorded where contamination has occurred, particularly from industrial sources, such situations are relatively rare, since there are a number of natural mechanisms that control the levels. The concentrations in finished water (i.e., water after treatment) prior to its distribution are generally lower than in source waters since lead is partially removed by most conventional water-treatment processes (*4*). The levels in drinking-water, however, can be much higher owing to the use of lead service pipes running from the street to a dwelling, from lead plumbing, and/or lead-lined storage tanks (*4, 5*). Particularly high lead levels can result when the water is aggressive, soft, or has a low pH. These conditions tend to produce the highest levels (*1, 4, 6*). Lead pipes have not been used extensively throughout the world, but in some towns and cities in certain countries lead is still being widely used. This occasionally results in undesirably high levels of lead in the tap-water (*1, 4*).

111

10.2 Routes of exposure

10.2.1 Drinking-water

In most countries, the levels of lead in domestic tap-water are relatively low, i.e., normally well below 10–20 μg/litre. In some places, however, they may be rather high. For example, in Scotland, the water is extremely soft, has a low pH, and lead plumbing and lead-lined storage tanks are common. As a result, it has been estimated that in over 10 % of houses in Scotland (population about 5 million), the first-draw water contains more than 300 μg of lead per litre (7). Values in excess of 2000 μg/litre have been recorded in some places in the world (8). It is very difficult, however, to define precisely the average exposure in terms of the concentration of lead in the water because of the very wide variation of levels produced at the tap. These levels depend critically on factors such as the stagnation time of the water in a lead service pipe or in household plumbing. Even in the same water supply area, there may be considerable home-to-home variations in the level of lead because of differences in the length of pipe, water-use patterns, and types of deposit that have built up.

Based on a water consumption of 2 litres per day, calculations show that the daily intake of lead from water varies from 10 to 20 μg to 1 mg or more. Such estimates are based on the assumption that all lead is consumed. It has been shown, however, that in the preparation of certain beverages, e.g., tea (9), not all of the lead in the water will appear in the beverages prepared. On the other hand, tap-water is also used for cooking and food preparation, which provides an additional opportunity for ingesting lead from domestic tap-water.

10.2.2 Food

Lead is present in a wide variety of foodstuffs. The amounts vary depending on the type of food. For instance, canned foods (3, 10, 11) tend to contain the highest levels if lead solders have been used in the manufacture of the can. Many fresh vegetables, cereals, and fruits contain small quantities of lead as a result of some limited absorption of the metal from the soil in which they are grown, and because of the deposition on surfaces from lead in the air. Lead is also present in milk and dairy products and in wine (10).

Because of the large differences in individual diets, precise calculations of lead intake are not possible. Estimates of typical daily intakes range from less than 100 to over 500 μg of lead (1, 5, 8, 10, 12); the worldwide average for adults is about 200 μg/day. Estimates of levels of lead in the diet appear to have been falling in recent years (8, 10). Women generally eat less than men and their intake of lead in food is consequently lower. It has been estimated that children aged 1–5 years ingest about 90 μg of lead per day (10). Additional lead in food can arise from contamination by cooking vessels, such as pots that have soldered joints and some glazed earthenware utensils. Some lead comes from tap-water used for preparing food. In general, the major source of ingested lead is food.

10.2.3 Air

In rural areas, average levels of lead of 0.1 μg/m^3 of air can be found (3, 10, 12). Average city levels are typically in the range 0.5–2 μg/m^3 (8, 10). The levels in any one particular area will depend on the type and the extent of the emission sources (e.g., traffic, industry) and the natural dispersion conditions of the area (prevailing weather conditions). At present, most of the airborne lead in non-industrialized cities comes from motor vehicle traffic. In general, people living near busy highways will be most exposed. In some industrialized areas, average ambient air levels as high as 6 μg/m^3 have been recorded (3, 8). Deposited lead from the air contaminates soil, and levels of 2 g/kg have been reported (8); in highly contaminated soils, levels of over 10 g/kg have been recorded (8).

Most of the lead in the air is in the form of fine particles. When these particles are inhaled, only 20–60 % will be deposited in the respiratory system (8). On the basis of a daily respired volume of air of 15–22.8 m^3 (1, 8) a typical daily intake for an urban dweller (exposed to 1 μg of lead per cubic metre of air and with 40 % retention) would be 6–9 μg. Much of this retained lead will ultimately be absorbed.

10.2.4 Occupational exposure

Levels of airborne lead in industrial work areas can be much higher than in the general environment. Levels of up to 100 μg/m^3 of air are not uncommon (8).

10.2.5 Smoking

Small quantities of lead are found in tobacco, but in general the exposure from this source is relatively small.

10.2.6 Ingestion of soil and dust and chewing of paint

Soil, dust, and particularly household paints (especially old paint) contain elevated levels of lead. Because of the "hand-to-mouth" activity (pica) of young children, they may be subject to significant exposure from such sources (1, 3–8). Despite a host of studies analysing the relative contribution of soil, dust, and paint to lead exposure in children, the precise contributions from these sources remain uncertain (6).

10.2.7 Relative significance of different routes of exposure

For different individuals and population groups the exposure to lead from water, food, air, etc., can vary significantly. Since the relative contributions of each of these sources can also vary widely it is not possible to provide comprehensive information for a wide range of circumstances. A range of some environmental situations has, however, recently been considered and numerous estimates provided (*10*). To give some idea of possible situations, a few simplified examples are given here. The last two columns of the tables give estimates of the relative contribution of water to the total intake of lead and to the uptake of lead by the body. An important group omitted in the following tables is that of infants (up to one year old). The available information about lead in their diet and its absorption is insufficient to allow reasonable estimates to be made. It is likely, however, that the contribution of daily lead via drinking-water in this group is as high, or perhaps higher, than that for the 1–5-year-old children, which is given in the tables. No account is taken in the calculations of the contributions made by cigarette smoking, occupational exposure, or various other sources.

10.2.7.1 *Estimates of weekly intake and uptake of lead by the body in adults*

The tabular examples below assume that an average person consumes 2 litres of water per day, and breathes air at the rate of 20 m^3/day (*10*). Lead absorption from individual sources is assumed not to be influenced by the uptake from other sources; absorption from food and water is assumed to be 10% of the intake and there is assumed to be complete absorption of the 40% retained from inhalation.

(a) *Daily intake of 100 μg of lead per day in food; 1.0 μg of lead per m^3 of air*

Lead concentration in water	Weekly intake of lead (mg)				Intake ratio: water/total (%)	Uptake ratio: water/total (%)
	Water only	Air only	Food only	Total		
20 μg/litre	0.28	0.14	0.70	1.12	25	18
50 μg/litre	0.70	0.14	0.70	1.54	47	35
100 μg/litre	1.40	0.14	0.70	2.24	63	52

(b) *Daily intake of 300 μg of lead per day in food; 1.0 μg of lead per m^3 of air*

Lead concentration in water	Weekly intake of lead (mg)				Intake ratio: water/total (%)	Uptake ratio: water/total (%)
	Water only	Air only	Food only	Total		
20 μg/litre	0.28	0.14	2.10	2.52	11	10
50 μg/litre	0.70	0.14	2.10	2.94	24	20
100 μg/litre	1.40	0.14	2.10	3.64	40	34

10.2.7.2 *Estimates of weekly intake and uptake of lead by the body in children aged 1–5 years*

The example below assumes that water is consumed at the rate of 1 litre per day and air is breathed at the rate of 4.7 m³/day (*10*). The basic assumptions for lead absorption are the same as for adults except that the absorption from food and water is 50% of the intake (*4, 8*).

Daily intake of 93 µg of lead per day in food; 1.0 µg of lead per m³ of air

Lead concentration in water	Weekly intake of lead (mg)				Intake ratio: water/total (%)	Uptake ratio: water/total (%)
	Water only	Air only	Food only	Total		
20 µg/litre	0.14	0.03	0.65	0.82	18	17
50 µg/litre	0.35	0.03	0.65	1.03	35	35
100 µg/litre	0.70	0.03	0.65	1.38	51	51

10.3 Metabolism

An important consideration is the proportion of the lead in drinking-water that is actually absorbed when ingested. Although little is known about the absorption of fine particles of lead present in tap-water, some information exists on the intestinal uptake of lead from aqueous solutions containing dissolved lead. Generally a figure of about 10% (*4, 5, 8*) is regarded as typical of the fraction absorbed for an adult,[a] but this value depends on whether the water is consumed on a full or an empty stomach. For example, humans who fast for 6 hours before and after an oral dose of lead ions have markedly increased absorption (e.g., 50% or more) (*13, 14*). This finding has been confirmed in mice (*15*). Other factors influence the absorption of lead from the gastrointestinal tract, such as the presence of elements such as calcium, phosphorus, iron, copper, and zinc in the diet, and the age and physical state of the subject (*1, 6, 8, 11*).

Pulmonary absorption depends on the size of lead particles and on the depth and rate of breathing (*5, 8*). Some large particles are deposited on the mucous lining of the respiratory tract, and some are ultimately swallowed (*8, 10*). Pulmonary retention is typically 40% (*5, 8, 10*).

Absorbed lead enters the blood and is distributed to soft tissues and bone. After prolonged exposure an equilibrium is reached between the blood and soft tissues. In contrast, bone has the ability to accumulate lead with time. Postmortem studies indicate that the skeletal burden of lead increases with age; in fact, about 90% of the total body burden lies

---

[a] For children, the absorption seems to be higher; values up to 50% have been regarded as typical for children under 5 years of age (*4, 8*).

in the bone (8, 11). The respective half-lives of lead in blood, soft tissues, and bone have been estimated to be 2–4 weeks (5, 8, 14), 4 weeks (14), and 27.5 years (10).

Lead passes through the placenta easily and fetal blood has almost the same lead concentration as maternal blood. Lead also passes the blood-brain barrier, although the brain does not accumulate lead (16).

Relationships between lead intake and blood lead level have been extensively studied and reviewed (6). It appears that there is a curvilinear relationship between air lead and blood lead; as lead exposure increases the corresponding blood lead increments become smaller (17). The interrelationship between lead exposure from water and blood level is also curvilinear, and this has been demonstrated in human beings drinking household tap-water where lead was present in high concentrations, i.e., exceeding 50 µg/litre (18, 19).

It should be stressed, however, that at the present time there is considerable uncertainty regarding the precise relationship between low-level exposure to lead and levels in the blood; consequently, great care needs to be exercised when using data derived from such relationships.

Lead is excreted in urine, faeces, sweat, hair, fingernails, and toenails. The metabolism of lead has been reviewed in two WHO publications (8, 16).

10.4 Health effects

Lead in high doses has been recognized for centuries as a cumulative general metabolic poison. Some of the symptoms of acute poisoning are tiredness, lassitude, slight abdominal discomfort, irritability, anaemia and, in the case of children, behavioural changes (8). Such symptoms are difficult to quantify and currently there is considerable interest in various possible subtle effects, including neurophysiological ones, caused perhaps by exposure to low levels of lead (5, 11). Lead at low levels can reduce the activity of an enzyme, porphobilinogen synthase (EC 4.2.1.24) (8). This enzyme is involved in normal haeme synthesis at the stage of conversion of aminolevulinic acid to porphobilinogen; a decrease in the activity of this enzyme may be used as an index of exposure to lead. Lead also has an affinity for amino acids containing sulfur. In addition, lead has a tendency to bind to mitochondria, leading to interference in the regulation of oxygen transport and energy generation (10). Significantly higher lead blood levels (> 400 µg/litre) have been found in mentally retarded children (20).

Many animal studies have been carried out on the effects of lead on the haemopoietic, nervous, renal, cardiovascular, and reproductive systems (8, 10, 11). While the results of these studies cannot be directly extrapolated to man, they do provide valuable dose-response data that are not available from epidemiological human studies.

A wide range of epidemiological and clinical studies have been conducted on human subjects, including retrospective studies to ascertain the possible causes of mortality and morbidity in lead-exposed populations and studies on the effects on specific organs and systems (8). Of particular relevance are the studies of levels of lead in water in relation to levels of lead found in the blood of both adults and children, and possible associated subtle behavioural effects in children (6). Many of these studies have shown that there is a small increase in lead in the blood where the levels in water are relatively high, but in general the blood levels attributable to water are not high in relation to acceptable levels of blood lead for individuals and population groups (6, 8). For example, for adults a mean increase of 25 μg of lead per litre of blood has been estimated to be associated with drinking-water containing an average concentration of 100 μg of lead per litre (6). Increments of about 40 and 50 μg of lead per litre of blood have been estimated for young children and pregnant women, respectively (6, 10). The interpretation of such figures must be considered with care, however, owing to the curvilinear relationship between lead intake and lead in the blood.

Population group mean exposure limits for lead from all environmental sources combined has been specified as 200 μg per litre of blood, but for individuals values are within the range 300–350 μg/litre. For individual children, the Centers for Disease Control in the USA and also the American Academy of Pediatrics (6, 10) have recommended an exposure limit value of 300 μg per litre of blood. Lower lead values may need to be considered in the future and 250–300 μg litre has been proposed for individuals (6, 10). Typical values for the lead levels in the general population are not known very accurately, but for non-occupationally exposed adults figures of below 200 μg per litre of blood are often quoted. In the USA (6) it has been estimated that 99.5% of children would have <300 μg of lead per litre of blood if the geometric mean level was maintained at 150 μg/litre.

Many of the studies on behavioural effects in children relate to situations where there are high levels of lead in the environment generally rather than from the lead in water specifically. In a few instances there is a tenuous suggestion of some association between adverse effects and levels of lead in water (8, 10).

In several studies, chromosomal aberrations were found in peripheral lymphocytes of lead-exposed populations whose blood lead levels ranged from 100 to 1000 μg/litre. Negative results were obtained in other studies in which blood lead levels ranged from 40 to 500 μg/litre (21).

Lead is not known to be essential for the functioning of biological systems and the general view is that where possible the exposure to lead should be kept as low as possible. Although in 1972, an FAO/WHO Expert Committee recommended that the maximum intake of lead should be 3 mg/week (0.05 mg/kg of body weight) for adults (12), no corresponding value was suggested for children. The situation will be different for children (including infants), because absorption of lead is higher than for adults (*12*); children also have a higher susceptibility, due in part to their rapid growth rate (*4*). Pregnant women and developing fetuses also appear to be more sensitive to lead because of increased maternal food intake and changes in hormonal status (*8*). Should the intake in the diet be greater than 220 μg of lead per day (using the assumptions in the tables on page 114), the 3-mg weekly limit for adults would be exceeded.

The health effects of lead have been reviewed by three WHO groups (*8, 16, 22*).

REFERENCES

1. *Guidelines for Canadian drinking water quality, 1978*. Quebec, Ministry of Supply and Services, 1979 (supporting documentation).
2. HARRISON, R. M. & LAXEN, D. P. H. Natural source of tetra-alkyl lead in air. *Nature*, **275**: 738 (1978).
3. *The hazards to health and ecological effects of persistent substances in the environment*: report of a working group. Copenhagen, WHO Regional Office for Europe, 1973.
4. NATIONAL RESEARCH COUNCIL. *Drinking water and health*. Washington, DC, National Academy of Sciences, 1977.
5. *Toxicology of water, vol. II*. Washington, DC, US Environmental Protection Agency, 1977 (Environmental Health Effects Research Series).
6. US Environmental Protection Agency. *Ambient water quality criteria for lead*. Washington, DC, Criteria and Standards Division, Office of Water Planning and Standards, 1980 (EPA 440/5-80-057).
7. *Lead in drinking water, a survey in Great Britain*. London, Department of the Environment, 1977 (Pollution paper No. 12).
8. *Lead*. Geneva, World Health Organization, 1977 (Environmental Health Criteria 3).
9. ZOETEMAN, B. C. J. & BRINKMANN, F. J. J. In: Amavis, R. et al., ed., *Hardness of drinking water and public health*, Oxford, Pergamon Press, 1975.
10. DRILL, S. ET AL. *The environmental lead problem. An assessment of lead in drinking water from a multi-media perspective*. Washington, DC, US Environmental Protection Agency, 1979.
11. UNDERWOOD, E. J. *Trace elements ih human and animal nutrition*. New York, Academic Press, 1977.
12. WHO Technical Report Series, No. 505, 1972 (*Evaluation of certain food additives and the contaminants: mercury, lead and cadmium*).
13. CHAMBERLAIN, A. C. ET AL. *Investigations into lead from motor vehicles*. Harwell, Oxfordshire, Atomic Energy Research Establishment, Environmental and Medical Sciences Division, 1978 (AERE-R9198).
14. RABINOWITZ, M. ET AL. Studies of human lead metabolism by use of stable isotope tracers. *Environmental health perspectives*, **7**: 145 (1974).
15. GARBER, B. T. & WEI, E. Influence of dietary factors on the gastrointestinal absorption of lead. *Toxicology and applied pharmacology*, **27**: 685 (1974).
16. WHO Technical Report Series, No. 647, 1980 (*Recommended health-based limits in occupational exposure to heavy metals*).
17. HAMMOND, P. B. & BELILES, R. P. Metals: lead. In: Doull, J. et al., ed., *Casarett and Doull's toxicology: the basic science of poisons*, 2nd ed. New York, Macmillan, 1980.
18. MOORE, M. R. ET AL. Contribution of lead in water to blood-lead. *Lancet*, **2**: 661 (1977).
19. THOMAS, H. F. ET AL. Relationship of blood lead in women and children to domestic water lead. *Nature*, **282**: 712 (1979).

20. MOORE, M. R. ET AL. A retrospective analysis of blood-lead in mentally retarded children. *Lancet*, 1: 717 (1977).
21. *Some metals and metallic compounds*. Lyon, International Agency for Research on Cancer, 1980 (IARC Monographs on the evaluation of the carcinogenic risk of chemicals to humans, vol. 23).
22. *Health hazards from drinking-water*: report of a Working Group. Copenhagen, WHO Regional Office for Europe, 1977 (ICP/PPE 005).

11. MERCURY

11.1 General description

11.1.1 Sources

The major source of mercury in the environment is the natural degassing of the earth's crust, quantities ranging between 25 000 and 150 000 tonnes of mercury per year being released. In addition, many industrial activities not directly related to mercury production or use contribute significant amounts of this element to the environment; these include burning of fossil fuels, smelting of various metals, cement manufacture, and waste disposal. In addition, mercury is used in chloralkali plants (producing chlorine and sodium hydroxide), in paints as preservatives or pigments, in electrical switching equipment and batteries, in measuring and control equipment (e.g., thermometers, medical equipment), in dentistry, and in agriculture (especially as seed dressings). Mercury can exist in the environment as the metal, as monovalent and divalent salts, and as organomercurials, the most important of which is methyl mercury. Methyl mercury may be produced from inorganic mercury by microorganisms found in aquatic sediments and sewage sludge; other microorganisms can demethylate mercury back to inorganic mercury. Fish and mammals absorb and retain methyl mercury to a greater extent than inorganic mercury; it is methyl mercury that accumulates along food chains (1).

11.1.2 Occurrence in water

Rainwater in Sweden is reported to contain approximately 300 ng of mercury per litre (1).

In most surface-waters, mercuric hydroxide and chloride are the predominant mercury species; levels are generally less than 0.001 mg/litre (2–5). In polluted rivers and lakes, levels of up to 0.03 mg/litre have been reported (6). Inland waters in the Federal Republic of Germany contain mercury at concentrations of about 400 ng/litre whereas values ranging between 100 and 1800 ng/litre were found in rivers (7).

Levels of mercury in drinking-water are usually very low (8). For example, median levels in several Canadian provinces are approximately 0.0002 mg/litre (9). Inorganic mercury can be controlled in water treatment by iron and alum coagulation (10). Seven hundred water samples collected from the drinking-water reservoirs in the Federal Republic of Germany indicate that the purest drinking-water contained less than 0.00003 mg/litre (*11*).

11.2 Routes of exposure

11.2.1 Drinking-water

Where there is no evidence of mercury contamination, levels of mercury in freshwater bodies are less than 0.0002 mg/litre. Levels of mercury in ambient waters may be significantly reduced by conventional treatment processes and further losses may occur during the preparation of beverages. Therefore, the estimate that the intake of mercury from drinking-water would not normally exceed 0.1 μg daily (*1*) may be on the high side (*4*).

11.2.2 Food

Food is the main source of mercury in non-occupationally exposed populations and fish and fish products account for most of the methyl mercury in food (*1*). The average daily intake of mercury from food is in the range 10–12 μg (*1, 12, 13*) but in regions where ambient waters have become contaminated with mercury and where fish comprises a high proportion of the diet, the intake from food may be much higher.

11.2.3 Air

Because metallic mercury and organomercurial compounds have relatively high vapour pressure, they can be released to the atmosphere by volatilization. Ambient air levels, except in polluted areas, appear to be of the order of 0.02 μg/m^3. Assuming an ambient air level of 0.05 μg/m^3, the average daily intake of metallic mercury vapour would amount to about 1 μg/day, of which about 80% is retained (*1*).

11.2.4 Industrial exposure

A large number of occupations or trades involve exposure to mercury (*14*). Especially significant are the mining industry, the chloralkali industry, and the manufacture of some scientific instruments; mercury levels in air may attain values as high as 5 mg/m^3 (*1*). Accepting the time-weighted average threshold limit value of 0.05 mg/m^3 proposed by the American Conference of Government Industrial Hygienists (ACGIH), the occupational exposure would lead to a daily average intake of 500 μg of mercury or less, assuming a pulmonary ventilation of 10 m^3/day at work (*1*).

11.2.5 Relative significance of different routes of exposure

For adults the relative significance of mercury in drinking-water is illustrated in the following table:

	Weekly intake (uptake) of mercury (μg)			
Food	Air	Water	Total	Ratio: water/total (%)
(a) 70 (5.6)	2.8 (2.2)	14 (2.1)	87 (9.9)	16 (21.0)
(b) 140 (11.2)	2.8 (2.2)	14 (2.1)	157 (15.5)	9 (13.6)
(c) 70 (5.6)	7.0 (5.6)	14 (2.1)	91 (13.3)	15 (15.8)
(d) 140 (11.2)	7.0 (5.6)	14 (2.1)	161 (18.9)	9 (11.1)

Assumptions: (1) Food intake 10 μg/day (a, c) or 20 μg/day (b, d) with 8% of ingested mercury absorbed.
(2) Air containing 0.02 μg/m³ (a, b) or 0.05 μg/m³ (c, d) with total ventilation of 20.0 m³/day and 80% retention.
(3) Drinking-water containing 0.001 mg/litre with intake of 2 litres per day and 15% of ingested mercury absorbed.

No reliable information is available for children.

11.3 Metabolism

Mercury serves no beneficial physiological function in man. The various physical and chemical states (metal, inorganic compounds, organomercurial compounds) each have intrinsic properties that dictate independent toxicological assessment (14).

Absorption of inorganic mercury compounds from food is about 7–8% of the ingested dose; in contrast, gastrointestinal absorption of methyl mercury is practically complete. Absorption of inorganic mercury compounds from water may be 15% or less (15) whereas the methyl mercury is almost completely absorbed.

Inorganic mercury compounds are rapidly accumulated in the kidney, the main target organ for these compounds (15). Absorbed methyl mercury rapidly appears in the blood where, in man, 80–90% is bound to red cells; demethylation of methyl mercury to inorganic mercury occurs at a slow but significant rate. The greater intrinsic toxicity of methyl mercury compared with inorganic mercury is due to its lipid solubility, which permits it to cross biological membranes more easily than inorganic mercury, especially into brain, spinal cord, and peripheral nerves, and across the placenta.

Mercury salts are excreted from the kidney, liver, intestinal mucosa, sweat glands, salivary glands, and through milk; the most important routes are via the urine and faeces (15).

11.4 Health effects

The major effects of mercury poisoning take the form of neurological and renal disturbances, which are primarily associated with organic and inorganic mercury compounds respectively (*16*). According to Krasovsky (*17*), in addition to producing general toxic effects, mercury causes gonadotoxic and mutagenic effects and disturbs the cholesterol metabolism. The toxicology of mercury compounds has been reviewed in depth in the WHO *Environmental health criteria* series (*1*). It was not possible to identify even approximate minimum effect values for inorganic, aryl, and alkoxyalkylmercurials. There exists no evidence that inorganic mercury is carcinogenic. Alkylmercurials are embryotoxic and teratogenic in laboratory animals (*9*).

REFERENCES

1. *Mercury*. Geneva, World Health Organization, 1976 (Environmental Health Criteria 1).
2. HOLDEN, A. V. Present levels of mercury in man and his environment. In: *Mercury contamination in man and his environment*, Vienna, International Atomic Energy Agency, 1972, p. 143 (Technical Report Series No. 137).
3. WIKLANDER, L. Mercury in ground and river water. *Grundfoerbaettring*, **21**: 151 (1968).
4. WERSHAW, R. L. Sources and behaviour of mercury in surface water. In: *Mercury in the environment*, Washington, DC, US Geological Survey, 1970 (Professional Paper No. 713).
5. VOEGE, F. A. Levels of mercury contamination in water and its boundaries. In: *Proceedings of the symposium on mercury in man's environment*, Ottawa, Royal Society of Canada, 1971, p. 107.
6. *Investigations of mercury in the St. Clair River – Lake Erie systems*. Washington, DC, US Department of the Interior, 1970 (Report of the Federal Water Quality Administration), p. 108.
7. SCHRAMEL P. ET AL. Some determinations of Hg, As, Se, Sb, Sn and Br in water, plants, sediments and fishes in Bavarian rivers. *International journal of environmental studies*, **5**: 37 (1973).
8. NATIONAL RESEARCH COUNCIL. *Drinking water and health*. Washington, DC, National Academy of Sciences, 1977.
9. *Guidelines for Canadian drinking water quality, 1978*. Quebec, Ministry of Supply and Services, 1979 (supporting documentation).
10. *Guidance for the issuance of variances and exemptions*. Washington, DC, Office of Drinking Water, US Environmental Protection Agency, 1979.
11. BOUQUIAUX, J. In: *Proceedings of an international symposium on the problems of contamination of man and his environment by mercury and cadmium*. Luxembourg, Commission of the European Communities, 1974, p. 23.
12. MERANGER, J. C. & SMITH, D. C. The heavy metal content of a typical Canadian diet. *Canadian journal of public health*, **63**: 53 (1972).
13. NEILSEN-KUDSK, F. Absorption of mercury vapour from the respiratory tract in man. *Acta pharmacologica et toxicologica*, **23**: 250 (1965).
14. KEY, M. M. ET AL., ed. *Occupational diseases–a guide to their recognition*. Washington, DC, US Department of Health, Education and Welfare, 1977 (NIOSH publication 77–181), pp. 370–373.
15. WHO Technical Report Series, No. 647, 1980 (*Recommended health-based limits in occupational exposure to heavy metals*).
16. SWEDISH EXPERT GROUP. Methyl mercury in fish. A toxicologic-epidemiologic evaluation of risks. *Nordisk hygienisk Tidskrift*, Suppl. 4 (1971).
17. KRASOVSKY, G. N. ET AL. [The need for revising the existing hygienic standard for mercury in water.] *Gigiena i sanitarija* (2): 20 (1981) (in Russian).

12. NICKEL

12.1 General description

12.1.1 Sources

Nickel is ubiquitous; typical soils contain between 10 and 100 mg of nickel per kg (*1*). The chief ores are mainly arsenides and sulfides. The processing of minerals as well as the production and use of nickel has caused environmental contamination. Nickel is used as a component in some alloys and for metal plating, for catalysts, for batteries, and in certain fungicides; the use of the metal in food processing equipment can give rise to some contamination of food.

12.1.2 Occurrence in water

Many nickel salts are water-soluble, therefore contamination of water can arise; significant problems are associated with industrial discharge to rivers of effluents containing nickel compounds. Levels as high as 1 mg/litre have been reported in surface-waters (*1*), although the levels are generally much lower, e.g., 5–20 μg/litre (*2*). In the USSR, underground water supplies have been found to contain nickel at levels up to 0.13 mg/litre (*3*).

12.2 Routes of exposure

12.2.1 Drinking-water

Some nickel is removed by conventional water treatment, so that the levels in treated water are generally lower than in untreated water (*2*). Few comprehensive surveys of nickel levels in tap-water have been identified. Limited data suggest that concentrations of 2–5 μg/litre are fairly typical (*1–5*); elevated levels may sometimes be found (*6*), especially where nickel-plated plumbing fittings are used. Exceptionally, levels in drinking-water up to 0.5 mg/litre have been reported (*4, 7*). Assuming a consumption of 2 litres of water per day, human exposure to nickel from this source would not normally exceed 10–20 μg/day.

12.2.2 Food

Nickel is present in most foodstuffs, but at levels below (and often well below) 1 mg/kg (*8*). Little is known about the chemical form of nickel in food, although it is probably partly complexed with phytic acid (8). Dietary contributions have been reported ranging from less than 200 to 900 μg/day (1, 2, 8–10). A typical diet might contribute about 400 μg/day. Nickel concentrations of 100 μg/litre and 50 μg/litre have been reported in wines and beers respectively (1).

12.2.3 Air

Few data have been reported concerning the nickel content of air. However, it would appear that levels in air are generally less than 0.5 $\mu g/m^3$. Higher levels were reported in the past, mainly associated with industrial areas (1, 2). A typical urban air concentration is 0.2 $\mu g/m^3$.

12.2.4 Other routes of exposure

12.2.4.1 Industrial exposure

Levels up to 400 μg of nickel per m^3 of air have been reported in the industrial environment (1), although generally exposure levels in industry are much lower. In some cases the major route of exposure of a worker could be his industrial environment.

12.2.4.2 Smoking

It has been reported that about 10–20% of the nickel content of cigarettes (typically around 3 μg per cigarette) can be inhaled (1, 2). This appears to be mainly as a volatile nickel compound, nickel carbonyl (2). A typical weekly intake for someone who smokes 20 cigarettes a day might be 40–80 μg of nickel.

12.2.5 Relative significance of different routes of exposure

Estimates of weekly human exposure to nickel for adults are given below:

(a) Weekly intake from food, air, and water

Nickel concentration in water	Weekly intake of nickel (μg)				Ratio: water/total (%)
	Water only	Air only	Food only	Total	
50 μg/litre	700	28	3150	3878	18.0
75 μg/litre	1050	28	3150	4228	25.0
100 μg/litre	1400	28	3150	4578	30.6
150 μg/litre	2100	28	3150	5278	40.0

Assumptions: Daily nickel intake from food, 450 μg; ventilation 20 m^3 of air daily, nickel content, 0.2 $\mu g/m^3$.

(b) Weekly absorption from food, air, and water

Nickel concentration in water	Weekly absorption of nickel (μg)				Ratio: water/total (%)
	Water only	Air only	Food only	Total	
50 μg/litre	7.0	14	31.5	52.5	13.3
75 μg/litre	10.5	14	31.5	56.0	18.8
100 μg/litre	14.0	14	31.5	59.5	23.5
150 μg/litre	21.0	14	31.5	66.5	31.6

Assumptions: Inhaled nickel 50% absorption; ingested nickel 1% absorption.

12.3 Metabolism

Nickel is almost certainly essential for animal nutrition, and consequentially it is probably essential to man (*11*). Absorption of nickel through the gastrointestinal tract seems to be very low, i.e., 1% or even less (*1*), although higher absorption values have been reported, i.e., 10% (*8*). There is little evidence of accumulation of nickel by various tissues (*12*). No significant accumulation of nickel was observed in rats fed nickel in drinking-water at a concentration of 5 mg/litre. It is clear that at least in the animal body a mechanism controls excessive intake of nickel.

Certain disease states in man do give rise to elevated nickel in tissue; the reasons, however, are not understood (*8*).

Nickel is readily excreted, mainly in the faeces, with smaller quantities in the urine. Significant amounts can also be discharged in sweat (*8*).

12.4 Health effects

Nickel is a relatively nontoxic element. The levels of nickel usually found in food and water are not considered a serious health hazard (*1, 8*); however, high doses (1600 mg/kg in the diet) were shown in early animal studies to cause minimal toxic effects (decreased number of mice pups weaned) (*8*). Such effects were not substantiated in later three-generation reproduction studies. Rats and mice were given drinking-water containing 5 mg of nickel per litre throughout their lifetime without adverse effects (*13*).

Certain nickel compounds have been shown to be carcinogenic in animal experiments (*1, 6*). However, soluble nickel compounds are not currently regarded as either human or animal carcinogens (*2*). As in the case of other divalent cations, nickel can react with DNA, and at high concentrations can result in DNA damage as shown in *in vitro* mutagenicity tests (IARC, personal communication).

Dermatitis is most commonly associated with industrial exposure. However, the same effects have been observed from dermal contact with coinage or jewellery. High-level occupational exposures have been associated with renal problems, and effects such as vertigo and dyspnoea have been observed (*1*).

Studies using large numbers of patients have investigated the role of contact dermatitis in eczema of the hands; between 4% and 9% of the patients were found to respond positively to nickel patch tests (*14*). Women appear to be more sensitive to nickel than men by a factor of ten (*15*). These studies are of limited value in that they examined eczema patients and cannot fully reflect the true incidence of sensitization or contact dermatitis in the general population (*16*).

REFERENCES

1. COMMISSION OF THE EUROPEAN COMMUNITIES. *Trace metals: exposure and health effects.* Oxford, Pergamon Press, 1979.
2. NATIONAL RESEARCH COUNCIL. *Drinking water and health.* Washington, DC, National Academy of Sciences, 1977.
3. SIDORENKO, G. I. & ITSKOVA, A. I. *Nickel.* Moscow, Medicina, 1980.
4. DURFOR, C. N. & BECKER, E. *Public water supplies of the 100 largest cities in the United States, 1962.* Washington, DC, Government Printing Office, 1964 (Geological Survey Water Supply Paper 1812).
5. KOPP, J. F. & KRONER, R. C. *Trace metals in waters of the United States.* Cincinnati, US Department of the Interior, 1967.
6. *Cadmium, nickel, some epoxides, miscellaneous industrial chemicals and general considerations on volatile anaesthetics.* Lyon, International Agency for Research on Cancer, 1976 (IARC Monographs on the evaluation of the carcinogenic risk of chemicals to humans, vol. 11).
7. KOPP, J. F. The occurrence of trace elements in water. In: Hemphill, D. P., ed. *Proceedings of the Third Annual Conference on Trade Substances in Environmental Health, 1969,* Columbia, University of Missouri, 1970, pp. 59–73.
8. UNDERWOOD, E. J. *Trace elements in human and animal nutrition.* New York, Academic Press, 1977.
9. HAMILTON, E. I. & MINSKI, M. J. Abundance of the chemical elements in man's diet and possible relations with environmental factors. *Science of the total environment,* **1**: 375 (1973).
10. MASIRONI, R. How trace elements in water contribute to health. *WHO chronicle,* **32**: 382 (1978).
11. WHO Technical Report Series, No. 532, 1973 (*Trace elements in human nutrition*: report of a WHO Expert Committee).
12. ENVIRONMENTAL PROTECTION AGENCY. Water quality criteria; availability. *Federal register,* **44**: 43684 (1979).
13. SCHROEDER, H. A. ET AL. Long-term effects of nickel in rats: survival, tumors, interactions with trace elements and tissue levels. *Journal of nutrition,* **104**: 239 (1974).
14. WILKINSON, D. S. ET AL. The role of contact allergy in hand eczema. *Transactions of the St. John's Hospital Dermatological Society,* **56**: 19–25 (1970).
15. FISHER, A. A. & SHAPIRO, A. Allergic eczematous contact dermatitis due to metallic nickel. *Journal of the American Medical Association,* **161**: 717–721 (1956).
16. KAALTER, K. ET AL. Low nickel diet in the treatment of patients with chronic nickel dermatitis. *British journal of dermatology,* **98**: 197–201 (1978).

13. NITRATE AND NITRITE

Nitrate and nitrite are considered together because conversion from one form to the other occurs in the environment. The health effects of nitrate are generally a consequence of its ready conversion to nitrite in the body.

Concentrations in water are expressed as mg/litre for nitrate-nitrogen (nitrate-N) and nitrite-nitrogen (nitrite-N).

13.1 General description

13.1.1 Sources

Nitrates are widely present in substantial quantities in soil, in most waters, and in plants, including vegetables (1). Nitrites also occur fairly widely, but generally at very much lower levels than nitrates (1). Nitrates are products of oxidation of organic nitrogen by the bacteria present in soils and in water where sufficient oxygen is present. Nitrites are formed by incomplete bacterial oxidation of organic nitrogen (2). One of the principal uses of nitrate is as a fertilizer; most other nitrogen-containing fertilizers will, however, be converted to nitrate in the soil (2). Nitrates are also used in explosives, as oxidizing agents in the chemical industry, and as food preservatives (2). The main use of nitrites is as food preservatives, generally as the sodium or the potassium salt (1). Some nitrates in the environment are produced in the soil by fixation of atmospheric nitrogen (bacterial synthesis). Some nitrates and nitrites are formed when oxides of nitrogen produced by the action of lightning discharge or via man-made sources are washed out by rain (2). Nitrates and some nitrites are also produced in the soil as a result of bacterial decomposition of organic material, both vegetable and animal.

Because nitrates and nitrites are widespread in the environment, they are found in most foods, in the atmosphere, and in many water sources.

13.1.2 Occurrence in water

Fertilizer use, decayed vegetable and animal matter, domestic effluents, sewage sludge disposal to land, industrial discharges, leachates from refuse dumps, and atmospheric washout all contribute to these ions in water sources (1, 3, 4). Changes in land use may also give rise to increased nitrate levels. Depending on the situation, these sources can contaminate streams, rivers, lakes, and groundwater, especially wells (3).

Contamination may result from a direct or indirect discharge, or it may arise by percolation over a period of time, sometimes after many years. The levels of nitrates in polluted water are almost invariably very much higher than the levels of nitrites (4).

Levels of nitrate in water are typically below 5 mg of nitrate-N per litre, but levels exceeding 10 mg/litre occur in some small water sources. In chlorinated supplies, levels of nitrite are often less than the limit of detection, i.e., < 0.005 mg of nitrite-N per litre (2–4), but relatively high levels may occur in unchlorinated water. Very high nitrite levels are usually associated with water of unsatisfactory microbiological quality.

A number of studies have revealed levels of nitrate in the range of 20 to over 200 mg of nitrate-N per litre (3), but this is rare. Most of the higher levels of nitrate are found in groundwater (3); nitrates in surface-waters tend to get depleted by aquatic plants (4). Increases in the levels of nitrates in water are associated with the application of nitrogen fertilizers. Marked seasonal variations can occur in concentration in rivers and high levels may occur, particularly after heavy rainfall following severe drought periods (5). The levels in groundwater tend to be much more steady during the year.

13.2 Routes of exposure

13.2.1 Drinking-water

Because none of the conventional water treatment and disinfection practices modify the levels of nitrate to any appreciable extent, and since nitrate concentration is not changed markedly in water distribution systems, the levels in tap-water are often very similar to those for source waters. For nitrite, the levels in tap-water are likely to be markedly lower than in source waters because of oxidation during water treatment, particularly when water is chlorinated (2). It is very difficult to define a range and average exposure to nitrate or nitrite in water because the concentrations vary widely depending on the water source. For the majority of the world's population the exposure is likely to be well below 5 mg of nitrate-N per litre (3). For small populations, often in remote areas, the levels may be up to 100 mg or more of nitrate-nitrogen per litre. Assuming a consumption of 2 litres of water per day, the exposure is typically under 20 mg of nitrate-N per day, but in rare circumstances it could be over five times this value (3).

13.2.2 Food

Considerable quantities of nitrates and lesser amounts of nitrites are present in certain foods and in general the major source of human intake of both nitrates and nitrites is food (4). In certain crops the nitrate levels may be as high as 100 mg/kg. Certain vegetables, including cabbage, celery, lettuce, potatoes, several root vegetables, and spinach, contain relatively high levels of nitrate, but only small quantities of nitrite (*4*). Nitrates and nitrites are added as preservatives to certain foods, particularly for certain meats and cheeses (*2*).

A further important source of ingested nitrates and nitrites is saliva (*4*); human beings secrete about 10 mg of nitrate-N per day, of which about 2 mg/day is reduced to nitrite (*4*); the nitrate in the saliva is essentially derived from food sources (*6*), mainly vegetables.

The variation in the quantities of nitrates and nitrites ingested from the diet is extremely high (*4*). For example, individuals who eat few vegetables and little cured meat will have a very much lower intake (*4*). Various typical dietary inputs have been estimated as ranging from about 120 mg to 230–300 mg of nitrate per day (no estimate for nitrites) (*4, 7, 8*). As infants are the most sensitive group of the population to nitrate, it is important to define their exposure; a daily intake from food alone for a two-month-old infant has been estimated to be approximately 25 mg of nitrate-N (*2*).

13.2.3 Air

Apart from natural sources of oxides of nitrogen and nitrates in the air, there are some important man-made sources, in particular the products of combustion of fossil fuels (coal, oil, gas) and from the chemical industry. Inorganic and organic forms of nitrates occur; these and the oxides of nitrogen will be inhaled and some will be absorbed by the respiratory system, producing a mixture of nitrates and nitrites in the body. It has been estimated, for areas with elevated levels of nitrogen compounds in the air, that if all these compounds were absorbed by an adult, it would amount to an intake of approximately 0.1 mg of nitrate-N per day.

13.2.4 Other routes of exposure

13.2.4.1 *Industrial exposure*

Oxides of nitrogen are fairly common industrially, and although aerosols of nitrates and nitrites may occur, their levels would generally be low. A maximum of 5 parts of nitrogen dioxide in 10^6 parts of air is permitted in the USA in industrial situations for an eight-hour shift (*9*); this could represent an exposure of up to about 30 mg of nitrate-N per day in extreme circumstances.

13.2.4.2 *Smoking*

Oxides of nitrogen and nitrate aerosols are produced in a burning cigarette, but in comparison with food and water, the quantities contributed by smoking tobacco are regarded as relatively small (*2*).

13.2.5 Relative significance of different routes of exposure

For different individuals, there can be a wide range of possible exposures to nitrates and nitrites in water, food, and air. General air pollution appears to be a relatively unimportant source, making food, saliva, and water the main normal sources (2). Two levels of food nitrate are considered and, ignoring contributions from saliva, the proportions of nitrate from water containing 10 mg of nitrate-N per litre have been calculated.

The estimates are based on 100% absorption from food and water, with a consumption of 2 litres of water per day for adults and 1 litre of water per day for children. Negligible intakes from inhalation, cigarette smoking, occupational exposure, and other sources are assumed.

	Daily intake of nitrate-N in food (mg)	Weekly uptake of nitrate-N (mg)			Ratio: water/total (%)
		Water only[a]	Food only	Total	
Adults—case 1:	20	140	140	280	50
Adults—case 2:	70	140	490	630	22
Children—case 3:	25	70	175	245	28

[a] 10 mg nitrate-N per litre.

13.3 Metabolism

The metabolism of ingested nitrate is not fully understood; it seems that absorption takes place in the upper portion of the small intestine and that excretion is primarily, if not exclusively, through the kidney (4). It is well known that nitrate is absorbed in the upper gastrointestinal tract and concentrated ultimately into saliva by the salivary glands (4). The nitrate metabolism in man has not been studied fully, and the results of animal experiments are not very reliable when extrapolated to man (4). Both nitrates and nitrites are very readily absorbed by the body.

A very important consideration is the fact that nitrate can be readily converted *in vivo* to nitrite as a result of bacterial reduction (4). Exposure to high levels of nitrate has been shown to give rise to large increases in the concentration of salivary nitrite (4). However, there can be marked variations between individuals; this can be a function of differences in their oral microflora and various constituents in their diet (4). Reduction of nitrate to nitrite also occurs elsewhere in the body, including the stomach; little conversion takes place unless the pH is greater than 4.6 (4). In infants, where the stomach acidity is normally very low, about pH 4 or higher (3, 4), a high yield of nitrite is obtained. In contrast, the acidity of an adult stomach is pH 1–5, and less conversion of nitrate occurs (4).

The formation of nitrite is especially important for two reasons. Firstly, it can oxidize haemoglobin to methaemoglobin, a pigment that is incapable of acting as a carrier of oxygen. Secondly, under certain conditions, nitrites may react in the human body with secondary and tertiary amines and amides (commonly derived from food and other sources) to form nitrosamines, some of which are considered to be carcinogenic (1). This process occurs in acidic solution within the pH range 1–5 (4) characteristic of the normal acidity range in a human stomach. The reaction rate is greatest at pH 3.5 or less.

13.4 Health effects

13.4.1 Methaemoglobinaemia

Normally, 1–2% of the body's haemoglobin is in the methaemoglobin form, but when the proportion is in excess of 10% (3), clinical effects are detectable (methaemoglobinaemia); 30–40% leads to anoxia.

It has been well documented that, in some countries, water supplies containing high levels of nitrate have been responsible for cases of infantile methaemoglobinaemia and death (2, 10). The extent of the worldwide problem has been reviewed in a WHO document.[a] It has been recommended that water supplies containing high levels of nitrate (> 100 mg of NO_3 per litre) should not be used for the preparation of infant foods; alternative supplies having a low nitrate content, even to the extent of using bottled water, have been recommended (5). The susceptibility of infants to nitrate has been attributed to their high intake relative to body weight (3), to the presence of nitrate-reducing bacteria in the upper gastrointestinal tract, and to the greater ease of oxidation of fetal haemoglobin (present in this form for the first few months of life) (11). The problem of methaemoglobinaemia does not arise in adults. Increased sensitivity may also occur when infants suffer from gastrointestinal disturbances, which increase the numbers of bacteria that can convert nitrate to nitrite (3, 12). The reconstitution of powdered milk, as opposed to other forms of milk, has also been regarded as increasing the sensitivity to the nitrate content in water. The stomach pH in infants, being about neutral, enables bacterial growth to occur in both the stomach and upper intestine. Infants, in contrast to adults, are also deficient in two specific enzymes that can convert methaemoglobin back to haemoglobin (3).

The most common cause of infantile methaemoglobinaemia is excessive levels of nitrate in water used for the reconstitution of baby food (3). Prolonged boiling of water may exacerbate the problem by increasing the nitrate levels owing to evaporation. The vast majority of

---

[a] *Cyanosis of infants produced by high nitrate concentration in rural wells.* WHO Expert Committee on Maternal and Child Health, 1949 (Unpublished document WHO/MCH/13:19).

cases of infantile methaemoglobinaemia have been associated with the use as a source of water of private wells that were microbiologically contaminated (3).

A large number of studies have been carried out on the levels of nitrate in water giving rise to methaemoglobinaemia, but there are conflicting conclusions regarding the threshold level for an effect (3, 13). Cases of infantile methaemoglobinaemia have not been reported in areas where the drinking-water consistently contains less than 10 mg of nitrate-N per litre (2). Many infants have consumed much higher levels than this without developing the disease (2); only 2.3 % of all cases appear to be associated with nitrate levels of between 10 and 20 mg of nitrate-N per litre of water (1). There is therefore some doubt about the effects at these concentrations. However, although clinical manifestations of infantile methaemoglobinaemia may not be apparent at these levels, undesirable increases in levels of methaemoglobin in the blood do occur (2).

There is also a suggestion that pregnant women are at greater risk than the general adult population (1), but further work is needed to confirm this.

Although methaemoglobinaemia is well recognized and is unlikely to be a problem in areas with adequate medical facilities, it may be more important in the developing areas where such facilities are lacking.

13.4.2 Carcinogenicity of nitrosamines

Since ingested nitrates can be readily converted to nitrites, either in the mouth or elsewhere in the body where the acidity is relatively low (high pH), it is possible that nitrosamines, some of which may be carcinogenic, will be produced. It has been shown that the formation of nitrosamines may be increased in individuals with bladder infections and people suffering from achlorohydria (a condition of low stomach acidity) (14). In the case of bladder infections, it is probable that the nitrosamines produced there would be absorbed into the blood (2).

Although tests on animals have shown that a number of nitrosamines are carcinogenic, there is no direct evidence of their carcinogenicity in man (1, 4, 15). There have been several such studies, but the overall evidence that nitrate in water might relate to cancer remains inconclusive (4, 16). Evidence of carcinogenicity from nitrate via the formation of nitrosamines rests with epidemiological studies, there being no appropriate animal studies that relate to humans. In a review of gastric cancer in China (17), the Putian Prefecture of the Fujian Province was found to have the highest mortality from this disease (120–147 per 100 000 males). Data showed that in this area the levels of both nitrate and nitrite in drinking-water and in vegetables were higher than in the low-risk areas. A study of the epidemiology and etiology of the disease is proceeding in both high- and low-risk areas.

REFERENCES

1. *Nitrates, nitrites and N-nitroso compounds.* Geneva, World Health Organization, 1978 (Environmental Health Criteria 5).
2. *Guidelines for Canadian drinking water quality, 1978.* Quebec, Ministry of Supply and Services, 1980 (supporting documentation).
3. Nitrates in water supplies. Report by the International Standing Committee on Water Quality and Treatment. *Aqua,* 1: 5–24 (1974).
4. NATIONAL RESEARCH COUNCIL. *Drinking water and health.* Washington, DC, National Academy of Sciences, 1977.
5. *Royal Commission on Environmental Pollution.* London, HM Stationery Office, 1979, Chapter 4.
6. TANNENBAUM, S. R. ET AL. Nitrite in human saliva. Its possible relationship to nitrosamine formation. *Journal of the National Cancer Institute,* **53**: 79 (1974).
7. WHITE, J. W. Relative significance of dietary sources of nitrate and nitrite. *Journal of agricultural and food chemistry,* **23**: 886 (1975).
8. PHILLIPS, W. E. J. Change in nitrate and nitrite content of fresh and processed spinach during storage. *Journal of agricultural and food chemistry,* **16**: 88 (1968).
9. *TLVs-Threshold limit values for chemical substances and physical agents in the workroom environment with intended changes for 1976.* Cincinnati, American Conference of Governmental Industrial Hygienists, 1976.
10. *Health effects of nitrates in water.* Cincinnati, US Environmental Protection Agency, 1977 (EPA-600/1–77–030).
11. BETKE, K. ET AL. Vergleichende Untersuchungen über die Spontanoxydation von Naberschnur- und Erwachsenenhämoglobin. *Zeitschrift für Kinderheilkunde,* **77**: 549 (1956).
12. SHUVAL, H. I. & GRUENER, N. Epidemiological and toxicological aspects of nitrates and nitrites in the environment. *American journal of public health,* **62**: 1045 (1972).
13. FRAZER, P. & CHILVERS, C. *Health aspects of nitrate in drinking water.* Netherlands, 1980 (paper presented at the International Symposium on Water Supply and Health).
14. HILL, M. J. ET AL. Bacteria, nitrosamines and cancer of the stomach. *British journal of cancer,* **28**: 562 (1973).
15. *Some N-nitroso compounds.* Lyon, International Agency for Research on Cancer, 1978 (IARC Monographs on the evaluation of the carcinogenic risk of chemicals to humans, Vol. 17).
16. FRAZER, P. ET AL. Nitrate and human cancer: a review of the evidence. *International journal of epidemiology,* **9**: 3 (1980).
17. XU GUANG-WEI. Gastric cancer in China: a review. *Journal of the Royal Society of Medicine,* **74**: 210 (1981).

14. SELENIUM

14.1 General description

14.1.1 Sources

As a result of geochemical differences, levels of selenium in soil and vegetation vary within broad limits (1). The chemical form of selenium, and thus its solubility, is another decisive factor both as regards its entry into the food chain and its presence in water. Environmental processes can decrease selenium solubility, converting soluble selenate compounds, or some selenites, into compounds of very low solubility, such as elemental selenium or selenides (or even selenites of certain metals) (2). Selenium is usually present in water as selenite or selenate, the chemical form being influenced by such factors as pH and by the presence of salts of certain metals such as iron (2).

14.1.2 Occurrence in water

Several reviews (1–7) of data from different parts of the world indicated that the selenium content in most surface-water samples analysed was well below 10 µg/litre. Only 2 of 535 samples from the major watersheds of the USA studied over a four-year period exceeded this value, 14 µg/litre being the maximum level found. Non-seleniferous regions of the USSR reported values ranging in general from a few tenths to several µg per litre, the maximum reported value being 5.1 µg/litre. The selenium level in 22 surface-waters in Argentina ranged from less than 2 to 19 µg/litre, with a median value of 3 µg/litre.

One study analysing 42 samples of surface-water from Colorado, USA, reported values ranging from less than 1 µg/litre to 400 µg/litre, with a median value of 1 µg/litre. Waters draining in the USSR Ural Mountain areas nearest to pyrite deposits were also found to contain similar high values. Irrigation drainage from seleniferous soils increases selenium levels in surface (river) waters. Water from some springs and shallow wells contains selenium at levels exceeding 100 µg/litre; levels as high as 330 µg/litre have been reported from some wells in a seleniferous area of South Dakota, USA (2,5).

14.2 Routes of exposure

With the exception of occupational exposure, where the exposure through air and dermal contact is of particular significance, the general population is exposed to selenium mainly through food.

14.2.1 Drinking-water

The level of selenium in tap-water samples from various public water supply systems in Canada and the USA, and some village supplies in Australia and the Federal Republic of Germany, only exceptionally exceeded 0.01 mg/litre *(2–8)*. Higher selenium levels are found in seleniferous areas, particularly in well-water. Systematic studies on selenium in drinking-water in high selenium areas of the world are scarce.

Thus, in countries like the USA or Canada, the contribution made by selenium in drinking-water would usually not exceed 5–10 % of the daily dietary intake and would generally, in most localities, be much lower than this value. Drinking-water in general does not represent the only or the main source of selenium exposure for the resident population in seleniferous areas.

14.2.2 Food

Dietary intake of selenium depends on food consumption patterns and selenium levels in foodstuffs, the latter being determined mainly by the character of the foodstuff and by geochemical conditions.

Vegetables and fruits generally represent a poor dietary source of selenium, in contrast to grain, grain products, meat (particularly internal organ meat), and seafood, which contain substantial selenium levels, usually well above 0.2 mg/kg on a wet weight basis. The chemical composition of the soil and its selenium content have a marked influence on the selenium content in grain from different countries, ranging from 0.04 mg/kg to 21 mg/kg *(4, 6–9)*.

Recently reviewed data from studies on daily dietary intakes of selenium in different countries range from 56 µg/day in New Zealand (low selenium region) to over 320 µg/day in Venezuela, a country with very high selenium levels in soils and vegetation *(7, 9)*. On the other hand, daily intakes of selenium as low as 20 µg/day have been recorded in some apparently healthy individuals in New Zealand *(10)*. Nutritional surveys indicate that, in countries such as the USA or Canada, "typical" diets would provide about 100–200 µg of selenium per day in the adult population *(7, 9)*.

14.2.3 Air

Available data on selenium levels in the ambient air and in tobacco indicate that respiratory exposure does not contribute significantly to the daily intake of selenium in the general population *(4, 7)*.

14.3 Metabolism

Soluble selenium salts, such as sodium selenite, are readily absorbed in the gastrointestinal tract of rats. Absorption exceeded 95 % whether the diet contained 20 μg or 4000 μg of selenium per kg (*11*). About 93% of selenium was absorbed by man when milligram doses of sodium selenite were administered in aqueous solutions. Thus people, like rats, exhibit no homoeostatic control limiting gastrointestinal absorption of large amounts of selenite (*12*).

Absorbed selenium is widely distributed in organs and tissues, with high levels present in the liver and kidneys. Selenium penetrates through the placenta and also into the milk, the extent depending on the chemical form (*4,6,13*). Within the body, two primary metabolic pathways predominate. One is direct incorporation into or binding by proteins. The other, reduction followed by methylation, is responsible for the production of dimethylselenide and trimethylselenonium ions. When its rate of formation exceeds the rate of further methylation to a urinary metabolite trimethylselenonium ion, the volatile dimethylselenide is exhaled. Under the conditions of exposure prevailing in the general population, urinary selenium excretion predominates (*3,13*).

The rate of selenium elimination depends on the chemical form in which selenium is administered and on the selenium nutritional status. Available human data indicate that selenium administered as selenite is excreted more rapidly from the body than when given in organic form, such as selenomethionine (*14*). In rats, the biological half-time of selenium decreased with increased dietary selenium levels (*15*).

14.4 Health effects and dose-response relationships

Selenium has been identified as an essential nutrient in several animal species (*8,16*). Certain endemic diseases of farm animals have been identified in areas with low selenium levels and selenium supplementation has been highly effective in preventing these diseases. Higher selenium doses are, of course, toxic, resulting in other diseases in farm animals (*1, 2*).

14.4.1 Studies on human populations

There is growing evidence that selenium is essential for human health. Manifest effects of selenium deficiency appear only under extreme conditions of long-term exposure to locally produced diets having extremely low selenium levels. Recent studies on Keshan disease suggest that this myocardial disease of children (*17–19*) could be induced by low-level selenium intake.

Smith et al. (*20*) and Smith & Westfall (*21*) studied a group of farmers living in seleniferous regions of the USA who were consuming essentially locally produced foodstuffs and were exposed to selenium levels as high as 200 μg/kg of body weight per day. Signs and symptoms observed were rather unspecific or vague. Nevertheless, in a group of 100 subjects exposed to high selenium intake, gastrointestinal disturbances, icteric discoloration of the skin, and bad teeth were observed in 31, 28, and 27 subjects, respectively.

Another study was conducted by Jaffé in Venezuela, a country with high selenium levels in the soil and vegetation (22, 23). In this study, children with very high selenium exposure levels were compared with a group of children from Caracas, where the blood selenium levels and urinary selenium excretion were lower. The blood selenium levels found in the first group of children were the highest reported so far in the general population worldwide (8). In this study, Jaffé recognized that the two Venezuelan groups differed not only in selenium intake, but in other variables, including nutritional status and parasite infestation. Nausea, dermatitis, and pathological changes in the nails were more frequent among the children in the high seleniferous area than in the group from Caracas. However, observed changes, in particular growth retardation and anaemia, may have been due to other factors.

Several studies have attempted to relate high dental caries prevalence in different population segments to high selenium exposure or, on the other hand, lower cancer incidence in certain areas to higher exposure levels to selenium (for review see 4, 24). However, these studies have not excluded the involvement of other variables and have been criticized also from other points, particularly as regards selenium exposure (25, 26).

14.4.2 Animal studies

In most animal species studied, basic dietary requirements approximate 0.04–0.10 mg/kg of food (8). However, vitamin E deficiency substantially increases the demand for selenium, and several other nutritional interactions need to be considered as well (4, 13).

Dietary selenium levels of 5 mg/kg of food or more may cause chronic intoxication, and in seleniferous areas this value has been considered as the dividing line between toxic and non-toxic feeds (4). This conclusion is based both on field experience with farm animals raised in seleniferous areas and on many available experimental animal studies. The main effects of excessive selenium intake in animals include reduced body growth, decreased survival, and damage to the liver and other organs; in some cases, there has also been damage to the myocardium, kidneys, and pancreas.

Some animal studies reported effects following long-term exposure to dietary selenium levels lower than those mentioned above (4). Proliferation of the hepatic parenchyma was reported to be more prevalent in rats fed a semipurified diet supplemented with 0.5–2.0 mg of selenium per kg of feed in comparison with control rats (27).

Increased concentration of glutathione in the blood, decreased activity of succinate dehydrogenase in the liver, and some impairment of the excretory function of the liver have been associated with long-term low-level exposure to sodium selenite. A number of behavioural effects were observed at this exposure level as well (28).

Certain discrepancies exist between two reports on the effects of high selenium levels in drinking-water. Water containing selenite at the level of 2 mg of selenium per litre resulted in 50% mortality after less than 3 months in male rats, with less pronounced toxic effects in females (29). On the other hand, sodium selenite at a level of 3 mg of selenium per litre had no effect on the survival of male rats (30). Selenium compounds have been shown to be less toxic to animals kept on a higher dietary intake of selenium (31, 32). Experiments with monkeys fed a cariogenic diet and exposed to drinking-water containing sodium selenite at a level of 2 mg of selenium per litre for 15 months followed by 1 mg/litre for 45 months indicated a cariogenic effect of selenium during tooth development, but not when the teeth were exposed post-eruptively (33).

There is insufficient evidence to support the claim of carcinogenic effects of high selenium intake in experimental animals (34). The criticism of deficiency in the design of the experiments or in the statistical evaluation of the results does not apply to a recent study revealing carcinogenic effects of long-term exposure to selenium disulfide in mice (5). On the other hand, several reports indicate that doses of selenium higher than nutritionally essential have a preventive effect on the development of cancer in experimental animals (4).

A review of the anticarcinogenic effects of selenium in experimental animals has recently been published (35).

REFERENCES

1. ROSENFELD, I. & BEATH, D. A. *Selenium: geobotany, biochemistry, toxicity and nutrition.* New York, Academic Press, 1964.
2. NATIONAL RESEARCH COUNCIL. *Drinking water and health.* Washington, DC, National Academy of Sciences, 1977.
3. MUTH, O. H., ed. *Selenium in biomedicine.* Westport, CN, Avi Publishing Company Inc., 1967.
4. *Selenium.* Washington, DC, National Academy of Sciences, 1976.
5. *Ambient water quality criteria for selenium.* Washington, DC, US Environmental Protection Agency, 1980.
6. ERMAKOV, V. V. & KOVALSKIJ, V. V. *Biological significance of selenium.* Moscow, Nauka Publishing House, 1974.
7. NATIONAL RESEARCH COUNCIL. *Drinking water and health,* vol. 3. Washington, DC, National Academy of Sciences, 1980.
8. WHO Technical Report Series, No. 532, 1973 (*Trace elements in human nutrition*).
9. LEVANDER, O. A. In: *Proceedings of the symposium on selenium – tellurium in the environment*, Pittsburgh, Industrial Health Foundation, Inc., 1976.
10. STEWART, R. D. H. ET AL. Quantitative selenium metabolism in normal New Zealand women. *British journal of nutrition,* **40**: 45 (1978).
11. BROWN, D. G. ET AL. Effect of dietary selenium on the gastrointestinal absorption of $^{75}SeO_3$ in the rat. *International journal for vitamin and nutrition research,* **42**: 588 (1972).

12. THOMSON, C. D. In: *Trace elements in human and animal health and disease in New Zealand.* Hamilton, Waikato University Press, 1977.
13. DIPLOCK, A. T. Metabolic aspects of selenium action and toxicity. *CRC critical reviews in toxicology,* **4**: 271 (1976).
14. GRIFFITHS, N. M. ET AL. The metabolism of (75 Se) selenomethionine in four women. *British journal of nutrition,* **35**: 373 (1976).
15. BURK, R. F., JR. ET AL. Influence of dietary and injected selenium on whole-body retention, route of excretion, and tissue retention of $^{75}SeO^{2-}$ in the rat. *Journal of nutrition,* **102**: 1049 (1972).
16. SCHWARZ, K. ET AL. Introduction. Symposium on nutritional significance of selenium (Factor 3). *Federation proceedings,* **20**: 665 (1961).
17. KESHAN DISEASE RESEARCH GROUP OF THE CHINESE ACADEMY OF MEDICAL SCIENCES. Observations on the effects of sodium selenite in the prevention of Keshan disease. *Chinese medical journal.,* **92**: 471 (1979).
18. KESHAN DISEASE RESEARCH GROUP OF THE CHINESE ACADEMY OF MEDICAL SCIENCES. Epidemiological studies on the etiologic relationship of selenium and Keshan disease. *Chinese medical journal,* **92**: 477 (1979).
19. XIAOSHU CHEN ET AL. The relations of selenium and Keshan disease. *Biological trace element research,* **2**: 91 (1980).
20. SMITH, M. I. ET AL. The selenium problem in relation to public health. A preliminary study to determine the possibility of selenium intoxication in the rural population living on seleniferous soil. *Public health reports,* **51**: 1496 (1936).
21. SMITH, M. I. & WESTFALL, B. B. Further field studies on the selenium problem in relation to public health. *Public health reports,* **52**: 1375 (1937).
22. JAFFÉ, W. G. ET AL. Estudio clínico bioquímico en niños escolares de una zona selinífera. *Archivos latinoamericanos de nutrición,* **22**: 595 (1972).
23. JAFFÉ, W. G. In: *Proceedings of the symposium on selenium – tellurium in the environment,* Pittsburgh, Industrial Health Foundation, Inc., 1976.
24. GLOVER, J. ET AL. In: Friberg, L. et al., ed. *Handbook on the toxicology of metals,* Amsterdam, Elsevier/North-Holland Biomedical Press, 1979.
25. ALLAWAY, W. H. An overview of distribution patterns of trace elements in soil and plants. *Annals of the New York Academy of Sciences,* **199**: 17 (1972).
26. SCHWARZ, K. In: Muth, O. H., ed. *Selenium in biomedicine.* Westport, CN, Avi Publishing Company Inc., 1967, pp. 225–226.
27. HARR, J. R. ET AL. In: Muth, O. H., ed. *Selenium in biomedicine.* Westport, CN, Avi Publishing Company Inc., 1967.
28. PLETNIKOVA, I. P. Biological effects and safe concentrations of selenium in drinking water. *Hygiene and sanitation,* **35**: 176 (1970).
29. SCHROEDER, H. A. & MITCHENER, M. Selenium and tellurium in rats. Effects on growth survival and tumors. *Journal of nutrition,* **101**: 1531 (1971).
30. PALMER, I. S. & OLSON, O. E. Relative toxicities of selenite and selenate in the drinking water of rats. *Journal of nutrition,* **104**: 306 (1974).
31. JAFFÉ, W. G. & MONDRAGON, M. C. Adaptation of rats to selenium intake. *Journal of nutrition,* **97**: 431 (1969).
32. JAFFÉ, W. G. & MONDRAGON, M. C. Effects of ingestion of organic selenium in adapted and non-adapted rats. *British journal of nutrition,* **33**: 387 (1975).
33. BOWEN, W. H. The effects of selenium and vanadium on caries activity in monkeys (*M. irus*). *Journal of the Irish Dental Association,* **18**: 83 (1972).
34. *Some aziridines, N-, S- and O-mustards and selenium.* Lyon, International Agency for Research on Cancer, 1975 (IARC Monographs on the evaluation of the carcinogenic risk of chemicals to humans, vol. 9), p. 245.
35. JACOBS, M. M. Effects of selenium on chemical carcinogens. *Preventive medicine,* **9**: 362–367 (1980).

15. SILVER

15.1 General description

15.1.1 Sources

Silver occurs naturally in elemental form and as various ores, such as argentite and horn silver; it is also associated with lead, gold, copper, and zinc ores. Silver is present in the earth's crust at a concentration of about 0.1 mg/kg (*1*). It is used as a component of various alloys and solders, in photography, electrical equipment, electroplating, for fungicides, silverware, jewellery, coins, and dentalware (*1*). Because of their bacteriostatic properties, silver salts are used for water disinfection and as prophylactic agents (*2*).

15.1.2 Occurrence in water

Levels of silver in natural waters are very low (*1*). There is insufficient information to provide precise levels in water, but published data (*1–3*) suggest that few water sources contain more than 1 µg of silver per litre and levels greater than 10 µg/litre are rare.

15.2 Routes of exposure

15.2.1 Drinking-water

A number of conventional water-treatment practices have been shown to be effective in removing silver from water and, consequently, many treated waters contain very low levels of silver (*3*). However, because some metals (such as lead and zinc) used in distribution systems may contain traces of silver, and also because in some countries silver oxide is used to disinfect water supplies, silver levels in tap-water may sometimes be elevated. Levels exceeding 50 µg/litre have been recorded on rare occasions, particularly when silver-containing, point-of-use water purifiers have been employed to obtain drinking-water (*2*). The average levels in tap-water are low and certainly less than 1 µg/litre. Assuming a consumption of 2 litres per day, the average daily exposure from drinking-water would thus not be likely to exceed 2 µg.

141

15.2.2 Food

Few published data are available on the silver content of foodstuffs or diet, although most foods seem to contain trace amounts (less than 1 mg/kg). The reported exception is mushrooms, which may contain several hundred mg/kg (3). Various diets have been estimated to provide from 1 to 80 μg of silver per day (3, 4), although, where silver utensils are used, the amounts ingested could be very much higher. Vegetables cooked in water containing silver absorb the metal very effectively (3). There is inadequate information to provide a precise average dietary intake, but estimated values within the range 20–80 μg/day might be reasonable.

15.2.3 Air

Little information has been published on the silver levels in air, but values up to 0.1 μg/m³ of air have been reported in the USA (5, 6). Industrial emissions of silver are controlled for economic reasons and, as the levels present in fossil fuels are reported to be very low, the silver content of ambient air will be correspondingly low. A typical level in urban air would not be expected to exceed 0.05 μg/m³ and the exposure from this source is negligible.

15.2.4 Other routes of exposure

15.2.4.1 *Industrial exposure*

Little information is available on industrial levels. Except in brazing operations and the manufacture of silver varnish, industrial exposures are considered to be very low.

15.2.4.2 *Use of pharmaceutical preparations*

The topical use of certain products containing silver can result in significant exposure (7, 8).

15.2.5 Relative significance of different routes of exposure

On the basis of the considerations in sections 15.2.1–15.2.4, the average daily intake of silver may be roughly assessed.

Assuming that the silver concentration in water is 1 μg/litre and the daily water consumption is 2 litres, the intake from water is 2 μg of silver daily. Assuming also that the silver intake from air is negligible and the daily diet contains 20–80 μg of silver, the average daily intake from all sources will be about 22–82 μg.

Using a silver-containing bacteriostatic point-of-use water purifier, producing a concentration of silver in the drinking-water of 50 μg/litre, together with a daily diet containing as much as 80 μg of silver, the daily intake may be as high as 180 μg per person.

Generally, the levels in drinking-water are low, usually not in excess of 1 μg/litre, and thus the intake of silver from this source, in relation to normal total daily intake from all sources, is unlikely to be more than 5%.

Precise data for the absorption of silver by human beings is not available and no figures for uptake can be provided.

15.3 Metabolism

Relatively little is known about the absorption and metabolism of silver in humans (1) except that individuals and individual organs absorb the metal selectively (3). Animals seem to absorb about 10% of any ingested silver (2). Silver can be detected in various organs; the liver and spleen especially seem to concentrate the metal (2). In humans, more than 50% of the body burden can be found in the liver 16 days after exposure (9). Inhaled silver is also absorbed to a slight extent (2). Silver combines with the sulfhydryl component of some enzyme systems and other biologically important chemical groups, thus influencing the precipitation of proteins and inactivating some enzyme systems (10). Animal experiments have also shown that silver interacts metabolically with copper and selenium (4). Most of the absorbed silver is excreted almost exclusively with the faeces, and only small quantities are permanently retained by the tissues, the exception being the skin, where larger amounts of silver can accumulate (3). Silver that is available for excretion has a biological lifetime in the body ranging from a few days to a few weeks (2).

15.4 Health effects

There is no evidence that silver is essential to the human organism. Cases of fatal poisoning have been recorded, but only with extremely high doses. The main effect of silver is discoloration of skin, hair, and fingernails (argyria). This has been detected when silver arsphenamine has been administered as a medication (3). A single dose of 1 g of silver, injected as silver arsphenamine, can produce this effect (11). The effect has also been observed in workers industrially exposed to silver; the condition is rarely encountered nowadays, however (3). It is possible that argyria may occasionally mask some mild systemic effects (2, 3). There is no evidence that ingested silver is carcinogenic (12).

Pathological changes have been observed in the kidneys and liver of rats consuming water with silver concentrations of 400 μg/litre and above. It is difficult, however, to extrapolate these results to man. If it is assumed that the first appearance of argyria has no significant health effect, then the discoloration could be used to estimate a safe exposure level. The minimum dose for human beings that might induce argyria is

1000 mg of silver (*13*). A lifetime (70 years) exposure to 1000 mg of silver would be equivalent to a continuous daily exposure to 40 μg of silver. However, because silver is continuously excreted and only approximately 10 % is absorbed, the daily exposure level needed to cause argyria over a lifetime could be as high as 400 μg.

REFERENCES

1. *Guidelines for Canadian drinking water quality, 1978*. Quebec, Ministry of Supply and Services, 1980 (supporting documentation).
2. *Toxicology of metals, vol. II*. Washington, DC, US Environmental Protection Agency, 1977 (Environmental Health Effects Research Series).
3. NATIONAL RESEARCH COUNCIL. *Drinking water and health*. Washington, DC, National Academy of Sciences, 1977.
4. UNDERWOOD, E. J. *Trace elements in human and animal nutrition*. New York, Academic Press, 1977.
5. GREENBERG, R. R. ET AL. Composition and size distributions of particles released in refuse incineration. *Environmental science and technology*, **12**: 566 (1978).
6. RAGAINI, R. C. ET AL. Environmental trace metal contamination in Kellogg, Idaho, near a lead smelting complex. *Environmental science and technology*, **11**: 733 (1977).
7. PARISER, R. J. Generalized argyria. Clinicopathologic features and histochemical studies. *Archives of dermatology*, **144**: 373 (1978).
8. MARSHALL, J. P. & SCHNEIDER, R. P. Systemic argyria secondary to topical silver nitrate. *Archives of dermatology*, **113**: 1077 (1977).
9. NEWTON, D. & HOLMES, A. A case of accidental inhalation of Zn-65 and Silver-110. *Radiation research*, **29**: 403 (1966).
10. GOODMAN, L. S. & GILMAN, A. *The pharmacological basis of therapeutics*, 5th ed. New York, Macmillan, 1975.
11. HILL, W. B. & PILLSBURY, D. M. *Argyria. The pharmacology of silver*. Baltimore, MD, Williams and Wilkins, 1939.
12. ENVIRONMENTAL PROTECTION AGENCY. Water quality criteria; availability. *Federal register*, **44**: 15 964 (1979).

16. SODIUM [a]

16.1 General description

16.1.1 Sources

Sodium is present in a number of minerals, the principal one being rock salt (sodium chloride). Seawater contains relatively high levels of sodium. Overall, sodium represents about 26 g/kg of the earth's crust ([1]).

Sodium and its salts are used for a wide variety of purposes, including the de-icing of roads, in the paper, glass and soap industries, in the pharmaceutical and general chemical industries, for the treatment of water, in the food industry, and for culinary purposes.

It is widely present, sometimes in substantial quantities, in soils, plants, water, and many foods. Most countries have significant mineral deposits of sodium. Considerable amounts are excreted by humans and it is a common constituent of domestic sewage.

16.1.2 Occurrence in water

The sodium ion is ubiquitous in water owing to the high solubility of its salts and the abundance of mineral deposits. Seawater contains about 10 g of sodium per litre. The highest freshwater levels are found in lowland rivers and in groundwater. Upland streams and associated reservoirs will tend to have a relatively low sodium content. Particularly elevated levels of sodium are associated with groundwater in areas where there is an abundance of sodium mineral deposits or where there has been contamination from saline intrusion (sea and estuarine sources) or other forms of pollution ([2]).

Near coastal areas, windborne sea spray can make an important contribution, either by fallout on to land surfaces where it drains to the water source or from washout by rain from the air to surface-water sources ([2]). Discharge of effluents (domestic, commercial, industrial wastes) to rivers is another important source of sodium in water. Levels in such rivers would be a function of many factors, including river flow rate and sodium concentration in the effluent. Substantial concentrations can sometimes be detected in some rivers, especially at times of low river flow.

---

[a] A discussion of some other aspects of the effect of sodium on water quality will be found in Part V, section 11, p. 286.

16.2 Routes of exposure

16.2.1 Drinking-water

In most countries, the majority of water supplies contain less than 20 mg of sodium per litre, but in some countries sodium levels can exceed 250 mg/litre. Apart from saline intrusion and natural contamination, salt used in de-icing roads, water-treatment chemicals, domestic water softeners, sea-salt spray and sewage effluents can all contribute significant quantities to water. Water-treatment chemicals, such as sodium fluoride, sodium silicofluoride, sodium hydroxide, sodium carbonate, sodium bicarbonate, and sodium hypochlorite provide, as individual chemicals, a relatively small contribution, but collectively the amounts can be significant; levels of 30 mg/litre might result in some instances (2). Domestic water softeners can provide levels of over 300 mg/litre, but in general the levels are much lower (2).

Most people are exposed to less than 50 mg of sodium per day by drinking tap-water (based on a consumption of 2 litres per day). As sodium salts are very soluble, virtually all the sodium present in water, whether consumed directly, in the preparation of beverages, or incorporated into food, will be absorbed.

16.2.2 Food

Sodium is naturally present in all foods. The way food is processed governs critically the final sodium content. For example, frozen peas contain much more sodium than fresh peas (2). Fresh fruit and vegetables contain from less than 10 mg/kg to 1 g/kg, in contrast with cereals and cheeses, which may contain 10–20 g/kg (2). Milk contains a relatively high proportion of sodium, i.e., 1.5 g/litre (2). Bottled waters can also sometimes contain similar levels of sodium (2). The estimation of the daily intake of sodium from food is difficult because of the wide variations of concentrations in foods and the fact that many people add salt to their food. In western Europe and North America, the overall dietary sodium chloride consumption is estimated to be 5–20 g/day (2–8 g of sodium per day), with an average of around 10 g/day (4 g of sodium) (2). For medical reasons, some people need a special low-sodium diet calling for a sodium intake of less than 2 g of sodium per day (3). In the case of artificial infant feeding, regulations requiring a reduction of the sodium content of infant food have been widely introduced.

16.2.3 Air

The level of inhaled sodium in ambient air and factory atmospheres is small relative to the amount of sodium absorbed into the body from the diet.

16.2.4 Relative significance of different routes of exposure

For different individuals there can be a fairly wide range of exposures to sodium. Generally, by far the major source is food. Since sodium is readily absorbed after ingestion, uptake can be considered to be equivalent to the exposure to the element. Uptakes for various categories of dietary and water sources have been calculated and are given in the tables below. The information for adults is based on published data on the dietary input of sodium (2). Calculations for infants (0–2 months) are based on an estimate of 250 mg of sodium per day and for children 1–5 years on an estimate of 2000 mg/day (2).

16.2.4.1 *Weekly uptake of sodium from water and diet in adults*

(a) Special restricted diet—500 mg of sodium per day

Sodium concentration in water	Weekly uptake of sodium (mg)			Ratio: Water/total (%)
	Water only	Food only	Total	
20 mg/litre	280	3 500	3 780	7
50 mg/litre	700	3 500	4 200	17
100 mg/litre	1 400	3 500	4 900	28
200 mg/litre	2 800	3 500	6 300	44

(b) Relatively low sodium diet—2000 mg per day

Sodium concentration in water	Weekly uptake of sodium (mg)			Ratio: water/total (%)
	Water only	Food only	Total	
20 mg/litre	280	14 000	14 280	2
50 mg/litre	700	14 000	14 700	5
100 mg/litre	1 400	14 000	15 400	9
200 mg/litre	2 800	14 000	16 800	17

(c) Typical sodium diet—5000 mg of sodium per day

Sodium concentration in water	Weekly uptake of sodium (mg)			Ratio: Water/total (%)
	Water only	Food only	Total	
20 mg/litre	280	35 000	35 280	1
50 mg/litre	700	35 000	35 700	2
100 mg/litre	1 400	35 000	36 400	4
200 mg/litre	2 800	35 000	37 800	7

(d) *Total intake restricted to 500 mg of sodium per day*

Sodium concentration in water	Weekly uptake of sodium (mg)			Ratio: water/total (%)
	Water only	Food only	Total	
20 mg/litre	280	3 220	3 500	8
50 mg/litre	700	2 800	3 500	20
100 mg/litre	1 400	2 100	3 500	40
200 mg/litre	2 800	700	3 500	80

All the above calculations are based on a water consumption of 2 litres per day.

16.2.4.2 *Weekly uptake of sodium from water and diet in children*

(a) *Infants aged 0–2 months—250 mg of sodium per day from food*

Sodium concentration in water	Weekly uptake of sodium (mg)			Ratio: water/total (%)
	Water only[a]	Food only	Total	
20 mg/litre	140	1 750	1 890	7
50 mg/litre	350	1 750	2 100	17
100 mg/litre	700	1 750	2 450	29
200 mg/litre	1 400	1 750	3 150	44

[a] Based on a water consumption of 1.0 litre per day.

(b) *Children aged 1–5 years—2000 mg of sodium per day from diet*

Sodium concentration in water	Weekly uptake of sodium (mg)			Ratio: water/total (%)
	Water only[a]	Food only	Total	
20 mg/litre	210	14 000	14 210	1
50 mg/litre	525	14 000	14 525	4
100 mg/litre	1 050	14 000	15 050	7
200 mg/litre	2 100	14 000	16 100	13

[a] Based on a water consumption of 1.5 litres per day.

16.3 Metabolism

The metabolism of sodium has been studied extensively because of its physiological properties and its importance to the body (3). Reference is made here only to the general aspects of the metabolism of sodium. It is the most abundant cation of plasma and extracellular fluid in man. It is present in bones, in cells, and in most tissues. The level of sodium in extracellular fluid is carefully maintained by the kidney under the influence of endocrine, cardiovascular, and autonomic regulatory mechanisms. The total amount of sodium in extracellular fluid thus determines the volume of these fluids (3).

Control of sodium balance is achieved through a complex interrelated system involving both nervous and hormonal systems (1). Increases in the plasma sodium concentration stimulate the osmoreceptors in the hypothalamic centre, regardless of fluid volume, with the resultant sensation of thirst (1). In hot climates and during heavy work, a substantial loss of sodium occurs by perspiration and additional salt may be needed to make up the loss (1).

The intake of sodium is not physiologically controlled. More than 90% of the amount in food is absorbed (2). Intake is normally dependent primarily on diet. The minimum sodium chloride requirement is about 120 mg/day (approximately 50 mg of sodium in this form) (2).

16.4 Health effects

16.4.1 Acute effects

In general, sodium salts are not acutely toxic substances because of the efficiency with which mature kidneys excrete sodium (1). Excessive intake of sodium chloride causes vomiting and the elimination of much of the salt. Acute effects may include convulsions, muscular twitching and rigidity, and cerebral and pulmonary oedema (1).

The effects on infants, in contrast to adults, are different because of the immaturity of infant kidneys (2). Acute effects and death have been reported in cases of accidental overdoses of sodium chloride (2).

Severe deterioration of chronic congestive heart failure can result from excessive salt intake, and ill effects due to high levels of sodium in drinking-water have been documented (2).

16.4.2 Hypernatraemia

Infants with severe gastrointestinal infections can suffer from fluid loss leading to dehydration and raised sodium levels in the plasma (hypernatraemia); permanent neurological damage is common under such conditions (2).

Evidence that a raised sodium intake is a factor in "sudden infant death" is limited and indirect and is not generally thought to be conclusive (2). However, for healthy infants and children, the total sodium intake should be kept as low as possible (2).

Modern infant feeding practice using cows' milk added to solid feed has been suggested as one of the causes of hypernatraemia (2). The situation could be exacerbated if tap-water containing high levels of sodium was also incorporated into the feed (2). The concentration of sodium in cows' milk is about three times that in human breast milk (2). The immature kidneys of infants are not as effective as those of adults in maintaining plasma osmolarity so that precautions are needed, such as the introduction of regulations requiring a reduction of sodium in infant foods.

16.4.3 Raised sodium intake and hypertension

There has been considerable scientific controversy for some time on this issue (2), and there are persuasive scientific arguments for the hypothesis that salt consumption affects the development and level of hypertension.

16.4.3.1 *Evidence from animal experiments*

Hypertension has been clearly demonstrated in different species of animals given high levels of sodium chloride in their diet (2). Despite the usual reservations about extrapolating animal results to humans, the consistency of the animal data suggests the validity of such a procedure.

16.4.3.2 *Evidence from human volunteer studies*

There is no conclusive evidence that elevated blood pressure is related to high-salt diets fed to volunteers. However, there is some doubt about these short-term studies in relation to hypertension. Most people in western societies ingest a high-salt diet from infancy, yet persistent hypertension is uncommon until the fourth decade (2).

16.4.3.3 *Evidence from epidemiological studies*

(a) *High-sodium diet population studies*

A particularly striking contrast is that between some non-westernized groups and western populations. Non-westernized groups have low-sodium diets and a very low prevalence of hypertension, with no increase in blood pressure with age (2). Although one may be tempted to conclude a causal relation, there are a number of other differences between the two populations that might account for such a contrast (2). However, the strong consistency between these results and those of other studies (2) gives further support to there being a direct link between raised sodium intake and hypertension.

(b) *Sodium intake studies via drinking-water*

Recently completed epidemiological studies in the USA and the Netherlands have demonstrated that schoolchildren (particularly girls) living in areas with moderate levels of sodium in the drinking-water (128–161 mg/litre) had slightly higher blood pressures (3–5 mmHg) than those living in areas with low levels of sodium (28 mg/litre) (4–6). A somewhat similar study in the USSR of people in the age range 16–60 years demonstrated a similar relationship between sodium in water and blood pressure (7).

In a study in the USA (*6*), children (10–11 years old) living in the same community where the drinking-water had a high sodium content (108 mg/litre) were grouped into triads and matched by systolic blood pressure. Two of the three groups were supplied with bottled water containing 108 mg of sodium per litre and the third with bottled water of low sodium content (8 mg/litre). The results for females were consistent with the results of the other two reported studies in the USA; girls provided with bottled water of low sodium content had lower blood pressures than those in the other two groups. The blood pressure differences for boys followed an identical trend for the first six weeks, but were not consistent for the remainder of the study.

16.4.4 Relationship between sodium in water and other diseases

Although there is an association between hypertension and some diseases, such as coronary heart diseases, genetic differences in susceptibility, possible protective minerals (potassium and calcium), and methodological weaknesses in experiments make it difficult to quantify a relationship. Sodium levels in drinking-water are generally only a small contributor to dietary sodium. No firm conclusions can at present be drawn on the importance of sodium in drinking-water and its possible association with disease.

The relevance of ingested sodium from all sources was reviewed by a WHO Working Group in 1978 (*2*). One of the recommendations of the group was that in areas where levels of sodium exceeded 20 mg/litre, public health authorities should be notified because certain people (patients with hypertension or congestive heart failure) need to restrict their overall dietary intake of sodium. As any action to be taken depends upon local conditions and policies, no specific level based on health considerations is recommended in the present guidelines (see Pàrt V, section 11.4, for a guideline value based on taste threshold).

REFERENCES

1. *Guidelines for Canadian drinking water quality, 1978.* Quebec, Ministry of Supply and Services, 1980 (supporting documentation).
2. *Sodium, chlorides and conductivity in drinking-water.* Copenhagen, WHO Regional Office for Europe, 1979 (EURO reports and studies, No. 2).
3. NATIONAL RESEARCH COUNCIL. *Drinking water and health.* Washington, DC, National Academy of Sciences, 1977.
4. CALABRESE, E. J. & TUTHILL, R. W. Elevated blood pressure and high sodium levels in the public drinking water. *Archives of environmental health*, **35**: 200 (1977).
5. TUTHILL, R. W. & CALABRESE, E. J. Elevated sodium levels in the public drinking water as a contributor to elevated blood pressure levels in the community. *Archives of environmental health*, **37**: 197 (1979).
6. TUTHILL, R. W. & CALABRESE, E. J. Drinking water sodium and blood pressure in children: a second look. *American journal of public health*, **71**: 722–729 (1981).
7. FATULA, M. I. The frequency of arterial hypertension among persons using water with an elevated sodium chloride content. *Sovetskaja medicina*, **30**: 123 (1967).

PART IV. HEALTH-RELATED ORGANIC CONSTITUENTS

1. CHLORINATED ALKANES

One of the major uses of chlorinated alkanes in the chemical industry is as an intermediate in the production of other organochlorine compounds. They are therefore produced in large quantities and consequently many are found in both raw and finished drinking-water. Of the large number of chlorinated ethanes known to be produced commercially, only 1,2-dichloroethane can be clearly labelled as a carcinogenic hazard on the basis of the available data.

1.1 Carbon tetrachloride

1.1.1 General aspects

Carbon tetrachloride (CCl_4) is a haloalkane with a wide range of industrial and chemical applications. At room temperature it occurs as a heavy, colourless liquid with a density of 1594 g/litre. It is relatively nonpolar, miscible with alcohol, acetone and most organic solvents, and soluble in water to the extent of 800 mg/litre at 25 °C. Approximately 423 000 tonnes (932.7×10^6 pounds) are produced at 11 plant sites in the USA.[a] The major portion of this production is used in the manufacture of fluorocarbons (95 % in 1973), which are used primarily as aerosol propellants. Accidental CCl_4 spills have occurred and as much as 63.6 tonnes were discharged into the Ohio River in February 1977, resulting in surface-water concentrations as high as 340 μg/litre (1). It is also frequently found in contaminated groundwater. Its general occurrence in raw water supplies results in reported levels of 2–3 μg/litre in finished drinking-water.

Hydrolytic decomposition, as a means of removal from water, appears to be insignificant as compared with evaporation.

1.1.2 Routes of exposure

1.1.2.1 *Water*

In the National Organic Reconnaissance Survey (NORS) study performed in the USA by the EPA (2), CCl_4 was found in 10 % of the drinking-water supplies at levels less than 2–3 μg/litre. In New Orleans,

---

[a] JOHNS, R. *Air pollution assessment of carbon tetrachloride*. Prepared under contract for the US Environmental Protection Agency. McLean, VA., Mitre Corp., 1976.

CCl_4 was found in both human blood plasma and drinking-water. CCl_4 has been found to be an occasional contaminant of chlorine used in the disinfection of drinking-waters, but it is not produced in drinking-water as a result of the chlorination process itself.

1.1.2.2 *Food*

CCl_4 has been detected in a variety of foodstuffs with levels ranging from 0.1 to 20 μg/kg. McConnell et al. (*3*) provide a summary of the different food categories that have been shown to be particularly susceptible to CCl_4 contamination. They noted that there is no evidence of significant bioaccumulation of CCl_4 via the food chain to higher trophic levels. The primary means of contamination of food with CCl_4 is through the use of CCl_4 as a fumigant.

1.1.2.3 *Air*

Carbon tetrachloride has been measured extensively in the atmosphere; consequently, there is a good understanding of its atmospheric distribution. The occurrence of CCl_4 in the atmosphere is due largely to the fact that it is such a volatile compound. The sources of CCl_4 have been found to be primarily man-made (*4–7*). The atmospheric distribution is approaching homogeneity. Some high concentrations have been reported in urban air; however, they are generally close to the background level of 0.00078–0.00091 mg/m^3 found in the continental air mass.

1.1.3 Metabolism

1.1.3.1 *Absorption*

CCl_4 is readily absorbed through the lungs and more slowly but still completely absorbed through the gastrointestinal tract (*8*). It can also enter the body by penetration through the skin. The rate and amount of absorption is enhanced by the simultaneous ingestion of fat (*8*) and alcohol (*8–10*). In an investigation of the absorption of CCl_4 from the gastrointestinal tract of dogs, Robbins (*11*) found that considerable quantities were absorbed from the small intestine, less from the colon, and little from the stomach.

1.1.3.2 *Distribution*

Nielsen & Larsen (*8*) found high concentrations of CCl_4 in animal testicular fatty tissues, liver, brain, bone marrow, and kidneys. Robbins (*11*) studied the distribution of CCl_4 in dogs after oral administration. The highest concentration of CCl_4 was found in the bone marrow. The amount found in the liver, pancreas, and spleen was one-fifth of the amount found in bone marrow. From work done by Recknagel &

Litteria (*12*), it would appear that the organ distribution of CCl_4 varies with the route of administration, its concentration, and the duration of exposure. At the cellular level, McLean et al. (*13*) found CCl_4 in all cell fractions, with higher concentrations in the ribosomes.

1.1.3.3 *Biotransformation*

When CCl_4 is administered to mammals, it is metabolized to a small extent, but the majority is excreted through the lungs. The metabolites include chloroform, hexachloroethane, and carbon dioxide. Research has revealed that these metabolites play an important role in the overall toxicity of CCl_4 (*14*).

1.1.4 Health effects

The toxicity of CCl_4 to humans is not often recognized (*15*). Acute and subacute toxicity has resulted from oral, dermal, and inhalation exposures, with adverse effects on the skin, circulation, respiration, blood, and the function of the kidneys, liver, eyes, and pancreas.

In many instances of acute poisoning, the patient develops signs of liver injury within a few days. The patient becomes jaundiced and the liver becomes enlarged and tender. As liver injury develops, and sometimes in its absence, injury of the kidneys may be observed; at times this may dominate the clinical picture and it is often responsible for early death (*16*). In general, hepatic complications are a more frequent response to CCl_4 toxicity than are renal complications. Changes in blood parameters, visual acuity, and the pancreas have also been noted. The clinical picture of chronic CCl_4 poisoning is much less characteristic than that of acute poisoning. Reports of pathological changes in persons dying from CCl_4 poisoning are generally limited to findings in the liver and kidneys.

There are very few reports on the mutagenesis of CCl_4. Kraemer et al. (*17*) found that CCl_4 was not mutagenic in the *Salmonella typhimurium* or *Escherichia coli* reversion tests; however, halogenated hydrocarbons are usually negative in the Ames test.

The available data appear to be sufficient to permit the conclusion that CCl_4 is a carcinogen to laboratory animals (*18–24*). The standard for a permissible level of human exposure is based on the study performed in the USA by the National Cancer Institute (*24*) on trichloroethene, in which CCl_4 was utilized as a positive control. In this work, CCl_4 was found to be carcinogenic in the B6C3-Fl mouse. Although other studies have been conducted, inadequate dose-response information was obtained or the experiments were of too short a duration to allow utilization of the data for risk estimates. Because there are doubts regarding the mechanism of tumorigenesis in the liver of this strain of mouse with agents that are known hepatotoxins (such as CCl_4), the appropriateness of a no-threshold model for extrapolation is questionable. Alternative extrapolation models do not exist for epigenetically induced cancer. The need for guidance on this frequent contaminant of drinking-water dictates a conservative approach. Consequently, a linear multistage extrapolation model was used to derive a tentative guideline value for CCl_4 of 3 μg/litre. This level should give rise to less than 1 additional cancer per 100 000 population for a lifetime of exposure, assuming a daily consumption of drinking-water of 2 litres.

REFERENCES

1. AMERICAN CHEMICAL SOCIETY. CCl_4 spill causes along Ohio River. *Chemical and engineering news*, **55**: 7 (1977).
2. *Ambient water quality criteria for carbon tetrachloride.* Washington, DC, US Environmental Protection Agency, 1980 (EPA 440/5-80-026).
3. McCONNELL, G. ET AL. Chlorinated hydrocarbons in the environment. *Endeavour*, **34**: 13 (1975).
4. ALTSCHULLER, A. P. Average tropospheric concentration of carbon tetrachloride based on industrial production, usage and emissions. *Environmental science technology*, **10**: 596 (1976).
5. LOVELOCK, J. E. ET AL. Halogenated hydrocarbons in and over the Atlantic. *Nature*, **247**: 194 (1974).
6. WILKNISS, P. E. ET AL. Atmospheric trace gases in the southern hemisphere. *Nature*, **245**: 45 (1973).
7. SINGH, H. B. ET AL. Atmospheric carbon tetrachloride: Another man-made pollutant. *Science*, **192**: 1231 (1976).
8. NIELSEN, V. K. & LARSEN, J. Acute renal failure due to carbon tetrachloride poisoning. *Acta medica Scandinavica*, **178**: 363 (1965).
9. FOLLAND, D. S. ET AL. Carbon tetrachloride toxicity potentiated by isopropyl alcohol. Investigation of an industrial outbreak. *Journal of the American Medical Association*, **23**: 1853 (1976).
10. MOON, H. D. Pathology of fatal carbon tetrachloride poisoning with special reference to histogenesis of the hepatic and renal lesions. *American journal of pathology*, **26**: 1041 (1950).
11. ROBBINS, B. H. The absorption, distribution, and excretion of carbon tetrachloride in dogs under various conditions. *Journal of pharmacology*, **37**: 203 (1929).
12. RECKNAGEL, R. O. & LITTERIA, M. Biochemical changes in carbon tetrachloride fatty liver: Concentration of carbon tetrachloride in liver and blood. *American journal of pathology*, **36**: 521 (1960).
13. McLEAN, A. S. M. ET AL. Cellular necrosis in the liver induced and modified by drugs and other agents. *International review of experimental pathology*, **4**: 127 (1965).
14. GORDIS, E. Lipid metabolites of carbon tetrachloride. *Journal of clinical investigation*, **48**: 203 (1969).
15. VON OETTINGEN, W. F. The halogenated hydrocarbons of industrial and toxicological importance. In: Browning, E., ed. *Elsevier monographs on toxic agents*. New York, Elsevier Publishing Co., 1964.
16. VON OETTINGEN, W. F. *The halogenated aliphatic, olefinic, cyclic, aromatic, and aliphatic-aromatic hydrocarbons including the halogenated insecticides, their toxicity and potential dangers.* Washington, DC, Department of Health, Education & Welfare, 1955.
17. KRAEMER, M. ET AL. *S. typhimurium* and *E. coli* to detect chemical mutagens. *Naunyn-Schmiedebergs archives of pharmacology*, **284**: 46R (Abstract).

18. *Some halogenated hydrocarbons.* Lyons, International Agency for Research on Cancer, 1979 (IARC Monographs on the evaluation of carcinogenic risk of chemicals to humans, vol. 20).
19. EDWARDS, J. Hepatomas in mice induced with carbon tetrachloride. *Journal of the National Cancer Institute*, **2**: 197 (1941).
20. EDWARDS, J. & DALTON, A. Induction of cirrhosis of the liver and hepatomas in mice with carbon tetrachloride. *Journal of the National Cancer Institute*, **3**: 19 (1942).
21. EDWARDS, J. ET AL. Induction of the carbon tetrachloride hepatoma in strain L mice. *Journal of the National Cancer Institute*, **2**: 297 (1942).
22. ESCHENBRENNER, A. B. & MILLER, E. Studies on hepatomas-size and spacing of multiple doses in the induction of carbon tetrachloride hepatomas. *Journal of the National Cancer Institute*, **4**: 385 (1943).
23. ESCHENBRENNER, A. B. & MILLER, E. Liver necrosis and the induction of carbon tetrachloride hepatomas in strain A mice. *Journal of the National Cancer Institute*, **6**: 325 (1946).
24. NATIONAL CANCER INSTITUTE. *Carcinogenesis bioassay of trichloroethylene.* Washington, DC, US Department of Health, Education & Welfare, 1976 (CAS No. 79-01-6, NCI-CG-TR-2).

1.2 1,2-Dichloroethane

1.2.1 General aspects

1,2-Dichloroethane ($CH_2 Cl$-$CH_2 Cl$) is a liquid with a relative density of 1.25, and a threshold odour limit of 2 mg/litre (*1*). It is used extensively as a solvent for a great many organic chemicals, as an intermediate in chemical synthesis, and as an insecticide. Production of 1,2-dichloroethane in the USA in 1976 was 3.63×10^6 tonnes (8×10^9 pounds) (*2*).

1.2.2 Routes of exposure

1.2.2.1 *Drinking-water*

As a result of the use of 1,2-dichloroethane in industry, it is a component of industrial effluents and has been detected in raw and finished drinking-water in the USA. 1,2-Dichloroethane was found in the water of 28 cities in the USA at levels up to 6 μg/litre (*3*).

1.2.2.2 *Food*

0.4 % of 1,2-dichloroethane exposure results from the consumption of aquatic organisms, which exhibit an average bioconcentration potential of 1.2-fold. The remaining 99.6 % of exposure results from drinking-water (*4*).

1.2.2.3 *Industrial exposure*

In the USA, the National Institute for Occupational Safety and Health (NIOSH) (*2*) estimated that 4.5 million workers are exposed to 1,2-dichloroethane by inhalation and dermal routes.

1.2.2.4 *Air*

1,2-Dichloroethane has been detected in urban air at levels between 0.04 and 38 $\mu g/m^3$ (*5*). As the result of manufacture, storage, and distribution it was calculated in the USA in 1974 that the emissions of 1,2-dichloroethane to the ambient air amounted to approximately 74×10^6 kg, i.e., about 1.8 % of the total production.

1.2.3 Metabolism

Few data are available concerning the metabolism of 1,2-dichloroethane. It is known that this substance is readily soluble in the lipids of the brain; this property promotes the influence of 1,2-dichloroethane on the nervous system (*1*).

1.2.4 Health effects

Data on the toxicity of 1,2-dichloroethane are related mainly to its inhalation exposure in occupational conditions; it acts as a narcotic and causes damage to the liver, kidneys, and cardiovascular system (*7*). Insignificant symptoms of intoxication were found at concentrations under 4 mg of 1,2-dichloroethane per m^3 in air (*1*).

On the basis of the available information, NIOSH recommended that occupational exposure to 1,2-dichloroethane should not exceed 20 mg/m^3 determined as a time-weighted average for up to a 10-hour working day with a 40-hour working week. Peak concentrations should not exceed 60 mg/m^3 as determined by a 15-minute sample. The exposure standard recommended by the Occupational Safety and Health Administration (OSHA) of the USA is 200 mg/m^3, while the standard for occupational conditions in the USSR is 10 mg/m^3.

The LD_{50} of 1,2-dichloroethane administered orally to white rats was 1120 ± 142 mg/kg of body weight (*7*).

Epidemiological studies have not disclosed a relationship between exposure to 1,2-dichloroethane and cancer. However the compound is carcinogenic in animal tests, inducing a statistically significant number of squamous cell carcinomas of the prestomach and haemangiosarcomas of the circulatory system in male rats, mammary adenocarcinomas in female rats and mice, and endometrial tumours in female mice (*6, 7*). The ambient water quality criterion for 1,2-dichloroethane in the USA was calculated by applying a linearized multistage model to the data from the appropriate bioassay of the National Cancer Institute.

1,2-Dichloroethane is a known mutagen. It was mutagenic in the Ames *Salmonella* assay for the strains TA 1530 and 1535, and was also mutagenic for the *E. coli* DNA polymerase-deficient system (*8*). It induced highly significant increases in somatic mutation frequencies in *Drosophila melanogaster* (*9*). Morphological and chlorophyll mutations in eight varieties of peas were induced by treatment of the seeds with 1,2-dichloroethane (*10*).

Chloroacetaldehyde, a postulated metabolite of 1,2-dichloroethane, is mutagenic in *Salmonella typhimurium* TA 100 (see *11*).

The guideline value for 1,2-dichloroethane is based on the induction of circulatory system haemangiosarcomas in male Osborne-Mendel rats given oral doses of 1,2-dichloroethane over a period of 78 weeks (*10*). The concentration of 1,2-dichloroethane in water, calculated to keep the lifetime cancer risk below 10^{-5}, is 9.4 µg/litre, i.e., approximately 10 µg/litre.

REFERENCES

1. ZOETEMAN, B. C. J. *Sensory assessment and chemical composition of drinking-water.* Leidschendam, Netherlands, Institute of Water Supply, 1978.
2. NATIONAL INSTITUTE FOR OCCUPATIONAL SAFETY AND HEALTH. *Ethylene dichloride (1,2-dichloroethane).* Washington, DC, Department of Health, Education and Welfare, 1978 (*Current Intelligence Bulletin 25* (NIOSH) publication No. 78149).
3. SYMONS, J. M. ET AL. National organics reconnaissance survey for halogenated organics. *Journal of the American Water Works Association*, **67**: 634 (1975).
4. *Ambient water quality criteria for chlorinated ethanes.* Washington, DC, Environmental Protection Agency, 1980 (440/5-80-029).
5. OKUNO, T. ET AL. [Gas chromatography of chlorinated hydrocarbons in urban air.] *Hyogo-ken kogai kenkyusho kenkyu hokoku,* **6**: 1–6 (*Chemical abstracts,* 87.72564 f) (1974).
6. NATIONAL CANCER INSTITUTE. *Bioassay of 1,2-dichloroethane for possible carcinogenicity.* Washington, DC, US Department of Health, Education & Welfare, 1978 ((NIH) 78-1305).
7. *Some halogenated hydrocarbons.* Lyon, International Agency for Research on Cancer, 1979 (IARC Monographs on the evaluation of the carcinogenic risk of chemicals to humans, vol. 20).
8. BREM, H. ET AL. The mutagenicity and DNA-modifying effect of haloalkanes. *Cancer research,* **34**: 2576 (1974).
9. NYLANDER P. O. ET AL. Mutagenic effects of petrol in *Drosophila melanogaster.* I. Effects of benzene and of 1,2-dichloroethane. *Mutation research,* **57**: 163 (1978).
10. KIRICHEK, Y. F. Effect of 1,2-dichloroethane on mutations in peas. *Uspehi himii mutageneza se.,* 232 (1974).
11. *Some monomers, plastics and synthetic elastomers, and acrolein.* Lyon, International Agency for Research on Cancer, 1979 (IARC Monographs on the evaluation of the carcinogenic risk of chemicals to humans, vol. 19).

2. CHLORINATED ETHENES[a]

This group of compounds is used widely in a variety of industrial processes as solvents, softeners, paint thinners, dry-cleaning fluids, intermediates, etc. Because of their wide use, they are often found in raw and treated drinking-water. They are known to occur in groundwater at concentrations of a few milligrams per litre. Because of their high volatility, they are usually lost to the atmosphere from surface-water and therefore generally occur at lower concentrations.

The compounds that are of interest within this group are those for which there are indications of carcinogenic activity in experimental animals. This group includes the well-known human carcinogen vinyl chloride. The occurrence of vinyl chloride in drinking-water seems to be primarily associated with the use of poorly polymerized poly(vinyl chloride) water-pipes, a problem that can be more appropriately controlled by product specification than by the setting of a guideline level.[b]

2.1 Vinyl chloride

2.1.1 General aspects

Vinyl chloride is mainly used for the production of poly(vinyl chloride) (PVC) resins which, in turn, form the most widely used plastics in the world. Minor uses (less than 5% of total production) are as an intermediate in the manufacture of methyl chloroform and as a comonomer with vinylidene chloride in the production of vinylidene chloride–vinyl chloride copolymers, which are widely used in food packaging and as coatings. Vinyl chloride was formerly used as an aerosol propellant and as a refrigerant, but these uses appear to have been discontinued (1).

The largest use of PVC is in the production of piping and conduits; other important uses are in floor coverings, in consumer goods, in electrical applications, and in transport applications (1).

Vinyl chloride is volatile and readily passes from solution into the gaseous phase under most laboratory and ecological conditions. Low concentrations have been detected in effluents discharged by chemical

---

[a] Formerly known as chlorinated ethylenes.

[b] US National Sanitation Foundation Standard No. 14. *Plastic piping components and related materials*, revised December 1980, permits 10 mg of vinyl chloride monomer per kilogram of pipe for drinking-water.

and latex manufacturing plants (*1*) and in drinking-water as a result of leaching from PVC pipes used in water distribution systems (*2*). A number of product standards exist which specify a quality of PVC water-pipe that limits the quantity of free vinyl chloride monomer (VCM) present. Provided pipes of this quality are used, the concentration of the vinyl monomer likely to be present in the drinking-water will be small compared with the value that would be derived by applying the same linear multistage extrapolation model that was used in the case of other carcinogenic organic substances (20 µg/litre based upon an acceptable risk of less than one additional cancer case per 100 000 population over a lifetime).

2.1.2 Routes of exposure

2.1.2.1 *Water*

Vinyl chloride in samples of wastewater from seven areas in the USA (associated with PVC-vinyl chloride manufacturing plants) ranged from 0.05 to 20 mg/litre (*3*). The highest concentration of vinyl chloride detected in finished drinking-water in the USA was 10 µg/litre (*4*). In a five-city survey in the USA, concentrations of vinyl chloride up to 1.4 µg/litre were detected in drinking-water taken from distribution systems constructed with PVC pipe (*2*).

2.1.2.2 *Food*

Small quantities of vinyl chloride are ingested as a result of migration into foods from PVC packaging materials. Studies by the Food and Drug Administration of the USA indicated that up to 20 mg of vinyl chloride per kg was present in alcoholic beverages packaged in PVC containers (*5*). Vinyl chloride has been found at concentrations up to 14.8 mg/kg in edible oils, butter, and margarine packaged and stored in PVC containers (*1*). Many countries now limit the content of entrained vinyl chloride monomer in PVC packaging materials and prohibit the use of such materials for products containing alcohol or edible oils.

2.1.2.3 *Air*

Vinyl chloride is a gas at normal atmospheric temperature and pressure and it occurs in the vicinity of vinyl chloride and PVC industries. Concentrations of up to 8.8 mg/m^3 have been detected in the air near vinyl chloride manufacturing plants (*1*). Vinyl chloride was formerly used as a propellant for many aerosol products, such as pesticides, hair sprays, and deodorants; consumers repeatedly using such products were undoubtedly exposed to moderately high concentrations. Vinyl chloride has been detected in concentrations of 1–3 mg/m^3 in the air in the interior of new automobiles (*1*).

2.1.3 Metabolism

2.1.3.1 *Absorption*

Vinyl chloride is readily absorbed following oral administration (*6, 7*) or inhalation (*7*).

2.1.3.2 *Distribution*

In studies on the distribution of vinyl chloride in rats, the greatest concentrations were detected in the liver, kidneys, and spleen (*6, 8*).

2.1.3.3 *Biotransformation*

Vinyl chloride is metabolized by microsomal mixed-function oxidases (predominantly through the P-450 system) to chloroethene oxide, which can rearrange spontaneously to chloroacetaldehyde. A major route of metabolism of chloroacetaldehyde involves oxidation to chloroacetic acid; this is either excreted as such or bound to glutathione which, after further enzymic degradation, is excreted. A number of other routes of metabolism of chloroacetaldehyde are also involved (*9*).

2.1.3.4 *Elimination*

The kinetic parameters and half-lives for the elimination of vinyl chloride after inhalation and intravenous injection have been reported (*7*). Rats given 250 µg of vinyl chloride per kg of body weight by the intragastric route eliminated more than 96 % within 24 hours (3.7 % exhaled as vinyl chloride, 12.6 % as carbon dioxide, 71.5 % as urinary metabolites, and 2.8 % in faeces) (*10*).

2.1.4 Health effects

2.1.4.1 *Acute and subacute toxicity*

The principal response to acute exposure to vinyl chloride is one of central nervous system depression. Pathological findings at necropsy include congestion and oedema of the lungs and hyperaemia of the liver and kidneys (*11*).

2.1.4.2 *Carcinogenicity*

The carcinogenicity studies in animals and epidemiological observations in man have been studied and reviewed. Carcinogenic effects have been demonstrated in rats, mice, hamsters, and rabbits following ingestion or inhalation; tumours were produced at several sites, including angiosarcomas of the liver. Vinyl chloride produces angiosarcomas in the liver of man, as well as tumours of the brain, lung, and haematolymphopoietic system (*1*). The International Agency for

Research on Cancer regards the evidence on vinyl chloride as sufficient to support a causal association between exposure and cancer (*12*).

2.1.4.3 *Mutagenicity*

The mutagenicity of vinyl chloride and several of its metabolites has been reviewed (*1*). It is mutagenic in a number of biological systems, including *Salmonella typhimurium, Escherichia coli* K12 bioauxotropic strain, several species of yeast, germ cells of *Drosophila*, and Chinese hamster V79 cells. The mutagenic action appears to be dependent on metabolic activation.

2.1.4.4 *Teratogenicity*

Skeletal abnormalities have been observed in mice and rats exposed during gestation (*13*).

REFERENCES

1. *Some monomers, plastics and synthetic elastomers, and acrolein.* Lyon, International Agency for Research on Cancer, 1979 (IARC Monographs on the evaluation of the carcinogenic risk of chemicals to humans, vol. 19).
2. DRESSMAN, R. C. & McFARREN, E. F. Determination of vinyl chloride migration from polyvinyl chloride pipe into water. *Journal of the American Water Works Association,* **70**: 29 (1978).
3. *Preliminary assessment of the environmental problems associated with vinyl chloride and polyvinyl chloride.* Springfield, VA, US Environmental Protection Agency, 1974 (EPA 560/4-74-001).
4. SAFE DRINKING WATER COMMITTEE. *Drinking water and health.* Washington, DC, National Academy of Sciences, 1977, p. 794.
5. ANON. FDA to propose ban on use of PVC for liquor use. *Food chemical news,* 14 May: 3–4 (1973).
6. WATANABE, P. G. ET AL. Fate of (^{14}C) vinyl chloride after single dose administration in rats. *Toxicology and applied pharmacology,* **36**: 339 (1976).
7. WITHEY, J. R. Pharmacodynamics and uptake of vinyl chloride monomer administered by various routes to rats. *Journal of toxicology and environmental health,* **1**: 381 (1976).
8. BOLT, H. M. ET AL. Disposition of [1,2-^{14}C] vinyl chloride in the rat. *Archives of toxicology,* **35**: 153 (1976).
9. PLUGGE, H. & SAFE, S. Vinyl chloride metabolism. A review, *Chemosphere,* **6**: 309 (1977).
10. GREEN, T. & HATHAWAY, D. E. The biological fate in rats of vinyl chloride in relation to its carcinogenicity. *Chemico-biological interactions,* **11**: 545 (1975).
11. PATTY, F. A., ed. *Industrial hygiene and toxicology.* Vol. II, New York, Interscience, 1963.
12. *Chemicals and industrial processes associated with cancer in humans.* Lyon, International Agency for Research on Cancer, 1979 (IARC Monographs on the evaluation of the carcinogenic risk of chemicals to humans, suppl. 1).
13. JOHN, J. A. ET AL. The effects of maternally-inhaled vinyl chloride on embryonal and foetal development in mice, rats and rabbits. *Toxicology and applied pharmacology,* **39**: 497 (1977).

2.2 1,1-Dichloroethene

2.2.1 General aspects

Of the three isomers of dichloroethene, 1,1-dichloroethene (1,1-DCE) is the most widely used in the chemical industry. It is an intermediate in the synthesis of methylchloroform and the production of poly(vinylidene chloride) (PVDC) copolymer. PVDC polymers are used as barrier coatings in the packaging industry and Saran, a 1,1-DCE containing polymer, is widely used in the food packaging industry. 1,1-DCE has a water solubility of 2500 mg/litre; its octanol/water partition coefficient has been reported as 5.37, indicating that it should not accumulate significantly in animals.

2.2.2 Routes of exposure

2.2.2.1 *Water*

In the USA, the National Organics Monitoring Survey of the EPA (*1*) reported detecting 1,1-DCE in drinking-water, but did not quantify its occurrence. One source of 1,1-DCE could be the decomposition of 1,1,1-trichloroethane which has occasionally been detected in drinking-water at concentrations of about 1 µg/litre (*2, 3*). Dichloroethene has been found in some European groundwaters.

2.2.2.2 *Food*

1,1-DCE copolymer food wrappers find extensive use but unfortunately no data are available on the extent to which the unreacted monomer migrates into the wrapped food. The possibilities of other human contact via the diet appear remote.

2.2.2.3 *Air*

The major exposure via the inhalation route is occupational. The threshold limit value (TLV) is 40 mg/m^3 of air and corresponds to an exposure of 280 mg/day for workers in an industry using or manufacturing DCE (*4*).

2.2.3 Metabolism

2.2.3.1 *Absorption*

On the basis of studies of related compounds, such as trichloroethene, it is assumed that virtually 100% of ingested 1,1-DCE may be systemically absorbed (*5, 6*).

2.2.3.2 *Distribution*

In studies on 1,1-DCE distribution in rats (*7*), the largest concentrations were found in the kidneys, followed by the liver, spleen, heart, and brain. Blood concentrations were high relative to tissue concentrations. Data from subcellular distribution studies suggest substantial binding of 1,1-DCE metabolites to macromolecules and associations with lipids.

2.2.3.3 *Biotransformation*

Liebman & Ortiz (*8*) identified the formation of chloracetic acid from 1,1-DCE. It appears that chloroethenes are metabolized through epoxide intermediates, which are reactive and may form covalent bonds with tissue macromolecules (*9*). In intact test animals, a large portion of systemically absorbed 1,1-DCE is metabolized. The relationship between 1,1-DCE metabolites and their toxicity is not well understood.

2.2.4 Health effects

2.2.4.1 *Acute, subchronic, and chronic toxicity*

1,1-DCE, like other chlorinated ethenes, possesses anaesthetic properties. Kidney and liver damage in rats and guinea-pigs exposed to air containing 1,1-DCE was reported by Prendergast (*10*). Differences were observed between intermittent and continuous exposures at similar 1,1-DCE concentrations and total exposure times. Continuous exposure, at lower concentrations than intermittent exposure, produced increased mortality. Oral administration of single doses of 200–400 mg of 1,1-DCE per kg of body weight had strong effects on liver enzyme activities. Only one epidemiological study has been published in which workers exposed to 1,1-DCE were examined (*11*). No abnormal findings could be associated with 1,1-DCE exposure in a population of 138 workers; measured concentrations in the work places ranged from 9 to 280 mg/m^3 (time-weighted averages).

2.2.4.2 *Mutagenicity*

1,1-DCE has been shown to be mutagenic in *Salmonella typhimurium* strains TA 1530 and TA 100 (*12*) and *E. coli* K12 (*13*). Henschler (*9*) and his associates have suggested that the mutagenic and presumably carcinogenic activities of the chloroethene series are related to the unsymmetrical chlorine substitution of the respective epoxide intermediates. Such substitution would result in less stable and more reactive intermediates than those derived from symmetrically substituted epoxides. The finding of increased mutation rates in bacterial systems has not yet been confirmed in mammalian systems.

2.2.4.3 *Teratogenicity*

Teratogenic effects of the DCEs do not appear to have been evaluated.

2.2.4.4 *Carcinogenicity*

Maltoni and his coworkers (*14, 15*) have reported the effects of inhalation exposures to 1,1-DCE. At concentrations of 100 mg/m³ of air, 25 of 300 Swiss mice developed kidney adenocarcinomas, while no adenocarcinomas were observed in control animals. A significant increase in mammary adenocarcinomas in Swiss mice inhaling 100 mg/m³ and in Sprague-Dawley rats exposed to 600 mg/m³ was also observed.

Lee et al. (*16*) observed a small increase in hepatic haemangiosarcomas in animals exposed to 1,1-DCE in a concentration of 220 mg/m³ for 4 hours a day, 5 days a week for 7–12 months.

Rampy et al. (*17*) exposed Sprague-Dawley rats to drinking-water containing 200 mg of 1,1-DCE per litre for 2 years and to 100 and 300 mg/m³ by inhalation. He found no evidence of increased tumours in animals treated with 1,1-DCE. In view of the demonstrated insensitivity of Sprague-Dawley rats in the Maltoni (*15*) study, however, these data are not considered to alter the interpretation of positive results in Swiss mice.

There is some evidence that rats are in general sensitive to the carcinogenic effects of low molecular weight chlorinated hydrocarbons (*18*). Epidemiological studies on workers exposed to vinylidine chloride are inadequate for evaluation (*19*).

2.2.5 Derivation of criterion

2.2.5.1 *Existing standards*

Existing standards in the USA are for occupational exposures via the inhalation route. The TLV as established by the American Conference of Governmental Industrial Hygienists (*4*) is 40 mg of 1,1-DCE per m³ of air in working-places. This value allows for a daily exposure of 286 mg of 1,1-DCE. This standard was established on the basis of the work of Prendergast et al. (*10*) described above.

2.2.5.2 *Carcinogenic risk limit*

1,1-DCE has been shown to produce mammary tumours in both mice and rats, and kidney adenocarcinomas in mice. Additionally, 1,1-DCE has been shown to be mutagenic in the Ames assay, a qualitative indicator of carcinogenic activity. Given this information, a linear multistage extrapolation model was applied to determine a limit that would give a calculated risk of less than one additional case of cancer per 100 000 population assuming a daily consumption of 2 litres of drinking-water by a 70-kg man. The value obtained was 0.3 μg/litre. This value is lower than that based on non-carcinogenic risks.

REFERENCES

1. *Statement of basis and purpose for an amendment to the national interim primary drinking water regulations on a treatment technique for synthetic organics.* Washington, DC, US Environmental Protection Agency, 1978.
2. *Preliminary assessment of suspected carcinogens in drinking water.* Washington, DC, US Environmental Protection Agency, 1975.
3. *List of organic compounds identified in US drinking water.* Cincinnati, OH, US Environmental Protection Agency, 1978.
4. *TLVs—Threshold limit values for chemical substances and physical agents in the workroom environment with intended changes for 1976.* Cincinnati, American Conference of Governmental Industrial Hygienists. 1976.
5. McKENNA, M. J. ET AL. The fate of (^{14}C) vinylidene chloride following inhalation exposure and oral administration in the rat. *Proceedings of the Society of Toxicology,* 206 (1977).
6. McKENNA, M. J. ET AL. Pharmacokinetics of vinylidene chloride in the rat. *Environmental health perspectives,* **21**: 99–106 (1977).
7. JAEGER, R. L. ET AL. 1,1-dichloroethylene hepatotoxicity: proposed mechanism of action of distribution and binding of ^{14}C radio-activity following inhalation exposure in rats. *Environmental health perspectives,* **21**: 113–120 (1977).
8. LEIBMAN, K. C. & ORTIZ, E. Metabolism of halogenated ethylenes. *Environmental health perspectives,* **21**: 91–98 (1977).
9. HENSCHLER, D. Metabolism and mutagenicity of halogenated olefins — A comparison of structure and activity. *Environmental health perspectives,* **21**: 61–64 (1977).
10. PRENDERGAST, J. A. ET AL. Effects on experimental animals of long-term inhalation of trichloroethylene, carbon tetrachloride, 1,1,1-trichloroethane, dichlorodifluoromethane, and 1,1-dichloroethylene. *Toxicology and applied pharmacology,* **10**: 270–289 (1967).
11. OTT, M. G. ET AL. A health study of employees exposed to vinylidene chloride. *Journal of occupational medicine,* **18**: 735 (1976).
12. BARTSCH, H. ET AL. Tissue-mediated mutagenicity of vinylidene chloride and 2-chlorobutadiene in *Salmonella typhimurium. Nature,* **255**: 641 (1975).
13. GREIM, H. ET AL. Mutagenicity *in vitro* and potential carcinogenicity of chlorinated ethylenes as a function of metabolic oxirane formation. *Biochemical pharmacology,* **24**: 2013 (1975).
14. MALTONI, C. ET AL. Carcinogenicity bioassays of vinylidene chloride. Research plan and early results. *Medicina de lavoro,* **68**: 241 (1977).
15. MALTONI, C. Recent findings on the carcinogenicity of chlorinated olefins. *Environmental health perspectives,* **21**: 1 (1977).
16. LEE, C. C. ET AL. Inhalation toxicity of vinyl chloride and vinylidene chloride. *Environmental health perspectives,* **21**: 25 (1977).
17. RAMPY, L. W. ET AL. Interim results of a two-year toxicological study in rats of vinylidene chloride incorporated in the drinking water or administered by repeated inhalation. *Environmental health perspectives,* **21**: 33 (1977).
18. NATIONAL CANCER INSTITUTE. *Bioassay of tetrachloroethylene for possible carcinogenicity.* Washington, DC, US Department of Health, Education and Welfare. 1977 (Technical Report Series No. 13. (NIH) 77–813).
19. *Some monomers, plastics and synthetic elastomers, and acrolein.* Lyon, International Agency for Research on Cancer, 1979 (IARC Monographs on the evaluation of the carcinogenic risk of chemicals to humans, vol. 19).

2.3 Trichloroethene

2.3.1 General aspects

Trichloroethene (1,1,2-trichloroethene; TCE) is a clear colourless liquid, with the empirical formula C_2HCl_3. It is used mainly as a degreasing solvent in metal industries. TCE has also been used as a household and industrial dry-cleaning solvent, an extractive solvent in foods, and as an inhalation anaesthetic during certain short-term surgical procedures (1).

The volatilization of TCE during production and use is the major source of environmental levels of this compound. TCE has been detected in air, in food, and in human tissues (2). Its detection in rivers, municipal water supplies, the sea, and aquatic organisms indicates that TCE is widely distributed in the aquatic environment (2–4). TCE is not expected to persist in surface-water because of its volatility. However, it has been found as a frequent contaminant of groundwater.

2.3.2 Routes of exposure

2.3.2.1 *Water*

The US National Organics Monitoring Survey observed TCE in drinking-water in 4 of 112 cities in March–April 1976, in 28 of 113 cities in May–July 1976, and in 19 of the cities in November 1976 to January 1977; the mean concentrations were 11 μg/litre, 21 μg/litre, and 1.3 μg/litre, respectively. TCE in water may occur as a result of direct contamination or from atmospheric contamination by rainfall (2). TCE may also be formed during the chlorination of water (5, 6).

2.3.2.2 *Food*

There is little information concerning the occurrence of TCE in foodstuffs. In England, TCE has been observed at concentrations up to 10 μg/kg in meats and up to 5 μg/kg in fruits, vegetables, and beverages (3). Packets of tea were found to contain 60 μg of TCE per kg. Little TCE would be expected in other foodstuffs, except in the case of ground and instant coffee and in spice extracts, when TCE has been used as a solvent.

2.3.2.3 *Air*

By far the most significant exposures of humans to TCE are confined to a relatively small industrial population (7). Other inhalation exposures are associated with the use of cleaning fluids containing TCE, but the hazards of such exposure would be acute.

2.3.3 Metabolism

2.3.3.1 *Absorption*

TCE is readily absorbed by all routes of exposure. This would be predicted on the basis of its physical and chemical properties (*8*). Most human data concerning TCE absorption have been obtained with inhalation as the route of exposure because of interest in the compound as an industrial toxicant and its use as an anaesthetic. Absorption of TCE following ingestion has not been studied in humans. In rats, 72–85% and 10–20% of the total orally administered dose could be accounted for in expired air and urine, respectively, with less than 0.5% appearing in the faeces (*9*). This indicates that at least 80% (and probably more) of ingested TCE is systemically absorbed.

2.3.3.2 *Distribution*

The distribution of TCE in the body is as would be expected on the basis of its chemical and physical properties (*8*). In guinea-pigs it was observed that the concentrations in the ovaries tend to be about 50%, and in other tissues about 25%, of the concentration observed in fat. Laham (*10*) demonstrated transplacental diffusion of TCE in humans. The ratio of fetal blood concentration to maternal blood concentration varies between 0.52 and 1.90.

2.3.3.3 *Biotransformation*

The metabolism of TCE appears to be central to its long-term deleterious effects. In a qualitative sense, the metabolism of TCE appears similar across species (*11–13*). The principal products of TCE metabolism measured in urine are trichloroacetaldehyde, trichloroethanol, trichloroacetic acid, and conjugated derivatives (glucuronides) of trichloroethanol (*14*). The metabolite trichloroethanol has been suggested as responsible for the long-term central nervous system (CNS) effects of TCE inhalation (*15*). In terms of reported carcinogenic and mutagenic effects of TCE, the metabolic pathway, rather than the final products of the pathway, is of paramount importance. The essential feature of the pathway is the formation of a reactive epoxide, trichloroethene oxide, which can alkylate nucleic acids and proteins (*9, 16–19*). Such covalent binding can be increased with epoxide hydrase inhibition (*16*).

2.3.3.4 *Elimination*

TCE and its metabolites are excreted in exhaled air, urine, sweat, faeces, and saliva (*12, 13*). TCE is lost from the body with a half-time of about 1.5 hours (*20*). Trichloroacetic acid, trichloroethanol, and the glucuronide of trichloroethanol are excreted more slowly. The biological half-life measured in urine in humans has ranged from 12 to 50 hours for trichloroethanol and 36 to 73 hours for trichloroacetic acid (*15, 21*).

2.3.4 Health effects

2.3.4.1 *Acute, subchronic, and chronic toxicity*

Classically, TCE is known as a central nervous system depressant. In fact, the compound has been used medically as a general anaesthetic (*22*). Although the predominant feature of the clinical picture is the direct CNS depression produced by high exposures to TCE, there is evidence of longer term CNS effects resulting from TCE exposure (*23*).

Fatal hepatic failure has been observed following use of TCE as an anaesthetic. Such failure has generally involved patients with complicating diseases, such as malnutrition, toxaemias, or burns, or who had received transfusions (*22*). Liver failure in experimental animals is marked by generalized binding of TCE metabolites to proteins and nucleic acids (*17*).

Renal failure has been an uncommon problem with TCE anaesthesia (*22*). Although depressed kidney function can be documented with TCE in experimental animals, it requires very high doses (*24*) and the effect is much less potent than that observed with chloroform or carbon tetrachloride. Renal damage has been reported in fatal cases involving TCE abuse (*1*).

2.3.4.2. *Mutagenicity*

TCE has been reported to possess mutagenic activity in a number of bacterial strains. Greim et al. (*25*) demonstrated reverse mutations in *E. coli* K12 when coupled with phenobarbital-induced mouse-liver microsomes at a concentration of 3.3 mmol/litre in the incubation media. In the presence of Aroclor-1254-induced rat-liver microsomes or B6C3-F1 mouse-liver microsomes, TCE increased the *S. typhimurium* revertant rate (*26*). Similar observations have been made in the yeast *Saccharomyces cerevisiae* (strain XV 185–14C) (*27*). There is some doubt as to the mutagenicity of TCE, however. On chemical analysis, technical grade TCE was found to contain epichlorohydrin and epoxibutane, two compounds that Henschler (*28*) observed to be more potent mutagens than TCE in *S. typhimurium* (TA 100). Pure TCE was weakly mutagenic. These investigators concluded that the mutagenic activity, formerly attributed to TCE, was probably due in part to mutagenic contaminants found in some samples of TCE. TCE has been uniformly negative in mutagenicity testing in the absence of metabolic activation (*25–27*), which would suggest that the two direct-acting compounds identified could not alone account for the activity.

2.3.4.3 *Teratogenicity*

Exposure of mice and rats to TCE in air at a concentration of 1600 mg/m³ on days 6–15 of gestation for 7 hours a day did not produce teratogenic effects (*29*). Although not statistically significant, there was evidence of haemorrhages in the cerebral ventricles (2/12 litters). A few cases of undescended testicles (2/12 litters) were observed in the TCE-treated mice but the incidence was very low. This appears to be the only teratogenesis study conducted with TCE.

2.3.4.4 Carcinogenicity

In the USA, the National Cancer Institute (30) observed increased incidence of hepatocellular carcinoma in mice (strain B6C3-F1) treated with TCE. The time-weighted doses administered for 5 days/week for 78 weeks were 1169 and 2339 mg/kg of body weight for males and 869 and 1739 mg/kg of body weight for females. Similar experiments with Osborne-Mendel rats failed to increase the incidence of tumours in this species. However, the rats also responded poorly in a control experiment with carbon tetrachloride, indicating that the B6C3-F1 mouse is a much more sensitive test animal than the rat to induction of carcinomas by chlorinated compounds. The data obtained from mice are summarized in Table 1. Some evidence of metastasis of hepatocellular carcinomas to the lung was observed in male mice treated with both low and high doses of TCE (4/50 and 3/48, respectively).

Table 1. Incidence of hepatocellular carcinoma in TCE-treated B6C3-F1 mice (30)

	Males	Females
Control	1/20	0/20
Low dose	26/50	4/50
High dose	31/48	11/47

TCE has been shown to induce transformation in a highly sensitive in vitro Fischer rat embryo cell system (F1706), which is used for identifying carcinogens. At a concentration of 1 mol/litre, TCE-induced transformation of rat embryo cells was characterized by the appearance of progressively growing foci made up of cells lacking contact inhibition, and by the growth of macroscopic foci when inoculated in semisolid agar. The transformed cells grew as undifferentiated fibrosarcomas at the site of inoculation in 100 % of newborn Fischer rats between 27 and 68 days after inoculation (31).

It has been pointed out that the TCE used in the National Cancer Institute bioassay (30) contained traces of the monofunctional alkylating agents epichlorohydrin and epoxibutane as stabilizers (32). However, TCE was also shown to induce cell transformation in the Fischer rat embryo cell system (31). In these latter experiments the TCE used was stated to be 99.9 % pure. TCE has also been shown to form covalent bindings with cellular macromolecules (16–19). In the majority of cases, covalent binding of TCE was disassociated from impurities by the use of radioactively labelled TCE (18). Such covalent binding properties are commonly associated with chemical carcinogens. In any event, it appears unlikely that epichlorohydrin or epoxibutane would account for liver tumours produced by TCE, since their instability in aqueous solution would militate against their responsibility for tumours remote from the site of application. The carcinogenicity of pure TCE was investigated by Henschler (33).

Although it is neither desirable, nor the general practice, to estimate carcinogenic potency for chemicals whose carcinogenic activity has been demonstrated in only one animal species, the NCI bioassay is the only long-term study of TCE toxicity by the oral route of exposure. In particular, the development of liver tumours following exposure to chemicals that are hepatotoxic raises questions as to the mechanism by which such tumours are produced—whether it is one of initiation or promotion. Models used to estimate potency are based on the somatic mutation theory of chemical carcinogenesis, which postulates a genetic basis for tumour initiation. There are no extrapolation models appropriate for tumour promoters. Because there is a genuine need for guidance relating to acceptable levels of TCE, especially in contaminated groundwater, a conservative approach was taken and the indicated carcinogenic hazards associated with exposure to TCE were estimated from its demonstrated carcinogenicity in B6C3-Fl mice by means of a linear extrapolation of risk using a multistage model (30). This calculation, based on a consumption of 2 litres of water per day for a 70-kg man and on an acceptable risk level of less than one additional case of cancer per 100 000 population for a lifetime of exposure, results in a *tentative* recommended value of 30 µg/litre for TCE. Epidemiological studies on workers exposed to TCE were inadequate for evaluation (34–37).

REFERENCES

1. HUFF, J. E. New evidence on the old problems of trichloroethylene. *Industrial medicine*, **40**: 25 (1971).
2. PEARSON, C. R. & McCONNELL, G. Chlorinated C_1 and C_2 hydrocarbons in the marine environment. *Proceedings of the Royal Society of London, Series B*, **189**: 305 (1975).
3. McCONNELL, G. ET AL. Chlorinated hydrocarbons and the environment. *Endeavour*, **34**: 13 (1975).
4. *Statement of basis and purpose for an amendment to the national primary drinking water regulations on a treatment technique for synthetic organics.* Washington, DC, US Environmental Protection Agency, 1978.
5. NATIONAL RESEARCH COUNCIL. *Drinking water and health.* Washington, DC, National Academy of Sciences., 1977.
6. BELLAR, T. A. ET AL. The occurrence of organohalides in chlorinated drinking waters. *Journal of the American Water Works Association*, **66**: 703 (1974).
7. FISHBEIN, L. Industrial mutagens and potential mutagens. I. Halogenated aliphatic derivatives. *Mutation research*, **32**: 267 (1976).
8. GOLDSTEIN, A. ET AL. The absorption, distribution, and elimination of drugs. In: *Principles of drug action: the basis of pharmacology.* 2nd ed. New York, John Wiley and Sons, 1974, pp. 129–154.
9. DANIEL, J. W. The metabolism of Cl-labelled trichloroethylene and tetrachloroethylene in the rat. *Biochemical pharmacology*, **12**: 795 (1963).

10. LAHAM, S. Studies of placental transfer of trichloroethylene. *Industrial medicine*, **39**: 46 (1970).

11. IKEDA, M. & OHTSUJI, H. A comparative study of the excretion of Fujiwara reaction-positive substances in urine of humans and rodents given trichloro- or tetrachloro-derivatives of ethane and ethylene. *British journal of industrial medicine*, **29**: 99 (1972).

12. KIMMERLE, G. and EBEN, A. Metabolism, excretion and toxicology of trichloroethylene after inhalation. 1. Experimental exposure on rats. *Archives of toxicology*, **30**: 115 (1973).

13. KIMMERLE, G. & EBEN, A. Metabolism, excretion and toxicology of trichloroethylene after inhalation. 2. Experimental human exposure. *Archives of toxicology*, **30**: 127 (1973).

14. IKEDA, M. ET AL. Urinary excretion of total trichloro-compounds, trichloroethanol, trichloroacetic acid, as a measure of exposure to trichloroethylene and tetrachloro-ethylene. *British journal of industrial medicine*, **29**: 46 (1970).

15. ERTLE, T. ET AL. Metabolism of trichloroethylene in man. I. The significance of trichloroethanol in long-term exposure conditions. *Archives of toxicology*, **29**: 171 (1972).

16. VAN DUUREN, B. L. & BANERJEE, S. Covalent interaction of metabolites of the carcinogen trichloroethylene in rat hepatic microsomes. *Cancer research*, **36**: 2419 (1976).

17. BOLT, H. M. & FILSER, J. G. Irreversible binding of chlorinated ethylenes to macromolecules. *Environmental health perspectives*, **21**: 107 (1977).

18. UEHLEKE, H. & POPLAWSKI-TABARELLI, S. Irreversible binding of ^{14}C-labelled trichloroethylene in mice liver constituents *in vivo* and *in vitro*. *Archives of toxicology*, **37**: 289 (1977).

19. ALLEMAND, H. ET AL. Metabolic activation of trichloroethylene into a chemically reactive metabolite toxic to the liver. *Journal of pharmacology and experimental therapeutics*, **204**: 714 (1978).

20. STEWART, R. D. ET AL. Observations on the concentrations of trichloroethylene in blood and expired air following exposure to humans. *American Industrial Hygiene Association journal*, **23**: 167 (1962).

21. IKEDA, M. & IMAMURA, T. Biological half-life of trichloroethylene and tetrachloro-ethylene in human subjects. *Internationales Archiv für Arbeitsmedizin*, **31**: 209 (1973).

22. DEFALQUE, F. J. Pharmacology and toxicology of trichloroethylene. A critical review of the world literature. *Clinical pharmacology and therapeutics*, **2**: 665 (1961).

23. GRANDJEAN, E. ET AL. Investigations into the effects of exposure to trichloroethylene in mechanical engineering. *British journal of industrial medicine*, **12**: 131 (1955).

24. KLAASEN, C. D. & PLAA, G. L. Relative effects of various chlorinated hydrocarbons on liver and kidney function in dogs. *Toxicology and applied pharmacology*, **10**: 119 (1967).

25. GREIM, H. ET AL. Mutagenicity *in vitro* and potential carcinogenicity of chlorinated ethylenes as a function of metabolic oxirane formation. *Biochemical pharmacology*, **24**: 2013 (1975).

26. SIMMON, V. F. ET AL. Mutagenic activity of chemicals identified in drinking-water. In: Scott, D. et al., ed., *Progress in genetic toxicology*. Amsterdam, Elsevier/North Holland Biomedical Press, 1977, pp. 249–258.

27. SHAHIN, M. & VON BARSTEL, R. Mutagenic and lethal effects of benzene hexachloride, dibutylphthalate and trichloroethylene in *Saccharomyces cervisiae*. *Mutation research*, **48**: 173 (1977).

28. HENSCHLER, D. Metabolism of chlorinated alkenes and alkanes as related to toxicity. *Journal of environmental pathology and toxicology*, **1**: 125 (1977).

29. SCHWETZ, B. A. ET AL. The effect of maternally inhaled trichloroethylene, perchloroethylene, methyl chloroform and methylene chloride on embryonal and fetal development in mice and rats. *Toxicology and applied pharmacology*, **32**: 84 (1975).

30. NATIONAL CANCER INSTITUTE, *Carcinogenesis bioassays of trichloroethylene*. Washington, DC, US Department of Health, Education and Welfare, 1976 (CAS No. 79-01-6, NCI-CG-TR-2).

31. PRICE, P. J. ET AL. Transforming activities of trichloroethylene and proposed industrial alternatives. *In vitro*, **14**: 290 (1978).
32. HENSCHLER, D. ET AL. Carcinogenicity of trichloroethylene: fact or artifact? *Archives of toxicology*, **37**: 233 (1977).
33. HENSCHLER, D. ET AL. Carcinogenicity study of trichloroethylene by long term inhalation in three animal species. *Archives of toxicology*, **43**: 237 (1980).
34. *Some halogenated hydrocarbons.* Lyon, International Agency for Research on Cancer, 1979 (IARC Monographs on the evaluation of the carcinogenic risk of chemicals to humans, vol. 20).
35. TOLA, S. ET AL. A short study on workers exposed to trichloroethylene. *Journal of occupational medicine*, **22**: 737 (1979).
36. BLAIR, A. Mortality among workers in the metal polishing and plating industry. *Journal of occupational medicine*, **22**: 158 (1980).
37. BLAIR, A. & MASON, T. J. Cancer and mortality in United States counties with metal plating industries. *Archives of environmental health*, **35**: 92 (1980).

2.4 Tetrachloroethene

2.4.1 General aspects

Tetrachloroethene (1,1,2,2-tetrachloroethene, perchloroethylene, PCE)[a] is a colourless, nonflammable liquid used primarily as a solvent in the dry-cleaning industries. It is used to a lesser extent as a degreasing solvent in metal industries (1).

Tetrachloroethene is widespread in the environment, and is found in trace amounts in water, aquatic organisms, air, foodstuffs, and human tissue (2). The highest environmental levels of PCE are found in the commercial dry-cleaning and metal-degreasing industries (3).

Although PCE is released into water via aqueous effluents from production plants, consumer industries, and household sewage, its level in ambient water is reported to be minimal owing to its high volatility. PCE has been detected at levels below 1 µg/litre in water surrounding chlorinated hydrocarbon production plants in England (4) and in surface-water in the USA (5). However, it is frequently found at high concentrations in contaminated groundwater.

2.4.2 Routes of exposure

2.4.2.1 *Water*

The US National Organics Monitoring Survey (6) detected PCE in 9 of 105 samples of drinking-water between November 1976 and January 1977 (range, 0.2–3.1 µg/litre). The mean concentration of the nine positive samples was 0.81 µg/litre. In Switzerland, PCE concentrations as high as 954 µg/litre have been found in contaminated groundwater (7). PCE was one of two halogenated compounds identified both in drinking-water and in the plasma of individuals living in New Orleans

---

[a] Perchloroethylene—abbreviated to PCE—is the old name for tetrachloroethene. Although the modern name has been adopted here, the abbreviation PCE is retained to avoid confusion with the abbreviation for trichloroethene.

(8). In the United Kingdom, municipal waters have been found to contain up to 0.38 μg of PCE per litre (4).

2.4.2.2 Food

PCE concentrations in seafood collected from the Liverpool Bay area in England ranged from 0.5 to 30 μg/kg (4, 9). PCE concentrations in foodstuffs ranged from non-detectable amounts (<0.01 μg/kg) in orange juice to 13 μg/kg in English butter (2).

2.4.2.3 Air

General environmental PCE concentrations tend to be low. Pearson & McConnell (4) observed concentrations in city atmospheres in Great Britain to range from less than 0.68 to 68 μg/m³. In a suburb of Munich, Loechner (10) found concentrations of 4 μg/m³ whereas air in the centre of Munich contained 6 μg/m³. Surveys at 8 locations in the USA indicated concentrations up to 6.7 μg/m³ in urban areas but less than 0.013 μg/m³ in rural areas (11). As with a number of related chlorinated hydrocarbon solvents of low relative molecular mass, by far the most significant exposure to PCE is in industrial environments (12). The major uses of PCE are in the textile and dry-cleaning industries (69%), metal cleaning (16%), and as a chemical intermediate (12%).

2.4.3 Metabolism

2.4.3.1 Absorption

Using inhalation exposure, Stewart et al. (13) found that PCE reached near steady-state levels in the blood of human volunteers after 2 hours of continuous exposure. Such results suggest a rapid attainment of steady-state levels of PCE within the body. Absorption of PCE by the oral route has not been specifically studied. However, there is every reason, based on analogy with other chemicals with similar properties (e.g., trichloroethene, chloroform), to believe that it would be completely absorbed from the gastrointestinal tract.

2.4.3.2 Biotransformation

The metabolism of PCE has been studied extensively in humans and experimental animals. In a qualitative sense, metabolic products appear to be similar in humans (14, 15) and experimental animals (16–18). Basically it is believed that PCE is metabolized through an epoxide (tetrachloroethene oxide) and an acid chloride intermediate (trichloroacetyl chloride) to form the end product of its metabolism, trichloroacetic acid. Ogata et al. (19) report that 1.8% of PCE retained by humans was converted to trichloroacetic acid and 1.0% to an unknown metabolite, in 67 hours.

2.4.3.3 *Elimination*

PCE itself is primarily eliminated from the body via the lungs (*13, 20, 21*). The half-time for respiratory elimination of PCE has been estimated at 65 hours (*20, 21*). Trichloroacetic acid, as a metabolite of PCE, is eliminated with a half-time of 144 hours via the urine (*21*).

2.4.4 Health effects

2.4.4.1 *Acute, subchronic, and chronic toxicity*

As with all other members of the chloroethene family, acute effects of PCE are very much dominated by central nervous system depression. The only indications of long-term effects on the central nervous system are changes in EEG patterns associated with increased electrical impedance of the cerebral cortex at exposures as low as 100 mg of PCE per m^3 of air for 4 hours a day for 15–30 days (*22, 23*). These effects were reported to be associated with sporadic swollen and vacuolized protoplasm in some cells (*23*). Although only limited information is available from experimental animals, it generally supports findings of acute central nervous system depression (*24*). As in the case of human clinical studies, little information is available concerning long-term exposures. That more serious central nervous system problems may be associated with chronic PCE exposure in humans is suggested by a few sporadic case reports (*25, 26*) and small-scale epidemiological and clinical studies (*27*) involving a group of men occupationally exposed to concentrations of 1890–2600 mg of PCE per m^3. However, these studies have often been complicated by exposures to other solvents (*28*).

Short-term PCE exposures at higher concentrations, and longer exposures at lower concentrations, can produce damage to the kidneys and liver in dogs (*29*). Kylin et al. (*30*) noted moderate fatty degeneration of the liver in mice following a single 240-minute exposure to 1340 mg of PCE per m^3. Exposure to this same concentration 4 hours daily, 6 days a week for up to 8 weeks was found to increase the severity of the lesions caused by PCE (*31*). Epidemiological studies on workers exposed to PCE are inadequate for evaluation (*32–34*).

2.4.4.2 *Mutagenicity*

Cerna & Kypenova (*35*) found that treatment with PCE resulted in elevated mutagenic activity in *Salmonella* strains sensitive to both base substitution and frameshift mutation. However, PCE had no effect on the spontaneous mutation rate in *E. coli* K12 in the presence of liver microsomes (*36*).

2.4.4.3 *Carcinogenicity*

PCE has been demonstrated to be a liver carcinogen in B6C3-F1 mice (*37*). Results in Osborne-Mendel rats were negative, but a high rate of early mortality precluded the use of rat data in evaluating the carcinogenicity of PCE. Furthermore, recent studies in which carbon tetrachloride was used as a positive control revealed that Osborne-Mendel rats have a low sensitivity to induction of hepatocellular carcinoma by chlorinated hydrocarbons (*38*).

Carcinogenic hazards associated with exposure to PCE may be estimated on the basis of its production of hepatocellular carcinoma in B6C3-F1 mice (*37*) using a linear multistage extrapolation model. Such a model is based on the somatic mutation theory of chemical carcinogenesis most appropriate for tumour initiators. However, chemicals that produce necrotic damage to tissues such as the liver are known to enhance tumorigenesis by epigenetic mechanisms (*39*). Extrapolation models for such effects do not exist. There was some reluctance to base an estimate of carcinogenic hazards on data from experiments utilizing hepatotoxic doses in the absence of other evidence of carcinogenicity. However, the lack of other suitable data and the genuine need for guidance for PCE-contaminated groundwater have prompted a conservative approach and PCE was treated as if it were a tumour initiator in order to derive a tentative recommended value. The results of the calculation indicate that a concentration in drinking-water of 10 µg of tetrachloroethene per litre would be predicted to increase the disease risk by less than one additional cancer per 100 000 population assuming a daily consumption of 2 litres of water.

REFERENCES

1. WINDHOLZ, M. ET AL. *The Merck index*. 9th ed. Rahway, NJ, Merck and Co., 1976.
2. McCONNELL, G. ET AL. Chlorinated hydrocarbons and the environment. *Endeavour*, **34**: 13–18 (1975).
3. NATIONAL INSTITUTE OF OCCUPATIONAL SAFETY AND HEALTH. *Criteria for a recommended standard—occupational exposure to tetrachloroethylene (perchloroethylene)*. Washington, DC, US Department of Health, Education & Welfare, 1976.
4. PEARSON, C. R. & McCONNELL, G. Chlorinated C_1 and C_2 hydrocarbons in the marine environment. *Proceedings of the Royal Society, Series B*, **189**: 305 (1975).
5. EWING, B & CHIAN, E. *Monitoring to detect previously unrecognized pollutants in surface waters*. Washington, DC, US Environmental Protection Agency, 1977.
6. *Statement of basis and purpose for an amendment to the national primary drinking water regulations on a treatment criteria for synthetic organics*. Washington, DC, US Environmental Protection Agency, 1978.
7. GIGER, W. & MOLNAR-KUBIEA, E. Tetrachloroethylene in contaminated ground and drinking waters. *Bulletin of environmental contamination and toxicology*, **19**(4): 475 (1978).
8. DOWTY, B. ET AL. Halogenated hydrocarbons in New Orleans drinking water and blood plasma. *Science*, **187**: 75 (1975).
9. DICKSON, A. G. & RILEY, J. P. The distribution of short-chained halogenated aliphatic hydrocarbons in some marine organisms. *Marine pollution bulletin*, **7**: 167 (1976).
10. LOECHNER, F. Perchloräthylen eine Bestandsaufnahme. *Umwelt*, **6**: 434 (1976).
11. LILLIAN, D. ET AL. Atmospheric fates of halogenated compounds. *Environmental science and technology*, **9**: 1042 (1975).

12. FISHBEIN, L. Industrial mutagens and potential mutagens I. Halogenated aliphatic hydrocarbons. *Mutation research*, **32**: 267 (1976).
13. STEWART. R. D. ET AL. Human exposure to tetrachloroethylene vapor. *Archives of environmental health*, **2**: 516 (1961).
14. IKEDA, M. ET AL. Urinary excretion of total trichloro-compounds, trichloroethanol and trichloroacetic acid as a measure of exposure to trichloroethylene and tetrachloro-ethylene. *British journal of industrial medicine*, **29**: 328 (1972).
15. IKEDA. M. Metabolism of trichloroethylene and tetrachloroethylene in human subjects. *Environmental health perspectives*, **21**: 239 (1977).
16. YLLNER, S. Urinary metabolites of ^{14}C-tetrachloroethylene in mice. *Nature*, **191**: 820 (1961).
17. DANIEL, J. W. The metabolism of ^{36}Cl-labelled trichloroethylene and tetrachloro-ethylene in the rat. *Biochemical pharmacology*, **12**: 795 (1963).
18. KEDA, M. & OHTSUJI, H. A comparative study of the excretion of Fujiwara – reaction-positive substances in urine of humans and rodents given trichloro- or tetrachloro-derivatives of ethane and ethylene. *British journal of industrial medicine*, **29**: 99 (1972).
19. OGATA, M. ET AL. Excretion of organic chlorine compounds in the urine of persons exposed to vapours of trichloroethylene and tetrachloroethylene. *British journal of industrial medicine*, **28**: 386 (1971).
20. STEWART, R. D. ET AL. Experimental human exposure to tetrachloroethylene. *Archives of environmental health*, **20**: 225 (1970).
21. IKEDA, M. & IMAMURA, T. Biological half-life of trichloroethylene and tetrachloro-ethylene in human subjects. *Internationales Archiv für Arbeitsmedizin*, **31**: 209 (1973).
22. DMITRIEVA, N. V. Maximum permissible concentrations of tetrachloroethylene in factory air. *Hygiene and sanitation*, **31**: 387 (1966).
23. DMITRIEVA, N. V. & KULESHOV, E. V. Changes in the bioelectric activity and electric conductivity of the brain in rats chronically poisoned with certain chlorinated hydrocarbons. *Hygiene and sanitation*, **36**: 23 (1971).
24. *Some halogenated hydrocarbons.* Lyon, International Agency for Research on Cancer, 1979 (IARC Monographs on the evaluation of the carcinogenic risk of chemicals to humans, vol. 20.)
25. GOLD, J. H. Chronic perchloroethylene poisoning. *Canadian Psychiatric Association journal*, **14**: 627 (1969).
26. MCMULLEN, J. K. Perchloroethylene intoxication. *British medical journal*, **2**: 1563 (1976).
27. COLER, H. R. & ROSSMILLER, H. R. Tetrachloroethylene exposure in a small industry. *Archives of industrial hygiene & occupational medicine*, **8**: 227 (1953).
28. TUTTLE, T. C. ET AL. *A behavioral and neurological evaluation of dry cleaners exposed to perchloroethylene.* Washington, DC, Department of Health, Education and Welfare, 1977 ((NIOSH) No. 77–214).
29. KLAASEN, C. D. & PLAA, G. L. Relative effects of chlorinated hydrocarbons on liver and kidney function in dogs. *Toxicology and applied pharmacology*, **10**: 119 (1967).
30. KYLIN, B. ET AL. Hepatotoxicity of inhaled trichloroethylene, tetrachloroethylene and chloroform. Single exposure. *Acta pharmacologica et toxicologica*, **20**: 16 (1963).
31. KYLIN, B. ET AL. Hepatotoxicity of inhaled trichloroethylene. Long-term exposure. *Acta pharmacologica et toxicologica*, **22**: 379 (1965).
32. BLAIR, A. ET AL. Causes of death among laundry and dry cleaning workers. *American journal of public health*, **69**: 508 (1979).
33. BLAIR, A. Mortality among workers in the metal polishing and plating industry. *Journal of occupational medicine*, **22**: 158 (1980).
34. BLAIR, A. & MASON, T. J. Cancer and mortality in the United States counties with metal plating industries. *Archives of environmental health*, **35**: 92 (1980).
35. CERNA, M. & KYPENOVA, H. Mutagenic activity of chloroethylenes analysed by screening system tests. *Mutation research*, **46**: 214 (1977).
36. GREIM, H. ET AL. Mutagenicity in vitro and potential carcinogenicity of chlorinated ethylenes as a function of metabolic oxirane formation. *Biochemical pharmacology*, **24**: 2013 (1975).

37. NATIONAL CANCER INSTITUTE. *Bioassay of tetrachloroethylene for possible carcinogenicity.* Washington, DC, Department of Health, Education and Welfare, 1977 (CAS No. 127-18-4, NCI-CG-TR-13 (NIH) 77-813).

38. NATIONAL CANCER INSTITUTE. *Carcinogenesis bioassay of trichloroethylene.* Washington, DC, Department of Health, Education and Welfare, 1976 (CAS No. 79-01-6, NCI-C6-TR-2 (NIH) 76-802).

39. SCHUMANN, A. ET AL. The pharmacokinetics and macromolecular interactions of perchloroethylene in mice and rats as related to oncogenicity. *Toxicology and applied pharmacology,* **55**: 207 (1980).

3. POLYNUCLEAR AROMATIC HYDROCARBONS

3.1 General description

Polynuclear aromatic hydrocarbons (PAH) are a large group of organic compounds consisting of two or more benzene rings with non-aromatic rings present in some instances. Adjacent rings share two carbon atoms.

Naphthalene

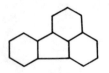

Fluoranthene

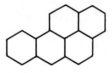

Benzo[a]pyrene

PAH are formed as a result of incomplete combustion of organic compounds but may also be synthesized by some bacteria, algae, and higher plants (1, 2). They are of relatively low solubility in water and are strongly adsorbed on to particulates and sediment clays (3), leading to higher concentrations in the water environment than would otherwise be possible on the basis of solubility considerations alone (1). PAH dissolved in water and adsorbed on particulate matter undergo photodecomposition provided that exposure to ultraviolet light from solar radiation of sufficient energy occurs (4, 5). Some microorganisms in soil can degrade PAH (6) and it is probable that at least some degradation of PAH in sediments also occurs (7).

3.2 Routes of exposure

PAH are present in the environment from both natural and anthropogenic sources. As a group, they are widely distributed in the environment, having been detected in animal and plant tissues, sediments, soils, air, and various water sources (8, 9).

3.2.1 Air

PAH concentrations in air vary with location and season. Industrial urban environments have higher PAH concentrations and these tend to be at a maximum in the colder months when more heating is used (*10*). Concentrations of PAH vary from about 0.001 μg to 2 μg per 1000 m^3 of air in rural areas to about 0.1 μg to 60 μg per 1000 m^3 in urban areas (*11*). In some polluted atmospheres, levels of benzo[*a*]pyrene varying from 20 to 400 μg per 1000 m^3 have been reported (*12*).

3.2.2 Food

PAH are found in substantial quantities in some foods, depending on the method of cooking, preservation, and storage, and are detected in a wide range of meat, fish, vegetables, and fruits (*13*). American sources indicate a total intake of PAH from food of the order of 1.6–16 μg/day (*14*).

3.2.3 Water

Levels of PAH in surface-waters are influenced by industrial discharges and in various German rivers have been shown to range from 0.12 to 3.1 μg/litre (*15*). Examination of a number of groundwater and drinking-water supplies for six PAH, namely fluoranthene, benzo[*b*]-fluoranthene, benzo[*k*]fluoranthene, benzo[*a*]pyrene, benzo[*ghi*]perylene and indeno[1,2,3-*cd*]pyrene, indicated that the collective concentrations generally did not exceed 0.05 μg/litre in groundwater and 0.1 μg/litre in drinking-water (*16*). Later work by Borneff (*17*) with data from a large number of analyses showed concentrations of these six PAH to be greater than 0.11 μg/litre in only 1% of samples, while for 90% of samples levels were between 0.001 and 0.01 μg/litre. Concentrations of benzo[*a*]pyrene were shown to range from 0.6 to 114 ng/litre in surface-waters and from 0.1 to 23.4 ng/litre in drinking-water (*18, 16*).

According to Borneff (*19*), two-thirds of the PAH in surface-waters are bound to particulate matter which is removed by sedimentation, flocculation, and filtration processes. Much of the remaining one-third of dissolved PAH may be removed by oxidation, the efficiency varying with the system used. Chlorination of water can remove 50–60% of benzo[*a*]pyrene, while filtration with activated carbon removed up to 99% of benzo[*a*]pyrene in field tests (*19*).

Contact with coal-tar-based pipe linings during distribution is known in some instances to lead to increases in PAH concentrations in the water. In such cases, an increase in the level of fluoranthene is particularly marked (*20*).

Estimates of annual intake of PAH are between 1 mg and 10 mg, with benzo[*a*]pyrene contributing 0.1–1.5 mg per person in developed countries (*21, 22*). However, it has been estimated that food accounts for about 99% of the daily oral intake and the typical intake from drinking-water is only 0.1% of the total (*23*).

3.3 Metabolism

PAH are highly lipid-soluble and are readily absorbed from the gut and lungs of mammals (24–26). They are initially cleared from the blood and liver very rapidly (27) and are distributed to a wide variety of tissues with a particular tendency to localize in body fat (26–29).

Metabolism of PAH is via the mixed-function oxidase system mediated by the cytochrome P450, with oxidation or hydroxylation as the first step. The resultant epoxides or phenols may then go through a detoxification reaction to produce glucuronides, sulfates, or glutathione conjugates. Some of the epoxides, however, may be metabolized to dihydrodiols, which in turn may undergo conjugation to form soluble detoxification products or oxidation to diol-epoxides (30, 31). These latter compounds are thought to be the ultimate carcinogens in cases where carcinogenicity has been demonstrated (32). The metabolites of PAH are primarily eliminated in the urine and faeces as water-soluble compounds (33, 34).

PAH can induce the synthesis of the enzymes responsible for its own metabolism (35). This in turn means that the efficiency of metabolism of PAH will increase with continued exposure. However, the effects of various exogenous and endogenous inducers, genetic factors, age, sex, and nutritional status, which have been shown to influence these enzyme systems, make it difficult to predict whether the carcinogenicity of PAH will be increased or decreased with time.

Numerous studies show that despite their high lipid solubility, PAH show little tendency to undergo bioconcentration in the fatty tissues of animals or man, primarily because PAH are rapidly and extensively metabolized (34, 36, 37).

3.4 Health effects

Little information is available on the acute, subacute, and chronic toxicity of PAH after ingestion. They have, however, been shown to cause hyperkeratosis, hyperplasia, and loss of sebaceous glands in the skin (38). Some, such as dimethylbenzanthracene, have been shown to exert marked effects on the bone marrow and lymphoid tissue in rats (39). However, overt signs of toxicity are not usually produced by carcinogenic PAH until the dose is sufficient to produce tumours.

Only very limited data are available on the teratogenic effects of PAH, but the data on benzo[a]pyrene indicate that effects occur only at comparatively high doses (40, 41).

A number of PAH have been shown to be mutagenic in bacterial systems, in vitro cell lines, and in vivo by sister chromatid exchange (42–45). The significance or impact of such changes on human populations is not clear, however, though there is a very good correlation between in vitro tests for mutagenesis and carcinogenic potential.

Many PAH are capable of producing tumours in skin and other

epithelial tissues in numerous test species (*13, 46*). Malignant tumours can often be induced by very small quantities of the test compound and with very short latency periods. The carcinogenesis of PAH by the oral route of administration has been less well studied but positive results have been obtained, particularly as regards gastrointestinal and oesophageal tumours (*13, 47*).

PAH are rarely encountered singly in the environment and many interactions can occur with mixtures of PAH whereby the potency of known carcinogenic PAH may be enhanced (*48–50*). These systems are not well understood, however, and their significance as regards environmental exposure to PAH is not yet clear.

Epidemiological studies of occupation-associated skin cancers in man provide strong presumptive evidence of the positive role of PAH in certain kinds of human cancer (*51–53*). However, this provides no quantitative data regarding environmental as opposed to occupational exposure or regarding the importance of mixtures of PAH.

3.4.1 Guideline value

Based on the premise that drinking-water should be comparable in quality with unpolluted groundwater, the drinking-water standards published by WHO in 1970 and 1971 set a limit of 200 ng/litre for the sum of six indicator PAH in drinking-water. These were fluoranthene, benzo[*a*]pyrene, benzo[*ghi*]perylene, benzo[*b*]fluoranthene, benzo[*k*]fluoranthene, and indeno[1,2,3-*cd*]pyrene. Concentrations of these indicator PAH were found to be 10–50 ng/litre in groundwater and 50–250 ng/litre in relatively unpolluted river-water, with higher levels in polluted rivers and effluents.

Subsequent studies showed that the levels of these PAH in drinking-water were considerably lower than the standard and that the concentration was markedly influenced by fluoranthene leached from coal-tar-lined distribution systems. In addition, the representative PAH chosen and the limit set were not based on any toxicological considerations.

In 1976, a limit of 5 ng/litre for benzo[*a*]pyrene in surface-water in the USSR was proposed and approved by the Ministry of Health of the USSR (*54*).

Though the relative contribution of PAH in drinking-water to the total intake of PAH is small, these are potentially hazardous substances and exposure should be minimized. Insufficient information is currently available, however, to set values for mixtures of PAH or individual compounds, except in the case of benzo[*a*]pyrene. It is probable that a guideline value for benzo[*a*]pyrene will affect the levels of other PAH in drinking-water, since methods to reduce benzo[*a*]pyrene to acceptable levels will result in the reduction of all PAH.

The proposed guideline value for benzo[*a*]pyrene is based on toxicological considerations and experimental data from Neal &

Rigdon (47) who fed benzo[a]pyrene at doses ranging between 1 and 250 mg/kg in the diet to strain CFW mice for about 110 days. There was a statistically significant, dose-related increase in the incidence of stomach tumours (papillomas and carcinomas) in this experiment. A linearized, multistage model for risk assessment, which defines acceptable risk as 1 in 100 000, was applied to these data (55).

The following is a summary of the recommendations made:

1. Based on the application of the multistage model to the available toxicological data for benzo[a]pyrene and taking account of the fact that this substance is associated in water with other PAH of known carcinogenicity, a guideline value of 0.01 µg/litre is proposed for benzo[a]pyrene in drinking-water.

2. Because of the close association of PAH with suspended solids, the application of treatment, when necessary, to achieve an acceptable level of turbidity[a] will ensure that minimum PAH levels are achieved.

3. No PAH should be added to water during water treatment and distribution; therefore, the use of coal-tar-based pipe linings *should be discontinued*. It is recognized that it may be impractical to remove coal-tar linings from existing pipes. However, research should be undertaken to develop methods for minimizing leaching of PAH.

4. Monitoring of PAH levels should continue with the objective of determining background levels against which any changes can be assessed and if necessary remedial action taken. For the purpose of monitoring PAH levels, the use of several specific compounds as indicators for the group as a whole is considered to be advantageous. The choice of indicator compounds will vary with the individual situation. Some authorities consider the compound benzo[a]pyrene can serve in some instances as an index of pollution by the whole of the PAH group of compounds.

5. The control of PAH in drinking-water should continue to be based on the concept that unpolluted groundwater represents a baseline contamination that should not be exceeded. This concept was embodied in the standards published by WHO in 1970 and 1971. This was a useful approach, but it is not applicable in all situations and is not based on toxicological considerations.

REFERENCES

1. ANDELMAN, J. G. & SNODGRASS, J. E. Incidence and significance of polynuclear aromatic hydrocarbons in the water environment. *CRC critical reviews in environmental control*, **4**: 69 (1974).
2. HARRISON, R. M. ET AL. Polynuclear aromatic hydrocarbons in raw, potable and waste waters. *Water research*, **9**: 331 (1975).
3. STROSHER, M. T. & HODGSON, G. W. *Polycyclic aromatic hydrocarbons in lake waters and associated sediments. Analytical determination by gas-chromatography mass*

---

[a] The level of acceptability is defined in Part V, section 16.4, p. 310.

spectrometry. Philadelphia, American Society for Testing and Materials, 1975 (Water Quality Parameters ASTM STP 573, pp 259–270).

4. McGINNES, P. R. & SNOEYINK, U. L. *Determination of the fate of PAH in natural water systems*. Urbana, IL, Illinois University, 1974 (Research Report No. 80, NTIS PB-232, 168).

5. *Water related environmental fate of 129 priority pollutants*. Washington, DC, EnvironmentaL Protection Agency 1979 (EPA-440/4-79-0296).

6. SHABAD, L. M. On the so-called MAC (Maximum Allowable Concentrations) for carcinogenic hydrocarbons. *Neoplasma*, **22**: 459 (1975).

7. NEFF, J. M. *Polycyclic aromatic hydrocarbons in the aquatic environment*. London, Applied Science Publishers, 1979.

8. RADDING, S. B. ET AL. *The environmental fate of selected polynuclear aromatic hydrocarbons*. Menlo Park, CA, Stanford Research Institute, 1976 (EDPA 560/5-75-009. NTIS PB-250 948).

9. SHACKELFORD, W. M. & KEITH, L. H. *Frequency of organic compounds identified in water*. Washington, DC, Environmental Protection Agency, 1976 (EPA-600/4-76-062).

10. HANGEBRAUK, R. P. ET AL. *Sources of PAH in the atmosphere*. Washington, DC, Department of Health, Education and Welfare, 1967 (999-AP-33).

11. SAWICKI, E. ET AL. Benzo[a]pyrene content of the air of American communities. *American Industrial Hygiene Association journal*, **21**: 443 (1960).

12. EGAN, H. & SAWYER, R. Contaminants in food: analytical problems and achievements. *Toxicology*, **4**: 245 (1975).

13. *Certain polycyclic aromatic hydrocarbons and heterocyclic compounds*. Lyon, International Agency for Research on Cancer, 1973 (IARC Monographs on the evaluation of the carcinogenic risk of chemicals to humans, vol. 3.)

14. SANTODONATO, J. ET AL. In: Bjorseth A. & Dennis A. J., ed. *Polynuclear aromatic hydrocarbons: Chemistry and biological effects*. Columbus, Ohio, Batelle Press, 1980.

15. ANDELMAN, J. B. & SUESS, M. J. Polynuclear aromatic hydrocarbons in the water environment. *Bulletin of the World Health Organization*, **43**: 479 (1970).

16. BORNEFF, J. & KUNTE, H. Kanzerogene Substanzen in Wasser und Boden. XXVI: Routinemethode zur Bestimmung von polyzyklischen Aromaten im Wasser. *Archiv für Hygiene und Bakteriologie*, **153**: 220 (1969).

17. BORNEFF, J. Drinking-water production from surface water. Example of the lake of Zurich with regard to polycyclics, metals and chlorinated hydrocarbons. *Gas, Wasser, Abwasser*, **55**: 467 (1975).

18. BORNEFF, J. & KUNTE, H. Kanzerogene Substanzen in Wasser und Boden. XVI. Nachweis von polyzyklischen Aromaten in Wasserproben durch direkte Extraktion. *Archiv für Hygiene und Bakteriologie*, **148**: 585 (1964).

19. *Ambient water quality criteria for polynuclear aromatic hydrocarbons*. Washington, DC, Environmental Protection Agency, 1980.

20. CRANE, R. I. ET AL. *Survey of polycyclic aromatic hydrocarbon levels in British waters*. Medmenham, England, Water Research Centre, 1981 (Technical Report TR 158).

21. BORNEFF, J. Trinkwassergewinnung aus Cber-flächenwasser am Beispiel des Zürichsees unter Berücksichtigung vor Polyzyklen, Metallen und Chlorkohlenwasserstoffen. *Gas, Wasser, Abwasser*, **55**: 467 (1975).

22. FRITZ, W. Unfang und Quellen der Kontamination unserer Lebensmittel mit Krebserzeugenden Kohlenwasserstoffen. *Ernährungsforschung*, **16**: 547 (1972).

23. SHABAD, L. M. & IL'NITSKII, A. P. [The perspectives of elaborating the problem of water bodies contamination with cancerogenic substances.] *Gigiena i sanitarija*, **8**: 84 (1970) (in Russian).

24. REES, E. O. ET AL. A study of the mechanism of intestinal absorption of benzo[a]pyrene. *Biochimica et biophysica acta*, **225**: 96 (1971).

25. VAINII, H. The fate of intratracheally instilled benzo[a]pyrene in the isolated perfused rat lung of both control and 20-methylcholanthrene pretreated rats. *Research communications in chemical pathology and pharmacology*, **13**: 259 (1976).

26. KOTIN, P. Distribution, retention, and elimination of ^{14}C-3,4-benzopyrene after administration to mice and rats. *Journal of the National Cancer Institute*, **23**: 541 (1969).

27. SCHLEDE, E. ET AL. Effects of enzyme induction on the metabolism and tissue distribution of benzo(alpha)pyrene. *Cancer research*, **30**: 2893 (1970).

28. BOCK, F. G. & DAO, T. L. Factors effecting the polynuclear hydrocarbon level in the rat mammary glands. *Cancer research*, **21**: 1024 (1961).

29. SCHLEDE, E. ET AL. Stimulatory effect of benzo(alpha)pyrene and phenobarbitol pretreatment on the biliary excretion of benzo(alpha)pyrene metabolites in the rat. *Cancer research*, **30**: 2898 (1970).

30. CONNEY, A. H. ET AL. Use of drugs in the evaluation of carcinogen metabolism in man. In: Montesano R. et al., ed. *Screening tests in chemical carcinogenesis.* Lyon, International Agency for Research on Cancer, 1976 (IARC Scientific Publications No. 12).

31. YANG, S. K. ET AL. In: Gelboin, H. V. & Ts'o, P.O.P., ed. *Polycyclic hydrocarbons and cancer. Vol. 1, Chemistry and metabolism.* New York, Academic Press, 1978.

32. LEHR, R. E. ET AL. In: Freudenthal, R. I. & Jones, P. W., ed. *Polynuclear aromatic hydrocarbons: 2nd International Symposium on Analysis, Chemistry and Biology.* New York, Raven Press, 1978.

33. DANIEL, P. M. ET AL. Metabolism of labelled carcinogenic hydrocarbons on rats. *Nature*, **215**: 1142 (1967).

34. WHITTLE, K. J. ET AL. Occurrence and fate of organic and inorganic contaminants in marine animals. *Annals of the New York Academy of Sciences*, **298**: 47 (1978).

35. MARQUARDT, H. In: Mohr, U. et al., ed. *Air pollution and cancer in man.* Lyon, International Agency for Research on Cancer, 1977 (IARC Scientific Publications No. 16).

36. STICH, H. F. ET AL. In: Rosenfeld, C. & Davis, W., ed. *Environmental pollution and carcinogenic risks.* Lyon, International Agency for Research on Cancer, 1976 (IARC Scientific Publications No. 13).

37. LEE, R. F. ET AL. Petroleum hydrocarbons: uptake and discharge by the marine mussel *Mytilus edulis. Science*, **177**: 344 (1972).

38. BOCK, F. G. Early effects of hydrocarbons on mammalian skin. *Progress in experimental tumor research*, **4**: 126 (1964).

39. CARWEIN, M. J. & SNYDER, K. L. Suppression of cellular activity in the reticuloendothelial system of the rat by 7,12-dimethylbenzo[a]anthracene. *Cancer research*, **28**: 320 (1968).

40. BULAY, O. M. The study of development of lung and skin tumours in mice exposed *in vitro* to polycyclic hydrocarbons. *Acta medica Turcica*, **7**: 3 (1970) (cited in *Registry of toxic effects of chemical substances*, 1978).

41. BULAY, O. M. & WATTENBERG, L. W. Carcinogenic effect of subcutaneous administration of benzo[a]pyrene during pregnancy on the progeny. *Proceedings of the Society for Experimental Biology and Medicine*, **135**: 84 (1970) (cited in *Registry of toxic effects of chemical substances*, 1978).

42. LA VOIE, E. J. ET AL. In: Jones, P. W. & Leber, P., ed. *Polynuclear aromatic hydrocarbons.* Ann Arbor, MI, Ann Arbor Science Publishers, 1979.

43. McCANN, J. ET AL. Detection of carcinogens as mutagens in the *Salmonella*/microsome test: assay of 300 chemicals. *Proceedings of the National Academy of Sciences*, **72**: 5135 (1975).

44. HUBERMAN, E. & SACHS, L. Mutability of different genetic loci in mammalian cells by metabolically activated carcinogenic polycyclic hydrocarbons. *Proceedings of the National Academy of Sciences*, **73**: 188 (1976).

45. BAYER, U. In: Freudenthal, R. I. & Jones, P. W., ed. *Polynuclear aromatic hydrocarbons: Second International Symposium on Analysis, Chemistry and Biology.* New York, Raven Press, 1978.

46. IBALL, J. Relative potency of carcinogenic compounds. *American journal of cancer*, **35**: 188 (1939).

47. NEAL, J. & RIGDON, R. H. Gastric tumours in mice fed benzo[a]pyrene: a quantitative study. *Texas reports on biology and medicine*, **25**: 553 (1967).

48. PFEIFFER, E. H. Investigations on the carcinogenic burden by air pollution in man. VII. Studies on the oncogenic interaction of polycyclic aromatic hydrocarbons.

Zentralblatt für Bakteriologie, Parasitenkunde, Infektionskrankheiten und Hygiene (Orig. B), **158**: 69 (1973).

49. PFEIFFER, E. H. Oncogenic interaction of carcinogenic and non-carcinogenic polycyclic aromatic hydrocarbons in mice. In: Mohr, U. et al., ed. *Air pollution and cancer in man.* Lyon, International Agency for Research on Cancer, 1977 (IARC Scientific Publications No. 16).

50. SCHMÄHL, D. ET AL. Syncarcinogenic action of polycyclic hydrocarbons in automobile exhaust gas condensates. In: Mohr, U. et al., ed. *Air pollution and cancer in man.* Lyon, International Agency for Research on Cancer, 1977 (IARC Scientific Publications No. 16).

51. POTT, P. In: *Chirurgical observations.* London, Hawes, Clark and Collins, 1975.

52. ECKHARDT, R. D. *Industrial carcinogens.* New York, Grune and Stratton, 1959.

53. SELIKOFF, I. J. Air pollution and asbestos carcinogenesis: investigation of possible synergism. In: Mohr, U. et al., ed. *Air pollution and cancer in man.* Lyon, International Agency for Research on Cancer, 1977 (IARC Scientific Publications No. 16).

54. [*Rules of surface waters protection from pollution by sewage and industrial wastes.*] Moscow, Ministry of Health, 1976.

55. *Ambient water quality criteria for polynuclear aromatic hydrocarbons.* Washington, DC, US Environmental Protection Agency, 1980 (EPA 440/5-70-069).

4. PESTICIDES

Pesticides that may be of importance to water quality include chlorinated hydrocarbons and their derivatives, persistent herbicides, soil insecticides, pesticides that are easily leached out from the soil, and pesticides systematically added to water supplies for disease vector control or other purposes. Of these compounds, only the chlorinated hydrocarbon insecticides occur frequently.

Chlorinated hydrocarbon pesticides persist in the environment and have become ubiquitous. For example, traces of DDT have been recovered from dust known to have drifted over thousands of kilometres and from water melted from antarctic snow. Traces of chlorinated hydrocarbon pesticides in water may progressively accumulate in different steps of a food chain; for example, DDT can bioaccumulate in fish at levels more than 10 000 times the concentrations present in the water of their habitat. Several of the pesticides in this group, including ones that have been used extensively in the past in agriculture and some that continue to be used for purposes such as disease control, have been shown to possess tumorigenic properties in animals.

Guideline values are recommended for several chlorinated hydrocarbon pesticides because of their known occurrence as adventitious residues in water. These values are derived from the acceptable daily intake (ADI) values set by the FAO/WHO Joint Expert Committee on Pesticide Residues; 1% of the ADI has been adopted as the basis for values in drinking-water. Application of the linear multistage extrapolation model for estimating potential carcinogenic risks to the recommended values suggests that water containing pesticide residues at these levels is unlikely to result in an incremental cancer risk per lifetime exceeding one per 100 000 of the population. Short-term excursions above these values resulting, for example, from vector control operations, may be permissible but require monitoring and assessment of health implications.

It is recognized that the pesticides for which values have been specified do not represent all of those that have been identified in water. Local circumstances may require the extensive use of a pesticide for which guidelines have not been developed. Monitoring for the presence of such substances in drinking-water may, therefore, be deemed desirable.

The recommended values are designed to protect human health. They may not be adequate for the protection of aquatic life.

190

4.1 DDT (total isomers)

4.1.1 General aspects

DDT is an acronym for dichlorodiphenyltrichloroethane. Its full chemical name is 1,1'-(2,2,2-trichloroethylidene)bis-[4-chlorobenzene]. This is also known as p,p'-DDT, since the structure of DDT permits several different isomeric forms. The term DDT is also applied to commercial products consisting predominantly of p,p'-DDT together with some o,p'-DDT and smaller amounts of p,p'-TDE, o,p'-TDE, p,p'-DDE, o,p'-DDE, and other unidentified compounds.[a]

DDT is highly insoluble in water but soluble in organic solvents. Its vapour pressure is 2.53×10^{-5} Pa $(1.9 \times 10^{-7}$ mmHg) at 20 °C.

DDT was first used for the protection of military areas and personnel, mainly against malaria, typhus, and certain other vector-borne diseases during the Second World War. Widespread agricultural use dates from 1946 in the USA and slightly later in most other countries. DDT was also applied extensively to forests. Its use has been restricted or even prohibited in some countries for ecological reasons and because of the increasing resistance of pests to the insecticide. A decrease in production has been observed particularly since 1970. DDT is still used extensively, both in agriculture and for vector control, in some tropical countries.

DDT is a persistent insecticide. It is stable under most environmental conditions and is resistant to complete breakdown by the enzymes present in soil microorganisms and higher organisms. Some of its metabolites, notably 1,1'-(2,2-dichloroethenylidene)bis[4-chlorobenzene] (DDE), have a stability equal to, or greater than that of the parent compound. WHO has issued an environmental health criteria document on DDT and its derivatives (1).

4.1.2 Routes of exposure

4.1.2.1 Air

Evaporation from fields treated with DDT can be detected for more than six months after application. Most of the DDT returns to the soil in the area of application with an almost straight-line inverse relationship to the logarithm of the distance from the source. However, a small proportion may undergo worldwide transportation and traces of DDT have been recovered from dust known to have drifted over 1000 km. Normally, the concentration in air in non-agricultural areas is in the range $1-2.36 \times 10^{-6}$ mg/m^3. In communities with anti-mosquito fogging programmes, concentrations of DDT may be much higher.

Knowledge of the circulation and fate of DDT and its analogues in the environment is generally lacking. Recently it has been demonstrated,

---

[a] p,p'-TDE (also known as DDD) is 1,1'-(2,2-dichloroethylidene)bis[4-chlorobenzene]; p,p'-DDE is 1,1'-(2,2-dichloroethenylidene)bis[4-chlorobenzene].

under laboratory conditions similar to those found in the upper atmosphere, that DDT breaks down to carbon dioxide and hydrochloric acid.

4.1.2.2 *Water*

Concentrations of DDT in rainwater have usually been of the same order of magnitude (1.8×10^{-5} to 6.6×10^{-5} mg/litre) in both agricultural and remote non-agricultural areas.

Concentrations of DDT in surface-water depend on levels in rainwater and in the soil and the nature of the soil. Concentrations in the USA are said to have reached a peak in 1966 and then dropped sharply in 1967 and 1968. The highest surface-water value for a DDT-related compound in those years was 0.84 μg/litre. By 1971, the average concentration in the Federal Republic of Germany was 0.01 μg/litre, never reaching as high as 1 μg/litre. In recent years, concentrations in potable water have been much less than 1 μg/litre and average concentrations have been similar to those for rainwater.

4.1.2.3 *Food*

The daily intake of DDT from food has been measured in several countries. In the USA during 1953 and 1954, average daily intakes of DDT and of total DDT (DDT + DDE + TDE) were 0.184 and 0.286 mg per person, respectively, most of which originated from foods of animal origin. Ten years later, following restrictions with regard to the application of DDT to livestock, intake had been reduced by more than 75%.

The US Market Basket Survey showed a gradual decrease in daily intake of DDT to 0.015 mg per person in 1970. Intake in Canada and the United Kingdom was slightly less for comparable periods. In many countries of Europe and in other countries with similar diets the intake of DDT has been judged to be about the same. Worldwide measurements of storage of DDT and its metabolites in human body fat indicate that the extremes of total exposure have varied by a factor of about 10, but that total exposure for most populations has varied by a factor of no more than 3.

Food is the major source of DDT for the general population and over 90% of the DDT stored in man is derived from food.

4.1.3 Metabolism

DDT is absorbed via both inhalation and ingestion. Absorption of small doses is virtually complete and is facilitated by the presence of fat in food. DDT is poorly absorbed through the skin. Levels of storage in adipose tissue increase rapidly at first and then more gradually until a steady state is reached. In man, the time necessary to reach storage equilibrium is at least one year. There is a gradual reduction in the amount of DDT stored in the tissue if exposure to the compound is discontinued.

Like most species, man converts some DDT to DDE. A small amount of TDE (DDD), an intermediate in the formation of the main excretory product 2,2-bis(4-chlorophenyl)acetic acid (DDA), may also be found in tissues.

Concentrations of total DDT in the blood of the general population of different countries lie within the range 0.01–0.07 mg/litre. In the blood and other tissues of the fetus or newborn infants, DDT concentrations are lower than in the corresponding tissues of the mother.

Levels of DDT in human milk have usually been reported to be in the range 0.01–0.10 mg/litre, with the concentration of DDT plus its metabolites, especially DDE, being about twice as high.

Among the general public, the average concentration of DDA in the urine is 0.014 mg/litre.

Animal studies indicate that the concentration in serum most accurately reflects the concentration in the brain, the critical tissue.

4.1.4 Health effects

The acute oral toxicity of DDT is affected by the solvent vehicle; administered in oil to the rat, a typical median lethal dosage is 250 mg/kg of body weight. DDT is poorly absorbed through the skin.

The main effect of DDT is on the nervous system. All parts, both central and peripheral are affected to some degree. It appears that its toxic action is associated with its effects on the membranes of the nervous system.

The liver is the only other organ significantly affected by DDT. Potentially fatal doses cause focal necrosis of liver cells in several species. DDT produced liver tumours in several strains of mice, non-metastasizing liver tumours in one study in rats, and no carcinogenic effect in hamsters (2–4). DDT induces microsomal enzymes in all species tested, but only in some rodents does the endoplasmic reticulum increase so much that the entire liver enlarges and granules are displaced to the margin of the cell. These changes are accompanied by a moderate increase in fat droplets, some of which form so-called lipospheres. In long-term feeding tests in mice and rats, the changes in the liver progress from hypertrophy margination and lipospheres to the formation of nodules of affected cells. The same series of changes can be produced in rodents by other inducers of microsomal enzymes. The entire continuum of changes from the prompt response in isolated cells to the eventual formation of tumours is peculiar to some rodents, and other species do not respond morphologically in the same way. Levels as high as 200 mg/kg of feed do not produce adverse effects on the reproduction of rats. In dogs receiving a dosage of 10 mg/kg of body weight per day, reproduction was also normal.

No teratogenic effects have been observed in the several animal species studied. DDT has not been found to be mutagenic in bacterial test systems. The evidence from mammalian test systems, *in vitro* and *in vivo*, is inconclusive.

Investigation of workers who have been exposed for as long as 25 years to higher levels of DDT than the general population has not revealed any evidence that DDT causes cancer in man. Although a large number of epidemiological studies were carried out over the years 1960 to 1981, involving workers exposed to DDT, the results were inadequate to permit an evaluation of its carcinogenicity to humans.

The toxicity data have been evaluated and in 1969 a conditional acceptable daily intake for man was estimated as 0–0.005 mg/kg of body weight (5).

REFERENCES

1. *DDT and its derivatives*. Geneva, World Health Organization, 1979 (Environmental Health Criteria 9).
2. *Some organochlorine pesticides*. Lyon, International Agency for Research on Cancer, 1974 (IARC Monographs on the evaluation of the carcinogenic risk of chemicals to humans, vol. 5).
3. ROSSI, L. ET AL. Long term administration of DDT or phenobarbital-Na in Wistar rats. *International journal of cancer*, **19**: 179 (1977).
4. National Cancer Institute Bioassay, Bethesda, MD, Department of Health, Education, and Welfare, 1978 (Technical Report No. 131).
5. *Evaluations of some pesticide residues in food*. Geneva, World Health Organization, 1970 (FAO/PL: 1969/M/17/1; WHO/Food Add./70. 38).

4.2 Aldrin and dieldrin

4.2.1 General aspects

The chemical names of these two related pesticides are as follows:

Aldrin   (HHDN):   1,2,3,4,10,10-hexachloro-1,4,4a,5,8,8a-hexahydro-*endo*-1,4-*exo*-5,8-dimethanonaphthalene.

Dieldrin (HEOD): 1,2,3,4,10,10-hexachloro-6,7-epoxy-1,4,4a,5,6,7,8,8a-octahydro-*endo*-1,4-*exo*-5,8-dimethanonaphthalene.

Aldrin and dieldrin are persistent insecticides and accumulate in the food chain. Dieldrin is formed from aldrin by metabolic oxidation in animals and by chemical oxidation in soils. Both insecticides were used for soil treatment against various soil insects, for seed treatment, and for foliar application on various agricultural crops. Over the years their use as foliar treatments has almost disappeared. Their use for the other purposes has also been gradually limited or prohibited; currently the largest use is for termite control. In some countries, it is necessary to treat the soil around fruit trees to form a barrier against termites attacking roots and trunks of trees.

The risk attached to the use of aldrin and dieldrin has been evaluated and reviewed by many groups since 1966 (*1–5*).

4.2.2 Routes of exposure

As aldrin is readily converted to dieldrin in plants and animals it is rarely found as such in soil, food, and water.

4.2.2.1 *Air*

Dieldrin has been detected in ambient air at very low concentrations. In the USA a maximum level of $29.7 \, \text{ng/m}^3$ has been reported; in London and its suburbs, levels up to $1.9 \times 10^{-4} \, \text{ng/m}^3$ of air have been found associated with airborne dust. In the USA, a threshold limit value of $0.25 \, \text{mg/m}^3$ has been established for an 8-hour, time-weighted average occupational exposure.

4.2.2.2 *Water*

Concentrations found in rainwater range from 5 to $42 \, \text{ng/litre}$ and in surface-water (mostly rivers and lakes in the USA) from 0 to $0.1 \, \mu\text{g/litre}$. Concentrations of $1–2 \, \text{ng/litre}$ were found in drinking-water.

4.2.2.3 *Food*

Dieldrin is stored in adipose tissues, liver, brain, and muscle of mammals, fish, and birds, and in other parts of the food chain. Fish can build up concentrations of dieldrin amounting to several mg/kg of body weight from concentrations of a few ng/litre in water. In the USA, the total dietary intake was found to range between 0.05 and $0.08 \, \mu\text{g/kg}$ of body weight per day during the period 1965–1970. In the United Kingdom, a total diet study from 1966 to 1967 showed the intake of dieldrin to be $0.09 \, \mu\text{g/kg}$ of body weight per day. This intake dropped to $0.03 \, \mu\text{g/kg}$ of body weight per day in 1970–71. An intake of $0.07 \, \mu\text{g/kg}$ of body weight per day has been estimated in Japan.

Levels of dieldrin in the body fat of the general population in the years 1961–68 in the USA and the United Kingdom and from 1964 to 1966 in six other countries ranged from 0.03 to $0.45 \, \text{mg/kg}$ of tissue. On the basis of a linear relationship between storage in body fat and exposure, the average intake of dieldrin can be estimated to have ranged from 0.01 to $0.35 \, \mu\text{g/kg}$ of body weight per day for a 70-kg man (*5*).

4.2.3 Metabolism

In all species examined (mouse, rat, rabbit, rhesus monkey, chimpanzee) 12-hydroxydieldrin and 4,5-aldrin-*trans*-dihydrodiol were the major metabolites. Regarding the ratios of these two metabolites, the rats and primates seemed to metabolize dieldrin mainly by direct oxidation resulting in 12-hydroxydieldrin, whereas in mice and rabbits the main metabolic process seemed to be the opening of the epoxide ring to the diol. This metabolic pathway, together with a high rate of metabolism compared with other animal species examined, could result in relatively high concentrations of 4,5-aldrin-*trans*-dihydrodiol in the mouse liver.

In humans, neither unchanged dieldrin nor the urinary metabolites found in rats were identified in workers occupationally exposed to dieldrin or in monkeys fed the insecticide.

Dieldrin is a powerful inducer of microsomal enzymes. Mice react rather anomalously and, therefore, may not be an appropriate model for humans. The results of prolonged enzyme induction are proliferation of the endoplasmic reticulum in the cells and hypertrophy and hyperplasia of the liver.

No measurable microsomal enzyme induction occurs in man after long exposure to approximately 0.01 mg/kg of body weight per day. The calculated half-life of dieldrin in workers who were removed from exposure was seven months. The mean dieldrin concentration in blood during the last half-year of exposure was 0.1 mg/litre, which corresponded to an average daily oral intake of about 0.17 mg/kg of body weight (*2, 4*).

4.2.4 Health effects

The primary site of action of dieldrin is the central nervous system. CNS stimulation is the cause of death in acute poisoning (*1*). Dieldrin was not proved to be mutagenic in any of the *in vitro* or *in vivo* mutagenicity studies. Teratogenicity studies in different animal species showed dieldrin not to be a teratogen. The administration of a single high dose of 15 mg of dieldrin per kg of body weight to mice or 30 mg/kg of body weight to hamsters resulted in minor malformations, which were attributed to maternal toxicity.

Results of various carcinogenicity tests in mice and in other mammalian species indicate that there is a species-specific effect of aldrin and dieldrin on the mouse liver resulting in an increased frequency of liver tumours that is peculiar to this animal species (*4*). A dose-response effect has been demonstrated in both sexes with an increased tumour incidence in females at the lowest dose tested (about 0.015 mg/kg of body weight per day). The available data in rats have not provided evidence of carcinogenicity at levels of up to 2.5 mg/kg of body weight per day. A carcinogenicity study on dieldrin in hamsters was negative (*6*).

An epidemiological study carried out on occupationally exposed workers did not allow any conclusions to be made concerning the existence of an excess risk of developing cancer (*5*). Epidemiological studies were inadequate for evaluation (*7, 8*).

Toxicological data available in 1977 supported the view that dieldrin and aldrin are not carcinogens; therefore, the previously estimated acceptable daily intake of 0.0001 mg/kg of body weight for aldrin and dieldrin residues, separately or for combined exposure, was reaffirmed (4).

REFERENCES

1. *Evaluations of some pesticide residues in food.* Geneva, World Health Organization, 1967 (FAO/PL: CP/15; WHO/Food Add./67.32).
2. *Evaluations of some pesticide residues in food.* Geneva, World Health Organization, 1971 (FAO/AGP/1970/M/12/1; WHO/Food Add./71.42).
3. WHO Technical Report Series, No. 612, 1976 (*Pesticide residues in food:* report of the 1976 Joint FAO/WHO Meeting).
4. *Evaluations of some pesticide residues in food. Report 1977.* Rome, FAO, 1978 (Plant Production and Protection Paper, No. 10).
5. *Some organochlorine pesticides.* Lyon, International Agency for Research on Cancer 1974. (IARC Monographs on the evaluation of carcinogenic risk of chemicals to humans vol. 5)
6. CABRAL, J. R. ET AL. A carcinogenicity study of pesticide dieldrin in hamsters. *Cancer letter*, 6: 241 (1979).
7. VAN RAALTE, H. G. Human experience with dieldrin in perspective. *Ecotoxicology and environmental safety*, 1: 201 (1977).
8. DEICHMANN, W. B. & MACDONALD, W. E. Organochlorine pesticides and liver cancer deaths in the United States, 1932–1972. *Ecotoxicology and environmental safety*, 1: 89 (1977).

4.3 Chlordane

4.3.1 General aspects

Pure chlordane is a pale yellow liquid having a molecular formula $C_{10}H_6Cl_8$ and a relative molecular mass of 409.8 (*1, 2*). The chemical name of chlordane is 1,2,4,5,6,7,8,8-octachloro-2,3,3a,4,7,7a-hexahydro-4,7-methanoindene (*1*). Pure chlordane is composed of a mixture of stereoisomers, with the cis- and trans- forms predominating and referred to as alpha- and gamma-isomers, respectively (*3*). It is soluble in water at concentrations that have been shown to be toxic to aquatic organisms. Brooks (*3*) reported the solubility of chlordane in water to be approximately 9 μg/litre at 25 °C.

Chlordane is a broad-spectrum insecticide of the group of polycyclic chlorinated hydrocarbons called cyclodiene insecticides. Chlordane has been used extensively over the past 30 years for termite control, as an insecticide for homes and gardens, and as a control for soil insects during the production of crops, such as corn. Both the uses and the production volume of chlordane have decreased extensively in recent years.

4.3.2 Routes of exposure

4.3.2.1 *Water*

A study of the persistence of technical chlordane in river-water showed 85% remaining after 8 weeks (*4*). Of the major components of technical chlordane, cis- and trans-chlordane were completely stable over the 8-week period. All but two of the remaining components were at least partially changed.

Schaefer et al. (*5*) examined over 500 grab samples from water supplies of the Mississippi and Missouri Rivers. Chlordane was detected in over 20% of the finished water samples, with the maximum concentration being 8 µg/litre. Chlordane has also been detected in rainwater (*6, 7*).

Although reports are occasionally received of individual household wells becoming contaminated after a house is treated with chlordane for termite control, only one report has been published of the contamination of a municipal water system (*8*). On 24 March 1976, a section of the public water system supplying 105 persons in Chattanooga, Tennessee, USA, became contaminated. Back siphonage apparently occurred while a chlordane concentrate was being diluted with a hose during a period of negative pressure. Of the 71 residents affected, 13 had mild symptoms of acute chlordane toxicity. None of the residents has had prolonged sequelae from the exposure.

4.3.2.2 *Food*

The Food and Drug Administration (FDA) of the USA has been systematically monitoring chlordane in the food supply of the USA since 1965. Chlordane was found infrequently during the first 11 years of the survey (*9*). The only quantifiable sample collected was found to contain 0.059 mg/kg, measured in a sample of grain in 1972 (*10*). In the most recent published results (for 1975) chlordane was not detected.

The National Academy of Sciences of the USA (*11*) in reviewing the results of Moore[a] reported that of 200 samples of milk collected in Illinois during the period 1971–73, 87% were positive for chlordane. The average concentration was 50 µg/litre. Cyclodienes, such as chlordane, are apparently ingested with forage and tend to concentrate in lipids. Oxychlordane, a major mammalian metabolite of chlordane and heptachlor, was found in 46% of 57 human milk samples collected during 1973–74 in Arkansas and Mississippi. The mean concentration was 5 µg/litre and the maximum was 20 µg/litre (*12*).

4.3.2.3 *Air*

In a survey of the extent of atmospheric contamination by pesticides in the USA, air was sampled at nine localities representative of both

---

[a] MOORE, S. III Proc. 27th Illinois Custom Spray Operators Training School. Urbana, IL, 1975.

urban and agricultural areas. At least one chlorinated pesticide was found at all locations, but chlordane was not found in any samples (*13*). In a larger survey summarized by Nisbet (*9*), 2479 samples were collected at 45 cities in 16 states of the USA. Chlordane was detected in only two samples, with concentrations of 84 and 204 ng/m^3.

4.3.2.4 *Other routes of exposure*

Chlordane can be absorbed through the skin and produce toxic effects (*14*). Dermal exposure would be expected to occur only with occupational manufacture or use of the pesticide. Absorption can range from negligible to that producing acute effects, depending on the degree of exposure. For the general population, dermal exposure would be negligible. In persons using chlordane, the pesticide may persist on the skin for long periods. In one study, hexane rinsings of the hands of a former pest-control operator contained chlordane two years after his last known exposure (*15*).

4.3.2.5 *Relative significance of routes of exposure*

For the non-occupationally exposed individual, fish and shellfish probably represent the most significant exposure to chlordane. The relative significance of seafood is due to the high bioconcentration factor whereby trace amounts of chlordane in the water may be concentrated to much higher levels in the seafood. Drinking-water and ambient air are relatively insignificant sources of chlordane. Occupational exposure, especially for persons who apply chlordane for pest-control purposes, undoubtedly represents the single greatest exposure to this chemical.

4.3.3 Metabolism

A single oral dose of chlordane administered to rats resulted in approximately 6% absorption (*16*). Small daily doses resulted in absorption of 10–15%. The levels of residues in the fat of the rats, after being fed diets containing 1, 5, and 25 mg/kg for 56 days, were approximately three times the concentration in the diet. Concentrations in the liver, kidney, brain, and muscle were 12, 10, 4 and 2%, respectively, of the concentration in the feed.

Once the chlordane was removed from the diet, all the residues declined steadily for 4 weeks, by which time the concentrations were reduced by about 60%. During the following 4 weeks, the residues declined only slightly.

In rats, most chlordane is excreted in the faeces. Only about 6% of the total intake is voided in the urine. Rabbits, however, show a different pattern. Urinary elimination of chlordane in rabbits is greater than excretion in the faeces.

Human half-life data were obtained when chlordane was accidentally ingested by a young boy (*17*). A whole-body half-life of 21 days was calculated, which is long compared with drugs used in therapy, but quite short when compared with other chlorinated insecticides. This half-life value compares with a half-life of about 23 days obtained by Barnett & Dorough (16) in studies with rats fed chlordane for 56 days. After the levels reached 60%, further reduction was slight. The serum half-life of chlordane in a young girl was found by Aldrich & Holmes (18) to be 88 days.

Polen et al. (19) and Street & Blau (20) found oxychlordane to be a mammalian metabolite of chlordane, and to persist in adipose tissue. Street & Blau (20) observed that the toxicity of oxychlordane was greater than that of the parent compound. Barnett & Dorough (16) tentatively identified several hydroxylated metabolites of chlordane in rat excreta, in addition to oxychlordane, and concluded that the metabolism of chlordane takes place via a series of oxidative enzyme reactions.

4.3.4 Health effects

4.3.4.1 *Toxicity*

Human toxicity data for chlordane are usually obtained after accidental exposure to the compound. Curley & Garrettson (17) reported that shortly after a 20-month-old boy accidentally drank an unknown amount of chlordane he vomited and began a series of convulsions lasting 3–5 minutes each. After being given phenobarbital in a dose of 14 mg/kg of body weight, the seizures stopped. Body temperature rose to 38.9 °C and then gradually decreased to normal. At no time was there evidence of pulmonary disease. Neurological examination at the time seizures were occurring revealed brisk deep tendon reflexes in all extremities. Cranial nerve function was intact and nystagmus was absent. An EEG taken 48 hours after exposure was normal. Three months after exposure, all tests appeared normal. Similar cases were reported by Dadey & Krammer (21) and Aldrich & Holmes (18).

A review of the literature by the National Institute for Occupational Safety and Health of the USA (22) indicated chlordane LD_{50} values ranging from 100 mg/kg of body weight for rabbits with oral administration to 700 mg/kg of body weight for rats with dermal administration. Chlordane fed to rats in a concentration of 2.5 mg/kg of feed caused slight liver damage (11).

4.3.4.2 *Teratogenicity*

Chlordane was found not to be teratogenic in rats when fed at concentrations of 150–300 mg/kg of feed during pregnancy (23).

4.3.4.3 *Mutagenicity*

Arnold et al. (24) administered chlordane to Charles River CD-1 male mice in a single dose of 50 or 100 mg/kg of body weight. The males were subsequently mated with untreated female mice. No dominant lethal changes were produced. Studies by Ahmed et al. (25) with SV-40 transformed human fibroblast cell line VA-4 showed chlordane-induced unscheduled DNA synthesis, indicating that chlordane is a potential mutagen. Metabolic activation eliminated the induction of unscheduled DNA synthesis. Simmon et al. (26) found that neither pure cis-chlordane nor trans-chlordane were mutagenic in the Ames *Salmonella* microsome assay. Technical grade chlordane, however, was found to be mutagenic in *Salmonella typhimurium* strains TA 1535, TA 98, and TA 100. An S-9 liver activation mix did not enhance the mutagenic activity.

4.3.4.4 *Carcinogenicity*

A bioassay of chlordane for possible carcinogenicity was conducted in the USA, by the Gulf South Research Institute under contract to the National Cancer Institute (27). Groups of 50 mice of each sex and 35 days of age were administered the test material at two concentrations for 80 weeks and then observed for 10 weeks. Hepatocellular carcinoma showed a highly significant dose-related trend for the mice.

Similar studies were conducted by the Gulf South Research Institute with chlordane using Osborne-Mendel strain rats. In contrast to the findings in mice, hepatocellular carcinomas failed to appear at a significant rate of incidence in rats administered chlordane.

Several epidemiological studies have been published on occupationally exposed persons. At Marshall, Illinois, where chlordane had been manufactured since 1946, no discernible hazard was uncovered among the workers engaged in the production of chlordane, and there was no evidence that chlordane was carcinogenic (28). Studies of personnel at three large pest control companies in the USA, including termite control operators, uncovered no evidence of increased mortality from cancer. There were no deaths attributable to cancer of the liver (28). Epidemiological studies on pesticide applicators exposed to chlordane were inadequate for evaluation (29).

In one report, 5 out of 14 children with neuroblastoma had had prenatal and/or postnatal exposure to chlordane, and in one epidemiological study, three persons with acute leukaemia were found to have been exposed to chlordane (which contained 3–7% heptachlor) (30).

As carcinogenicity has been conclusively demonstrated in only one animal, the mouse, a limit for chlordane should be based on toxicity. An acceptable daily intake for man has been estimated as 0.001 mg/kg of body weight (31). For a 70-kg man, this gives an acceptable daily intake of 0.07 mg. Allocating 1% of this figure to water, and assuming an average daily water intake of 2 litres, the value for chlordane in drinking water can be calculated as 0.35 µg/litre, rounded off to 0.3 µg/litre.

REFERENCES

1. WINDHOLZ, M. *The Merck index*. Rahway, NJ, Merck and Co., 1976.
2. WHETSTONE, R. R. *Kirk-Othmer encyclopedia of chemical technology*. New York, John Wiley and Sons, 1972.

3. BROOKES, G. T. *Chlorinated insecticides*. Cleveland, OH, Chemical Rubber Company Press, 1974.
4. EICHELBERGER, J. W. & LICHTENBERG, J. J. Persistence of pesticides in river water. *Environmental science and technology*, **5**: 541 (1971).
5. SCHAEFER, M. L. ET AL. Pesticides in drinking water. *Environmental science and technology*, **3**: 1261 (1969).
6. BEVENUE, A. ET AL. Organochlorine pesticides in rainwater, Oahu, Hawaii, 1971–72. *Bulletin of environmental contamination and toxicology*, **8**: 238 (1972).
7. ENVIRONMENTAL PROTECTION AGENCY. Consolidated heptachlor/chlordane hearing. *Federal register*, **41**: 7552 (1976).
8. HARRINGTON, J. M. ET AL. Chlordane contamination of a municipal water system. *Environmental research*, **15**: 155 (1978).
9. NISBET, I. C. T. *Human exposure to chlordane, heptachlor, and their metabolites*. Washington, DC, US Environmental Protection Agency, 1976.
10. MANSKE, D. D. & JOHNSON, R. D. Pesticide residues in total diet samples (VIII). *Pesticides monitoring journal*, **9**: 94 (1975).
11. NATIONAL RESEARCH COUNCIL *Drinking water and health*. Washington, DC, National Academy of Sciences, 1977.
12. STRASSMAN, S. C. & KUTZ, F. W. Insecticide residues in human milk from Arkansas and Mississippi, 1973–74. *Pesticides monitoring journal*, **10**: 130 (1977).
13. STANLEY, C. W. ET AL. Measurement at atmospheric levels of pesticides, *Environmental science and technology*, **5**: 430 (1971).
14. GOSSELIN, R. E. ET AL. *Clinical toxicology of commercial products*, 4th ed. Baltimore, MD, Williams and Wilkins Co., 1976.
15. KAZEN, C. ET AL. Persistence of pesticides on the hands of some occupationally exposed people. *Archives of environmental health*, **29**: 315 (1974).
16. BARNETT, J. R. & DOROUGH, H. W. Metabolism of chlordane in rats. *Journal of agricultural and food chemistry*, **22**: 612 (1974).
17. CURLEY, A. & GARRETTSON, L. K. Acute chlordane poisoning. *Archives of environmental health*, **18**: 211 (1969).
18. ALDRICH, F. D. & HOLMES, J. H. Acute chlordane intoxication in a child. *Archives of environmental health*, **19**: 129 (1969).
19. POLEN, P. N. ET AL. Characterization of oxychlordane, animal metabolite of chlordane. *Bulletin of environmental contamination and toxicology*, **5**: 521 (1971).
20. STREET, J. E. & BLAU, S. E. Oxychlordane: accumulation in rat adipose tissue on feeding chlordane isomers or technical chlordane. *Journal of agricultural and food chemistry*, **20**: 395 (1972).
21. DADEY, J. L. & KRAMMER, A. G. Chlordane intoxication. *Journal of the American Medical Association*, **153**: 723 (1953).
22. NATIONAL INSTITUTE FOR OCCUPATIONAL SAFETY AND HEALTH. *Suspected carcinogens–subfile of the NIOSH registry of toxic effects of chemical substances*. 2nd ed. Washington, DC, NIOSH, 1976 (NISH-77-149).
23. INGLE, L. Chronic oral toxicity of chlordane to rats. *Archives of industrial hygiene and occupational medicine*, **6**: 357 (1952).
24. ARNOLD, D. W. ET AL. Dominant lethal studies with technical chlordane, HCS-3260, and heptachlor: heptachlor epoxide. *Journal of toxicology and environmental health*, **2**: 547 (1977).
25. AHMED, F. E. ET AL. Pesticide induced DNA damage and its repair in cultured human cells. *Mutatation research*, **42**: 161 (1977).
26. SIMMON, V. F. ET AL. Mutagenic activity of chemicals identified in drinking water. *Developments in toxicology and environmental science*, **2**: 249 (1977).
27. NATIONAL CANCER INSTITUTE, *Bioassay of chlordane for possible carcinogenicity*. Bethesda, MD, Department of Health, Education and Welfare, 1977 (NCI-CG-TR-8).
28. ENVIRONMENTAL PROTECTION AGENCY. Consolidated heptachlor/chlordane hearing. *Federal register*, **41**: 7552–7572, 7584–7585 (1976).
29. WANG, H. H. & MACMAHON, B. Mortality of pesticide applicators. *Journal of occupational medicine*, **21**: 741 (1979).

30. *Some halogenated hydrocarbons.* Lyon, International Agency for Research on Cancer, 1979 (IARC Monographs on the evaluation of the carcinogenic risk of chemicals to humans, vol. 20).
31. *Evaluations of some pesticide residues in food.* (FAO/PL 1967/M/11/1; WHO Food Add./68.30.)

4.4 Hexachlorobenzene

4.4.1 General aspects

Hexachlorobenzene (HCB) is a white solid with a melting point of 230 °C. It is of low solubility in water (6 μg/kg) (*1*) but is soluble in organic solvents. The pure material is produced commercially, principally for use as a fungicide. The largest input to the environment, however, is its generation as a by-product in the manufacture of chlorine and other chlorinated chemicals, particularly solvents.

4.4.2 Routes of exposure

4.4.2.1 *Air*

People are exposed to HCB in air, water, and food. It is disseminated in the air as dust particles and as a result of volatilization from sites having a high HCB-concentration. Airborne HCB-laden dust particles appear to have been a major cause of increased blood concentrations of HCB in the general public living near an industrial site in Louisiana, USA (*2*).

4.4.2.2 *Water*

HCB has been found in 4 river-water samples, 8 finished drinking-water samples, 1 sample from a sewage-treatment plant, and in effluent water from 7 chemical plants in various locations in Europe and the USA (*3*). It has also been detected in urban rainwater runoff in the USA, at levels of 0–339 ng/litre (*4*); in the Rhine River (*5*); in 108 samples of surface-water in Italy, at average levels of 2.5 ng/litre (*6*); and in most river-water residues in an industrialized region of the USA, generally at levels of less than 2 μg/litre, but as high as 90 μg/litre in one sample (*7*).

4.4.2.3 *Food*

HCB occurs in a wide variety of foods, in particular, terrestrial animal products, including dairy products and eggs (*8*). The average intake of HCB from food in the USA was estimated as 0.4 μg/day in 1973 and 0.07 μg/day in 1974 (*9*). The dietary intake in Japan has been estimated to be 0.5 μg/day (*10*) and in Australia 35 μg/day (*11*). Breast-fed infants In Australia and Norway may consume up to 40 μg/day (*11, 12*). Average HCB levels in human adipose tissue have ranged from 0.02 to

8.2 mg/kg (13) and in human blood samples from 0.004 to 0.06 mg/kg (13).

The levels of HCB in the body fat of swine and sheep were, respectively, sixfold and eightfold greater than the dietary level (14). If these comparisons are valid when applied to man, it would appear that some adults have been exposed to levels of HCB of several mg/kg of body weight per day. A similar conclusion is reached by extrapolating the values for human blood. The HCB levels in the blood of rats are about one-tenth the dietary level (15).

Current evidence would indicate that food intake may be the primary source of the body burden of HCB for the general population, although inhalation and dermal exposure may be more important in selected groups, e.g., industrial workers.

4.4.3 Metabolism

Hexachlorobenzene (16) administered orally to rats was absorbed slowly from the gut, mainly via the lymphatic system, and was stored extensively in the fat after 48 hours (17). The quantitative recovery of intraperitoneally and orally administered [^{14}C]-HCB in rats was dose-dependent, but more ^{14}C was recovered from the faeces than from the urine. The major urinary metabolites were pentachlorophenol, tetra-chlorohydroquinone, and pentachlorothiophenol. The other urinary metabolites were tetrachlorobenzene, pentachlorobenzene, 2,4,5- and 2,4,6-trichlorophenols, and 2,3,4,6- and 2,3,5,6-tetrachlorophenols; 2,3,4-trichlorophenol and other tetrachlorophenols were present in traces. These metabolites were excreted as conjugates or in free form in the urine. Unchanged HCB was found in the faeces and in fat (18–21).

When [^{14}C]-HCB was given orally in a dosage of 110 μg/day to *Macaca mulatta* monkeys for 11–15 months, 50 % of the radioactivity found in the urine was in pentachlorophenol and 25 % in pentachlorobenzene, the remainder being in unidentified metabolites and unchanged HCB. In the faeces, 99 % of the radioactivity was in unchanged HCB. During the last 10 days of the experiment, males excreted 7.2 % of the administered dose in the urine and 52 % in the faeces; females excreted 4.6 % and 42.2 %, respectively (22).

4.4.4 Health effects

4.4.4.1 *Toxicity*

Although HCB has a low acute toxicity for most species (> 1000 mg/kg of body weight), it has a wide range of biological effects at prolonged moderate exposure.

Subacute toxic effects were examined in rats fed with HCB for 15 weeks. Histopathological changes were confined to the liver and spleen. In the liver, there was an increase in the severity of centrilobular liver lesions with as little as 2 mg of HCB per kg of food per day. It would appear that 0.5 mg of HCB per kg of body weight per day is the no-effect level in the rat (15). In a similar study, it was found that 0.05 mg of HCB per kg of feed per day is the no-effect level for swine (23).

In rats given 50 mg of HCB per kg of body weight every other day for 53 weeks, an equilibrium between intake and elimination was achieved after nine weeks. In general, the changes observed in the long-term studies resembled those described for short-term studies. When the administration of HCB was discontinued, elimination of the xenobiotic continued slowly for many months (24).

An epidemic of HCB-induced porphyria cutanea tarda occurred in Turkey during the period 1955 to 1959 (25). More than 600 patients were observed during a 5-year period, and it was estimated that a total of 3000 people were affected. The outbreak was traced to the consumption of wheat as food after it had been prepared for planting by treatment with HCB. The syndrome involves blistering and epidermolysis of the exposed parts of the body, particularly the face and hands. It was estimated that the subjects ingested 50–200 mg of HCB per day for a relatively long period before the skin manifestations became apparent. The symptoms were seen mostly during the summer months, having been exacerbated by intense sunlight. The disease subsided and symptoms disappeared 20–30 days after discontinuing the intake of HCB-contaminated bread. Relapses were often seen, either because the subjects were eating HCB-containing wheat again, or because of redistribution of HCB stored in body fat.

A disorder called pembe yara was described in infants of Turkish mothers who either had HCB-induced porphyria or had eaten HCB-contaminated bread (26). The maternal milk contained HCB. At least 95% of these infants died within a year. On the basis of toxicological considerations, FAO/WHO (27) suggested a human ADI of 0.6 μg per kg of body weight, but this has now been withdrawn.

4.4.4.2 Teratogenicity

Placental transfer of HCB has been reported in mice and rats (28, 29). A minimal teratogenic effect of HCB observed in Wistar rats could not be reproduced in the same laboratory with doses up to 120 mg/kg of body weight administered during organogenesis (30). In other studies with hexachlorobenzene and with pentachloronitrobenzene (PCNB) contaminated by hexachlorobenzene at a dose level of 100 mg/kg of body weight, cleft palate and some kidney malformations were found in mice (31).

In a 4-generation test, groups of 10 male and 20 female Sprague-Dawley rats were treated with HCB in dietary concentrations of 0, 10, 20, 40, 80, 160, 320 or 640 mg/kg from weaning. Suckling pups in the Fl generation were particularly sensitive, and many died prior to weaning when the mothers were fed dietary concentrations of 320 or 640 mg/kg of feed. No gross abnormalities were found (32).

4.4.4.3 *Carcinogenicity*

Two studies have been conducted that indicate that HCB is a carcinogen. The carcinogenic activity of HCB in hamsters fed 4, 8, or 16 mg/kg of body weight per day for life was assessed (*33*). HCB appears to have multipotential carcinogenic activity; the incidence of hepatomas, haemangioendotheliomas, and thyroid adenomas was significantly increased. Whereas 10 % of the unexposed hamsters developed tumours, 92 % of the hamsters fed 16 mg of HCB per kg of body weight per day developed tumours. The incidence of tumour-bearing animals was dose-related: 56 % for hamsters fed 4 mg of HCB per kg of body weight per day and 75 % for 8 mg per kg of body weight per day. Thyroid tumours, hepatomas, or liver haemangioendotheliomas were not detected in the unexposed group. An intake of 4–16 mg of HCB per kg of body weight per day in hamsters is near the exposure range estimated for Turkish people who accidentally consumed HCB-contaminated grain (*33*).

The carcinogenic activity of HCB in mice fed 6.5, 13, or 26 mg/kg of body weight per day for life was assessed. The incidence of hepatomas was increased significantly in mice fed 13 or 26 mg of HCB per kg of body weight per day. None of the hepatomas occurred or metastasized in the untreated control groups. The results presented by Cabral et al. (*34, 35*) confirm their earlier conclusion that HCB is carcinogenic. However, the incidence of lung tumours in strain A mice treated three times a week for a total of 24 injections of 40 mg/kg of body weight each was not significantly greater than the incidence in control mice (*36*). HCB is also carcinogenic in rats (*37*).

On the basis of the 10^{-5} risk level, the guideline value for HCB in drinking-water is recommended as 0.01 µg/litre.

REFERENCES

1. LU, P. Y. & METCALF, R. L. Environmental fate and biodegradability of benzene derivatives as studied in a model aquatic ecosystem. *Environmental health perspectives*, **10**: 269 (1975).
2. BURNS, J. E. & MILLER, F. M. Hexachlorobenzene contamination: its effects in a Louisiana population. *Archives of environmental health.*, **30**: 44 (1975).
3. SHACKELFORD, W. M. & KEITH, L. H. *Frequency of organic compounds identified in water.* Athens, GA, US Environmental Protection Agency, 1976 (EPA-600/4-76-062, p. 69).
4. DAPPEN, G. *Pesticide analysis from urban storm runoff.* Springfield, VA, Department of the Interior, 1974 (Report No. PB-238 593).
5. GREVE, P. A. Potentially hazardous substances· in surface water. I. Pesticides in the river Rhine. *Science of the total environment*, **1**: 173 (1972).
6. LEONI, V. & D'ARCA, S. U. Experimental data and critical review of the occurrence of hexachlorobenzene in the Italian environment. *Science of the total environment*, **5**: 253 (1976).
7. LASKA, A. L. ET AL. Distribution of hexachlorobenzene and hexachlorobutadiene in water, soil, and selected aquatic organisms along the lower Mississippi River, Louisiana. *Bulletin of environmental contamination and toxicology*, **15**: 535 (1976).
8. *Environmental contamination from hexachlorobenzene.* Washington, DC, US

Environmental Protection Agency, 1976 (EPA 560/6-70-014).

9. US FOOD & DRUG ADMINISTRATION. *Compliance program evaluation total diet studies (7320.08)*. Washington, DC, US Government Printing Office, 1977.

10. USHIO, F. & DOGUCHI, M. Dietary intakes of some chlorinated hydrocarbons and heavy metals estimated on the experimentally prepared diets. *Bulletin of environmental contamination and toxicology*, **17**: 707 (1977).

11. MILLER, G. J. & FOX, J. A. Chlorinated hydrocarbon pesticide residues in Queensland human milks. *Medical journal of Australia*, **2**: 261 (1973).

12. BAKKEN, A. F. & SEIP, M. Insecticides in human breast milk. *Acta paediatrica scandinavica*, **65**: 535 (1976).

13. *Ambient water quality criteria for chlorinated benzenes*. Washington, DC, US Environmental Protection Agency, 1980 (Document No. 440/5-80-028).

14. HANSEN, L. G. ET AL. Effects and residues of dietary hexachlorobenzene in growing swine. *Journal of toxicology and environmental health*, **2**: 557 (1977).

15. KUIPER-GOODMAN, T. ET AL. Subacute toxicity of hexachlorobenzene in the rat. *Toxicology and applied pharmacology*, **40**: 529 (1977).

16. *Some halogenated hydrocarbons*. Lyon, International Agency for Research on Cancer, 1979 (IARC Monographs on the evaluation of the carcinogenic risk of chemicals to humans, vol. 20.).

17. IATROPOULOS, M. J. ET AL. Absorption, transport and organotropism of di-chlorobiphenyl (DCB), dieldrin, and hexachlorobenzene (HCB) in rats. *Environmental research*, **10**: 384 (1975).

18. ENGST, R. ET AL. The metabolism of hexachlorobenzene (HCB) in rats. *Bulletin of environmental contamination and toxicology*, **16**: 248 (1976).

19. KOSS, G. ET AL. Studies on the toxicology of hexachlorobenzene. II. Identification and determination of metabolites. *Archives of toxicology*, **35**: 107 (1976).

20. MEHENDALE, H. M. ET AL. Metabolism and effects of hexachlorobenzene on hepatic microsomal enzymes in the rat. *Journal of agricultural and food chemistry*, **23**: 261 (1975).

21. RENNER, G. & SCHUSTER, K. P. 2,4,5-Trichlorophenol, a new urinary metabolite of hexachlorobenzene. *Toxicology and applied pharmacology*, **39**: 355 (1977).

22. ROZMAN, K. ET AL. Long-term feeding study of hexachlorobenzene in rhesus monkeys. *Chemosphere*, **6**: 81 (1977).

23. DEN TONKELAAR, E. M. ET AL. Hexachlorobenzene toxicity in pigs. *Toxicology and applied pharmacology*, **43**: 137 (1978).

24. KOSS, G. ET AL. Studies on the toxicology of hexachlorobenzene. III Observations in a long-term experiment. *Archives of toxicology*, **4**: 285 (1978).

25. CAM, C. & NIGOGOSYAN, G. Acquired toxic porphyria cutanea tarda due to hexachlorobenzene. *Journal of the American Medical Association*, **183**: 88 (1963).

26. CAM, C. Une nouvelle dermatose épidémique des enfants. *Annales de dermatologie et de syphiligraphie*, **87**: 393 (1960).

27. *1973 Evaluations of some pesticide residues in food*. Geneva, World Health Organization, 1974 (FAO/AGP/1973/M/9/1; WHO Pesticide Residue Series, 3 p. 291).

28. ANDREWS, J. E. & COURTNEY, K. D. Inter-and intralitter variation of hexachloro-benzene (HCB) deposition in fetuses. *Toxicology and applied pharmacology*, **37**: 128 (1976).

29. VILLENEUVE, D. C. & HIERLIHY, S. L. Placental transfer of hexachlorobenzene in the rat. *Bulletin of environmental contamination toxicology*, **13**: 489 (1975).

30. KHERA, K. S. Teratogenicity and dominant lethal studies on hexachlorobenzene in rats. *Food and cosmetics toxicology*, **12**: 471 (1974).

31. COURTNEY, K. D. ET AL. The effects of pentachloronitrobenzene, hexachlorobenzene, and related compounds on fetal development. *Toxicology and applied pharmacology*, **35**: 239 (1976).

32. GRANT, D. L. ET AL. Effect of hexachlorobenzene on reproduction in the rat. *Archives of environmental contamination and toxicology*, **5**: 207 (1977).

33. CABRAL, J. R. P. ET AL. Carcinogenic activity of hexachlorobenzene in hamsters. *Nature*, **269**: 510 (1977).

34. CABRAL, J. R. P. ET AL. Carcinogenesis study in mice with hexachlorobenzene. *Toxicology and applied pharmacology*, **45**: 323 (1978).
35. CABRAL, J. R. P. ET AL. Carcinogenesis of hexachlorobenzene in mice. *International journal of cancer*, **23**: 47 (1979).
36. THEISS, J. C. ET AL. Test for carcinogenicity of organic contaminants of United States drinking waters by pulmonary tumor response in strain A mice. *Cancer research*, **37**: 2717 (1977).
37. SMITH, A. G. & CABRAL, J. R. P. Liver cell tumours in rats fed hexachlorobenzene. *Cancer letter*, **11**: 169 (1980).

4.5 Heptachlor and heptachlor epoxide

4.5.1 General aspects

Pure heptachlor ($C_{10}H_5Cl_7$; relative molecular mass 373.35) is a white crystalline solid with a camphor-like odour. The chemical name for heptachlor is 1,4,5,6,7,8,8-heptachloro-3a,4,7,7a-tetrahydro-4,7-methano-indene. It has a vapour pressure of 4×10^{-2} Pa (3×10^{-4} mmHg) at 25 °C, a solubility in water of 0.056 mg/litre at 25–29 °C, and is readily soluble in relatively nonpolar solvents (*1*).

Heptachlor is a broad-spectrum insecticide of the group of polycyclic chlorinated hydrocarbons called cyclodiene insecticides. From 1971 to 1975 the most important use of heptachlor was to control agricultural soil insects (*1*).

Since 1975, both the uses and the production volume have declined extensively because of a voluntary restriction by the sole producer and because of the subsequent issuance (2 August 1976) of a registration suspension notice by the EPA for all food crop and home use of heptachlor. However, significant commercial use of heptachlor for termite control and non-field crops continues.

Heptachlor persists for prolonged periods in the environment. It is converted to the more toxic metabolite, heptachlor epoxide, in the soil (*2–5*), in plants (*6*), and in mammals (*7*). Heptachlor, in solution or thin films, undergoes photodecomposition to photoheptachlor (*8*) which is more toxic than the parent compound to insects (*9*) and aquatic invertebrates (*10, 11*).

Heptachlor and its epoxide undergo bioconcentration in numerous species and will accumulate in the food chain (*1*).

4.5.2 Routes of exposure

4.5.2.1 *Water*

Various investigators have detected heptachlor and/or heptachlor epoxide in the major river basins of the USA at a mean concentration for both of 0.0063 µg/litre (*12*). Levels of heptachlor ranged from 0.001 to 0.035 µg/litre (*13*).

4.5.2.2 *Food*

In their Market Basket Study (1974–75) for 20 different cities, the FDA showed that 3 of 12 food classes contained residues of heptachlor epoxide ranging from 0.0006 to 0.003 mg/kg (*14*). Heptachlor epoxide residues greater than 0.03 mg/kg have been found in 14–19% of red meat, poultry, and dairy products sampled from 1964 to 1974.[a] Heptachlor and/or heptachlor epoxide were found in 32% of 590 fish samples obtained nationally, with whole fish residues from 0.01 to 8.33 mg/kg (*15*).

Human milk can be contaminated with heptachlor epoxide. A nationwide survey indicated that 63.1% of 1936 mothers' milk samples contained heptachlor epoxide residues (*16*).

4.5.2.3 *Air*

Heptachlor volatilizes from treated surfaces, plants, and soil. Heptachlor and to a lesser extent heptachlor epoxide are widespread in ambient air, with typical mean concentrations of approximately 0.5 ng/m^3. On the basis of these data, typical human exposure was calculated to be 0.01 μg per person per day.[a] Thus, it appears that inhalation is not a major route for human exposure to heptachlor.

4.5.3 Metabolism

Heptachlor is readily metabolized by mammals to heptachlor epoxide. This metabolite is stored mainly in adipose tissue, but also in liver, kidney, and muscle (*17*). Both the rat and the dog rapidly metabolize ingested heptachlor to heptachlor epoxide and accumulate heptachlor epoxide primarily in adipose tissue. A positive relationship is found between the amount of heptachlor in the diet and the amount of heptachlor stored in the fatty tissue, female rats accumulating approximately six times as much heptachlor epoxide in their fat as males (*7, 18*).

Although there is no direct evidence showing the conversion of heptachlor to its epoxide in humans, there is little doubt that the epoxide that has been found in human tissues is derived from heptachlor. Various levels of heptachlor epoxide have been found in the blood, fat, and milk of humans. Because of its high lipid content, milk is one of the major excretion routes for organohalogenated compounds, including heptachlor epoxide. An extensive survey carried out in the USA indicated that women who had lactated following several births had lower pesticide levels in milk than primiparae. Heptachlor epoxide, dieldrin, and oxychlordane were the most common

[a] NISBETT, I. C. T. *Human exposure to chlordane, hepatochlor and their metabolites* (unpublished review).

pesticides found in human milk. Only 2% of human milk samples showed heptachlor residues, but 63.1% of the samples showed heptachlor epoxide residues, the levels ranging from 15 to 2050 μg/litre on a fat-adjusted basis, with a mean concentration of 91 μg/litre. Of the high-residue group of women, 11% were either occupationally exposed or lived in households where a household member was occupationally exposed (16).

4.5.4 Health effects

4.5.4.1 Toxicity

The mammalian LD_{50} of heptachlor and its metabolites has been reported in a variety of species as ranging from 6 mg/kg to 531 mg/kg of body weight (1).

Little information on chronic effects is available. When administered to rats in small daily doses over a prolonged period of time, heptachlor induced alterations in glucose homoeostasis, which were thought to be related to an initial stimulation of the cyclic AMP-adenylate cyclase system in liver and kidney cortex (19–21).

A joint FAO/WHO committee (22) has established a maximum acceptable daily intake for heptachlor plus heptachlor epoxide of 0.5 μg per kg of body weight.

4.5.4.2 Teratogenicity

In long-term feeding studies with heptachlor, cataracts developed in parent rats and in the offspring shortly after their eyes opened (23).

4.5.4.3 Other reproductive effects

In long-term feeding studies in rats, heptachlor caused a marked decrease in litter size and a decreased lifespan in suckling rats (23).

4.5.4.4 Mutagenicity

Heptachlor has been reported to be mutagenic in mammalian assays but not in bacterial assays. Heptachlor caused dominant lethal changes in male rats as demonstrated by the number of resorbed fetuses in intact pregnant rats (24). Bone marrow cells of the treated animals showed increases in the incidences of abnormal mitoses, chromatid abnormalities, pulverization, and translocation Both heptachlor and heptachlor epoxide induced unscheduled DNA synthesis in SV-40 transformed human cells (VA-4) in culture with metabolic activation (25). Neither heptachlor nor heptachlor epoxide was mutagenic for Salmonella typhimurium in the Ames test (26).

4.5.4.5 Carcinogenicity

Heptachlor and/or heptachlor epoxide have induced hepatocellular carcinomas in mice during three chronic feeding studies. Heptachlor epoxide has produced the same response in rats in one study (27).[a]

4.5.4.6 Guideline value

On the basis of the maximum acceptable daily intake of 0.5 µg per kg of body weight recommended by a joint FAO/WHO committee (22), the guideline value in drinking-water for a 70-kg man consuming 2 litres of water per day can be calculated as 0.1 µg/litre. As a carcinogenic response has been well documented in only one species of animal, it seems justified to recommend the application of this guideline value.

REFERENCES

1. *Heptachlor: ambient water quality criteria.* Washington, DC, US Environmental Protection Agency, 1980 (Document No. 440/5-80-052).
2. LICHTENSTEIN, E. P. Insecticidal residues in various crops grown in soils treated with abnormal rates of aldrin and heptachlor. *Journal of agricultural and food chemistry*, **8**: 448 (1960).
3. LICHTENSTEIN, E. P. ET AL. Degradation of aldrin and heptachlor in field soils. *Journal of agricultural and food chemistry*, **18**: 100 (1970).
4. LICHTENSTEIN, E. P. ET AL. Effects of a cover crop versus soil cultivation on the fate of vertical distribution of insecticide residues in soil 7 to 11 years after soil treatment. *Pesticides monitoring journal*, **5**: 218 (1971).
5. NASH, R. G. & HARRIS, W. G. Chlorinated hydrocarbon insecticide residues in crops and soil. *Journal of environmental quality*, **2**: 269 (1973).
6. GANNON, N. & DECKER, G. C. The conversion of aldrin to dieldrin in plants. *Journal of economic entomology*, **51**: 8 (1958).
7. DAVIDOW, B. & RADOMSKI, J. L. Isolation of an epoxide metabolite from fat tissues of dogs fed heptachlor. *Journal of pharmacology and experimental therapeutics*, **107**: 259 (1953).
8. BENSON, W. R. ET AL. Photolysis of solid and dissolved dieldrin. *Journal of agricultural and food chemistry*, **19**: 66 (1971).
9. KHAN, M. H. ET AL. Insect metabolism of photoaldrin and photodieldrin. *Science*, **164**: 318 (1969).
10. GEORGACKAKIS, E. & KHAN, M. A. Q. Toxicity of the photoisomers of cyclodiene insecticides to freshwater animals. *Nature*, **233**: 120 (1971).
11. KHAN, M. A. Q. ET AL. Toxicity-metabolism relationship of the photoisomers of certain chlorinated cyclodiene insecticide chemicals. *Archives of environmental contamination and toxicology*, **1**: 159 (1973).
12. *Chlordane and heptachlor in relation to man and the environment.* Washington, DC, US Environmental Protection Agency, 1976 (EPA 540/476005).
13. BREIDENBACH, A. W. ET AL. Chlorinated hydrocarbon pesticides in major river basins, 1957–65. *Public health reports*, **82**: 139 (1967).
14. JOHNSON, R. D. & MANSKE, D. D. Pesticide and other chemical residues in total diet samples (XI). *Pesticides monitoring journal*, **11**: 116 (1977).
15. HENDERSON, C. ET AL. Organochlorine insecticide residues in fish (National Pesticide Monitoring Program). *Pesticides monitoring journal*, **3**: 145 (1969).

[a] US Environmental Protection Agency. *Risk assessment of chlordane and heptachlor.* Carcinogen Assessment Group, Washington, DC, 1977 (unpublished report).

16. SAVAGE, E. P. *National study to determine levels of chlorinated hydrocarbon insecticides in human milk*. Washingdon, DC, US Environmental Protection Agency, 1976 (EPA/540/9-78/005).
17. *Evaluations of some pesticide residues in food*. Geneva, World Health Organization, 1967.
18. RADONSKI, J. L. & DAVIDOW, B. The metabolic of heptachlor, its estimation, storage and toxicity. *Journal of pharmacology and experimental therapeutics*, **107**: 266 (1953).
19. KACEW, S. & SINGHAL, R. L. The influence of p,p-DDT, and chlordane, heptachlor and endrin on hepatic and renal carbohydrate metabolism and cyclic AMP-adenyl cyclase system. *Life sciences*, **13**: 1363 (1973).
20. KACEW, S. & SINGHAL, R. L. Effect of certain halogenated hydrocarbon insecticides on cyclic adenosine 3′,5′-monophosphate-^{3}H formation by rat kidney cortex. *Journal of pharmacology and experimental therapeutics*, **188**: 265 (1974).
21. SINGHAL, R. L. & KACEW, S. The role of cyclic AMP in chlorinated hydrocarbon-induced toxicity. *Federation proceedings*, **35**: 2618 (1976).
22. *1971 evaluations of some pesticide residues in food*. (AGP: 1971/M/9/1; WHO Pesticide Residues Series, No. 1, p. 314).
23. MESTITZOVA, M. On reproduction studies on the occurrence of cataracts in rats after long-term feeding of the insecticide heptachlor. *Experientia*, **23**: 42 (1967).
24. CEREY, K. ET AL. Effect of heptachlor on dominant lethality and bone marrow in rats. *Mutation research*, **21**: 26 (1973).
25. AHMED, F. E. ET AL. Pesticide-induced DNA damage and its repair in cultured human cells. *Mutation research*, **42**: 161 (1977).
26. MARSHALL, T. C. ET AL. Screening of pesticides for mutagenic potential using *Salmonella typhimurium* mutants. *Journal of agricultural and food chemistry*, **24**: 560 (1976).
27. EPSTEIN, S. S. Carcinogenicity of heptachlor and chlordane. *Science of the total environment*, **6**: 103 (1976).

4.6 Lindane

4.6.1 General aspects

Lindane (gamma-hexachlorocyclohexane), also known as γ-HCH or γ-BHC, is a white solid with a melting point of 112.5 °C. It is fairly soluble in water (10 mg/litre) (*1*), but more soluble in organic solvents. Lindane is a broad-spectrum insecticide of the group of cyclic chlorinated hydrocarbons called organochlorine insecticides and is used in a wide range of applications, including treatment of animals, buildings, man (for ectoparasites), clothes, water (for mosquitos), plants, seeds, and soils (*2*).

It is slowly degraded by soil microorganisms (*3*) and can be isomerized to the alpha and/or delta isomers by microorganisms and plants (*2*).

4.6.2 Routes of exposure

4.6.2.1 *Water*

Contamination of water has occurred from direct application of technical hexachlorocyclohexane (HCH) or lindane to water for control of mosquitos, from the use of HCH in agriculture and forestry, and, to a lesser extent, from occasional contamination of wastewater from manufacturing plants (2).

In a survey of finished drinking-water in the USA, the highest reported level of lindane was 0.1 µg/litre (4). It is a ubiquitous contaminant of surface waters at levels up to 100 ng/litre (5, 6), presumably owing to its volatilization and precipitation in rain (7). In the Federal Republic of Germany, lindane was present in all surface-water samples taken, at levels ranging from 0.005 to 7.1 µg/litre.

4.6.2.2 Food

The daily intake of lindane has been reported to be 1–5 µg/kg of body weight and the daily intake of all other HCH isomers to be 1–3 µg/kg of body weight (8). The chief sources of HCH residues in the human diet are milk, eggs, and other dairy products (2).

In the USA, the EPA (2) has estimated the weighted average bioconcentration factor for lindane at 780. This estimate is based on the measured steady-state bioconcentration in the bluegill.[a]

4.6.2.3 Air

Traces of HCH have been detected in the air of central and suburban London (2). The uptake of lindane by inhalation is estimated at 0.002 µg per kg of body weight per day (9).

4.6.3 Metabolism

The rapidity of lindane absorption is enhanced by lipid-mediated carriers. Compared with other organochlorine insecticides, HCH and lindane are fairly soluble in water, which contributes to rapid absorption and excretion (2, 10). Lindane is absorbed after oral or dermal exposure (2).

After administration to experimental animals, lindane was detected in the brain at higher concentrations than in other organs (11–13). At least 75 % of an intraperitoneal dose of ^{14}C-labelled lindane was consistently found in the skin, muscle, and fatty tissue (14). Lindane enters the human fetus through the placenta; higher concentrations were found in the skin than in the brain and never exceeded the corresponding values for adult organs (15, 16).

Lindane is metabolized to an intermediate hexachlorocyclohexene; further degradation yields 2,3,4,5,6-pentachloro-2-cyclohexene-1-ol, two tetrachlorophenols, and three trichorophenols (17). These are commonly found in the urine as conjugates (18). Both free and conjugated chlorophenols are far less toxic than the parent compounds (19).

---

[a] A species of freshwater sunfish common in North America.

4.6.4 Health effects

The oral LD_{50} of lindane for rats is 125–230 mg/kg of body weight (20).

In chronic studies with rats given lindane in oil, liver cell hypertrophy (fat degeneration and necrosis) and nephritic changes were noted at high doses (21–23). Rats inhaling lindane (0.78 mg/m³ of air) for 7 hours, 5 days a week, for 180 days showed liver cell enlargement, but showed no toxic symptoms or other abnormalities (24). The addition of lindane to the diet of rats in a concentration of 10 mg/kg for 1 or 2 years decreased body weight after 5 months of treatment and altered ascorbic acid levels in urine, blood, and tissues (25). Dogs chronically exposed to lindane in the diet had slightly enlarged livers (26).

Irritation of the central nervous system with other toxic side-effects (nausea, vomiting, spasms, weak respiration with cyanosis, and blood dyscrasia) were reported after prolonged or improper use of a preparation containing 10 g of lindane per kg for the treatment of scabies on humans (27). Production workers exposed to technical HCH exhibited symptoms including headache, vertigo, irritation of the skin, eyes, and respiratory tract mucosa. In some instances, there were apparent disturbances of carbohydrate and lipid metabolism and dysfunction of the hypothalamo-pituitary-adrenal system (28, 29). A study of persons occupationally exposed to HCH for 11–23 years revealed biochemical manifestations of toxic hepatitis (30).

A maximum acceptable daily intake for lindane of 10 μg/kg of body weight has been estimated by a joint FAO/WHO meeting (31).

4.6.4.1 Teratogenicity

Lindane given in the diet during pregnancy at levels of 12 or 25 mg per kg of body weight per day did not produce teratogenic effects in rats (32).

4.6.4.2 Other reproductive effects

Chronic lindane feeding in a study of four generations of rats increased the average duration of pregnancy, decreased the number of births, increased the proportion of stillbirths, and delayed sexual maturation in F_2 and F_3 females. In addition, some of the F_1 and F_2 animals exhibited spastic paraplegia (25).

In rats and rabbits, lindane given in the diet during pregnancy increased postimplantation death of embryos (32, 33).

4.6.4.3 Mutagenicity

Evidence for the mutagenicity of lindane is equivocal. Some alterations in mitotic activity and the karyotype of human lymphocytes cultured with lindane at 0.1–10 g/litre have been reported (34). Lindane was not mutagenic in a dominant-lethal assay[a] or a host-mediated assay (35). It was found to be mutagenic in microbial assays using *Salmonella typhimurium* with metabolic activation, the host-mediated assay, and the dominant lethal test in rats. Other reports indicate that it does not have significant mutagenic activity (2).

4.6.4.4 *Carcinogenicity*

An increased incidence of liver tumours was reported in male and female mice of various strains fed lindane (gamma-HCH) (36–40). Epidemiological data were inadequate for evaluation (40, 41).

4.6.4.5 *Guideline value*

Based on the maximum acceptable daily intake of lindane for a 70-kg man recommended by an FAO/WHO meeting (31) and assigning 1% of the ADI to water, a guideline value of 3 μg/litre can be calculated for a water consumption of 2 litres per person per day.

REFERENCES

1. ULMANN, E., ed. *Lindane: monograph of an insecticide.* Freiburg, Verlag K. Schillinger, 1972.
2. *Hexachlorocyclohexane: ambient water quality criteria.* Washington, DC, US Environmental Protection Agency, 1979.
3. MATHUR, S.P. & SAHA, J. G. Microbial degradation of lindane-C-14 in a flooded sandy loam soil. *Soil science,* **120**: 301 (1975).
4. US Environmental Protection Agency. *Preliminary assessment of suspected carcinogens in drinking-water.* Report to Congress. Washington, DC, 1975, p. II-4.
5. CROLL B. T. Organo-chlorine insecticides in water–Part 1. *Water treatment examination,* **18**: 255–274 (1969).
6. GREVE, P. A. Potentially hazardous substances in surface waters. I. Pesticides in the R. Rhine. *Science of the total environment.* **1**: 173–180 (1972).
7. TARRANT, K. R. & TATTON, J. O'G. Organo-chlorine pesticides in rainwater in the British Isles. *Nature,* **219**: 725–727 (1968).
8. DUGGAN, R. E. & DUGGAN, M. B. Residues of pesticides in milk, meat and foods. In: Edwards, L. A., ed., *Environmental pollution from pesticides.* London, 1973, p. 334.
9. BARNEY, J. E. Pesticide pollution of the air studied. *Chemical and engineering news,* **47**: 42 (1969).
10. HERBST, M. & BODENSTEIN, G. Toxicology of lindane, In: Ulmann, E., ed. *Lindane: monograph of an insecticide.* Freiburg, Verlag K. Schillinger, 1972, p. 23.
11. LANG, E. P. Tissue distribution of a toxicant following oral ingestion of the gamma-isomer of benzene hexachloride by rats. *Journal of pharmacology and experimental therapeutics,* **93**: 277 (1948).
12. DAVIDOW, B. & FRAWLEY, J. P. Tissue distribution, accumulation and elimination of isomers of benzene hexachloride. *Proceedings of the Society for Experimental Biology and Medicine,* **76**: 780 (1951).
13. HUNTINGDON RESEARCH CENTRE. In: Ulmann, E., ed. *Lindane: monograph of an insecticide.* Freiburg, Verlag K. Schillinger, 1972, p. 97.

---

[a] US Environmental Protection Agency. *BHC-Lindane.* Washington, DC, Criteria and Evaluation Division (unpublished report).

14. KORANSKY, S. ET AL. Absorption, distribution and elimination of alpha- and beta-benzene hexachloride. *Archiv für experimentelle Pathologie und Pharmakologie*, **244**: 564 (1963).
15. PORADOVSKY, R. ET AL. Transplacental permeation of pesticides during normal pregnancy. *Ceskoslovenská gynekologie*, **42**: 405 (1977).
16. NICHIMURA, H. ET AL. Levels of polychlorinated biphenyls and organochlorine insecticides in human embryos and fetuses. *Pediatrician*, **6**: 45 (1977).
17. CHADWICK, R. W. ET AL. Dehydrogenation, a previously unreported pathway of lindane metabolism in mammals. *Pesticide biochemistry and physiology*, **6**: 575 (1975).
18. CHADWICK, R. W. & FREAL, J. J. The identification of five unreported lindane metabolites recovered from rat urine. *Bulletin of environmental contamination and toxicology*, **7**: 137 (1972).
19. NATIONAL RESEARCH COUNCIL. *Drinking water and health.* Washington, DC, National Academy of Sciences, 1977, p. 939.
20. *1966 Evaluations of some pesticide residues in food.* Geneva, WHO, 1967 (WHO/Food Add./67.32) pp. 126–147.
21. FITZHUGH, O. G. ET AL. Chronic toxicities of benzene hexachloride, and its alpha, beta, and gamma isomer. *Journal of pharmacology and experimental therapeutics*, **100**: 59 (1950).
22. LEHMAN, A. J. Chemicals in food: A report to the Association of Food and Drug Officials. *US Association of Food and Drug Officials quarterly bulletin*, **16**: 85 (1952).
23. LEHMAN, A. J. Chemicals in food: A report to the Association of Food and Drug Officials on current development. Part II, Pesticides. Section V: Pathology. *US Association of Food and Drug Officials quarterly bulletin*, **16**: 126 (1952).
24. HEYROTH, F. F. In: Leland, S. J., *Chemical Specialities Manufacturers Association. Proceedings of the annual meeting*, **6**: 110 (1952).
25. PETRESCU, S. ET AL. Studies on the effects of long term administration of chlorinated organic pesticides (lindane, DDT) on laboratory white rats. *Revue médico-chirurgicale (Jassy)*, **78**: 831 (1974).
26. RIVETT, K. F. ET AL. Effects of feeding lindane to dogs for periods of up to 2 years. *Toxicology*, **9**: 237 (1978).
27. LEE, B. ET AL. Suspected reactions to gamma benzene hexachloride. *Journal of the American Medical Association*, **236**: 2846 (1976).
28. KAZAHEVICH, R. L. State of the nervous system in persons with a prolonged professional contact with hexachlorocyclohexane and products of its synthesis. *Vrachebnoe delo*, **2**: 129 (1974).
29. BESUGLYI, V. P. ET AL. State of health of persons having prolonged occupational contact with hexachlorocyclohexane. *Zdravookhranenie Belorussii*, **19**: 49 (1973).
30. SASINOVICH, L. M. ET AL. Toxic hepatitis due to prolonged exposure to BHC. *Vrachebnoe delo*, **10**: 133 (1974).
31. *Pesticide residues in food. Report 1977.* FAO Plant production and protection paper, Rome, 1978.
32. MAMETKULIEV, C. H. Study of embryotoxic and teratogenic properties of the gamma isomer of HCH in experiments with rats. *Zdravookhranenie Turkmenistana*, **20**: 28 (1978).
33. PALMER, A. K. ET AL. Effect of lindane on pregnancy in the rabbit and rat. *Toxicology*, **9**: 239 (1978).
34. TSONEVA-MANEVA, M. T. ET AL. Influence of diazinon and lindane on the mitotic activity and the karyotype of human lymphocytes cultivated *in vitro*. *Bibliotheca haematologia*, **38**: 344 (1971).
35. BUSELMAIR, W. ET AL. Comparative investigation on the mutagenicity of pesticides in mammalian test systems. *Mutation research*, **21**: 25 (1973).
36. GOTO, M. ET AL. Ecological chemistry. Toxizitat von a-HCH in mausen. *Chemosphere*, **1**: 153 (1972).
37. HANADA, M. ET AL. Induction of hepatoma in mice by benzene hexachloride. *Japanese journal of cancer research*, **64**: 511 (1973).
38. NATIONAL CANCER INSTITUTE. *Bioassay of lindane for possible carcinogenicity.* Washington, DC, 1977 (Technical Report Series, No. 14; Department of Health, Education, and Welfare Publication No. (NIH) 77-814). *Federal register*, **42**: (1977).

39. THORPE, E. & WALKER, A. I. The toxicology of dieldrin (HEOD). II. In mice with dieldrin, DDT, phenobarbitone, beta-BCH, and gamma-BCH. *Food and cosmetics toxicology*, **11**: 433 (1973).

40. *Some halogenated hydrocarbons*. Lyon, International Agency for Research on Cancer, 1979 (IARC Monographs on the evaluation of the carcinogenic risk of chemicals to humans, vol. 20).

41. ERIKSSON, M. ET AL. Soft-tissue sarcomas and exposure to chemical substances: a case-referent study. *British journal of industrial medicine*, **38**: 27 (1981).

4.7 Methoxychlor

4.7.1 General aspects

Methoxychlor refers to 1,1′(2,2,2-trichloroethylidene)bis(4-methoxy-benzene). It is practically insoluble in water, but readily soluble in most aromatic organic solvents. Technical methoxychlor consists of about 88% of the *p,p′*-isomer, the remainder being the *o,p′*-isomer.

Methoxychlor is an insecticide used for the treatment of agricultural crops and livestock. Reviews of this compound are available (*1–4*) and relevant data are summarized below.

4.7.2 Routes of exposure

4.7.2.1 *Water*

The half-life of methoxychlor in water is about 46 days. Residues of methoxychlor have occurred in river-water at levels of 2.9–89.1 µg/litre, in lake water at levels up to 0.1 µg/litre, and in effluent from a biological sewage-treatment plant at levels of up to 106 µg/litre. It has been found in tributary streams of Lake Michigan at levels of 2.9–89.1 ng/litre. In finished drinking-water from the Mississippi and Missouri rivers no residues of methoxychlor were detected in 500 samples analysed.

4.7.2.2 *Food*

On the basis of measured residues in food, an average daily intake of 0.5 µg/day was calculated in the USA for the period 1965–70. Methoxychlor showed very little tendency to be stored in tissues or to be excreted in milk (*3, 4*).

4.7.3 Metabolism

Methoxychlor is rapidly metabolized in rats by the liver, its metabolic products being excreted mainly in the faeces and to a lesser extent in the urine. In mice, 98% of labelled methoxychlor given orally was eliminated within 24 hours. Methoxychlor is mainly degraded by hydrolysis of the methylether group leading to a polar phenol which is rapidly excreted.

Dose-dependent storage of methoxychlor occurred in fatty tissues of rats. At a dose of 500 mg/kg of diet, an equilibrium was reached within

4 weeks; methoxychlor disappeared from the fatty tissue within 2 weeks after the end of exposure.

From experiments in animals it may be concluded that the high rate and completeness of methoxychlor metabolism accounts for its low storage and accumulation (*1, 2, 4*).

4.7.4 Health effects

Methoxychlor is a compound of relatively low acute toxicity. Its oral LD_{50} in rats is 3460 mg/kg of body weight.

Methoxychlor did not exhibit teratogenicity in rats. It was not mutagenic in bacteria, yeast, or *Drosophila melanogaster*. Cytogenic and dominant lethal tests in mice were also negative.

Methoxychlor was tested for carcinogenicity in mice and in several experiments in rats by oral exposure. The study in mice gave negative results. A suggestion that methoxychlor was hepatocarcinogenic was not confirmed in three studies in rats. The available data did not provide evidence that methoxychlor is carcinogenic in experimental animals (*4*).

An acceptable daily intake for humans of 0.1 mg/kg of body weight, established in 1965, was reaffirmed in 1977 (*1, 2*). On this basis, the maximum daily intake for a 70-kg man would be 7 mg. Assigning 1 % of this value to water and assuming a water consumption of 2 litres per person per day, a guideline value of 30 µg/litre can be recommended.

REFERENCES

1. *Evaluation of the toxicity of pesticide residues in food.* (FAO/PL/1965:10/1; WHO/Food Add./27.65).
2. *Pesticide residues in food. Report 1977.* FAO Plant Production and Protection Paper, Rome, 1978.
3. *Some organochlorine pesticides.* Lyon, International Agency for Research on Cancer, 1974 (IARC Monographs on the evaluation of the carcinogenic risk of chemicals to humans, vol. 5).
4. *Some halogenated hydrocarbons.* Lyon, International Agency for Research on Cancer, 1979 (IARC Monographs on the evaluation of the carcinogenic risk of chemicals to humans, vol. 20).

4.8. 2,4-Dichlorophenoxyacetic acid

4.8.1 General aspects

2,4-D (2,4-dichlorophenoxyacetic acid) is used as a herbicide for the control of broad-leafed plants and as a plant growth regulator. It is produced commercially by the chlorination of phenol to form 2,4-dichlorophenol, which is reacted with monochloroacetic acid to form 2,4-D.

Commercial formulations are generally composed of salts or esters of the acid. Analyses have shown that dioxins are generally absent.

2,4-D is chemically quite stable, but its esters are readily hydrolysed to the free acid. The herbicide may be rapidly broken down in water. Residues are infrequently found in the soil as the substance is broken down by soil microorganisms and there is reportedly no accumulation. 2,4-D has been detected in streams following aerial application on adjacent forestland, while smaller concentrations (< 0.1 μg/l) have been reported in potable water sources prior to treatment.

4.8.2 Routes of exposure

4.8.2.1 Drinking-water

Information on the levels of 2,4-D in drinking-water is not available. Very low concentrations would be expected; microbial breakdown rapidly reduces the level of 2,4-D in contaminated surface-water.

4.8.2.2 Food

Some foods have been shown to be contaminated with low levels of 2,4-D (0.021–0.16 mg/kg) (1).

It has been reported that plants treated with 2,4-D produce increased amounts of nitrate. Although this is of concern with regard to the use of the herbicide on edible crops, there is no indication that foodstuff exposed to 2,4-D has become toxic in this manner.

4.8.2.3 Industrial exposure

Adverse health effects as the result of industrial exposure to the chemical have been reported (2).

4.8.3 Metabolism

2,4-D is rapidly excreted virtually unchanged in the urine of man and animals. Where the chemical is absorbed, it is distributed in various tissues but not stored. In rats that received 1–10 mg of 2,4-D, there was almost complete excretion in the urine and faeces in 48 hours; at higher doses some accumulation occurred in the tissues.

The effect of ingested phenoxyacid herbicides on muscular function may be related to interference with carbohydrate metabolism.

2,4-D is eliminated in the milk of cows maintained on pastures treated with 2,4-D or its esters.

4.8.4 Health effects

Individuals exposed to 2,4-D through use or manufacture have complained of fatigue, headache, liver pains, loss of appetite, etc.

However, these claims are subjective. Cases of arterial hypertension and liver dysfunction have also been encountered.

Workers exposed to 2,4-D at 0.43–0.57 mg/kg of body weight per day over a period of 0.5–22 years showed no differences in comparison with an unexposed human population.

Studies on the carcinogenic properties of this compound have proved inconclusive, because of either inadequate reporting or the small number of animals used. However, indications are that 2,4-D is not a potential carcinogen. In a case-control study on the association between malignant lymphomas and exposure to chlorophenols and phenoxyacids, 7 cases and 1 control were said to have been exposed only to 2,4-D (relative risk 14.6 with 95% confidence interval 2.9–29.9) (3).

The LD_{50} of 2,4-D has been estimated to be over 90 mg/kg of body weight.

4.8.4.1 Guideline value

The acceptable daily intake of 2,4-D has been established by a joint FAO/WHO meeting (4, 5) at 0.3 mg/kg of body weight. The guideline value for 2,4-D in drinking-water is 0.1 mg/litre, based on toxicity data. However, some individuals may be able to detect 2,4-D by taste or odour at levels around 0.05 mg/litre.

REFERENCES

1. NATIONAL RESEARCH COUNCIL. *Drinking water and health*. Washington, DC, National Academy Press, 1977.
2. *Some fumigants, the herbicides 2,4-D and 2,4,5-T, chlorinated dibenzodioxins and miscellaneous industrial chemicals*. Lyon, International Agency for Research on Cancer, 1977 (IARC Monographs on the evaluation of carcinogenic risk of chemicals to humans, vol. 15).
3. HARDELL, L. ET AL. Malignant lymphoma and exposure to chemicals, especially organic solvents, chlorophenols and phenoxy acids: a case-control study. *British journal of cancer*, **43**: 169 (1981).
4. *1974 Evaluations of some pesticide residues in food*. Geneva, World Health Organization, 1975 (Pesticide Residues Series, No. 4).
5. *1975 Evaluations of some pesticide residues in food*. Geneva, World Health Organization, 1976 (Pesticide Residues Series, No. 5).

5. CHLOROBENZENES

Monochlorobenzene is widely used as a solvent and in the manufacture of several chemicals, such as insecticides and phenols. Dichlorobenzenes are important intermediates for dyestuffs. 1,2-dichlorobenzene is used as a solvent and pesticide and 1,4-dichlorobenzene as a moth repellant and deodorant. 1,2,4-trichlorobenzene is used as a solvent, dielectric fluid, heat transfer medium and insecticide. Of the tetrachlorobenzenes, 1,2,4,5-tetrachlorobenzene is used as an intermediate in chemical syntheses, e.g., for the production of 2,4,5-trichlorophenol (*1*). Pentachlorobenzenes are not widely used. Hexachlorobenzene has been considered in the section on pesticides and is therefore not included here. Some lower chlorobenzenes are formed as by-products of chlorination of water.

Monochlorobenzene has been detected in ground water, surface water, and drinking-water at levels up to 10 μg/litre (*1, 2*). Dichlorobenzenes are frequently found in raw water sources at levels of 1–10 μg/litre or more (*3*). The main dichlorobenzenes in the aquatic environment are 1,2- and 1,4-dichlorobenzene. In drinking-water these compounds have been found at levels between 0.01 and 1 μg/litre. Of the trichlorobenzenes, 1,2,4-trichlorobenzene is detected most frequently; levels in drinking-water have varied between 0.01 and 1 μg/litre (*2*). The absence of data on the occurrence of the other trichlorobenzenes and of the tetrachlorobenzenes indicates that their concentrations are probably below 0.1 μg/litre, the limit of detection of the techniques used for their analysis.

On the basis of the anticipated maximum levels of chlorobenzenes in drinking-water and an examination of the data on toxicity and odour threshold concentrations (OTC), a selection can be made of the compounds for which the development of a guideline value is required (*4, 5*). A summary of these data is given in Table 2, page 222.

Considering the low levels detected in drinking-water compared with the organoleptic and toxicological properties of these compounds and the paucity of toxicological data, only the following three compounds were selected for further consideration: chlorobenzene; 1,2-dichlorobenzene; 1,4-dichlorobenzene.

As Table 2 shows, the organoleptic and the toxicological limits for chlorobenzene are of the same order of magnitude. For the dichlorobenzenes, the odour thresholds are somewhat lower than the toxicologically derived values.

Table 2. Criteria for selecting chlorinated benzenes of toxicological concern

Compound	Indication of maximum levels		Indication of OTC (μg/litre)	Indication of toxicity based on no-adverse-effect level (μg/litre)
	Raw water (μg/litre)	Drinking water (μg/litre)		
Chlorobenzene	10	10	20–100	5–50
1,2-dichlorobenzene	10	1	2–10	5–50
1,3-dichlorobenzene	1	0.1	20	5–50
1,4-dichlorobenzene	10	1	0.3–30	5–50
1,2,3-trichlorobenzene	—	—	10	*
1,2,4-trichlorobenzene	10	1	5–30	*
1,3,5-trichlorobenzene	—	—	50	*
1,2,3,4-tetrachlorobenzene	—	—	20	*
1,2,3,5-tetrachlorobenzene	—	—	400	*
1,2,4,5-tetrachlorobenzene	—	—	130	2–20

— no data available, which indicates that the levels are probably below 0.1 μg/litre.
* no data available on chronic toxicity.

5.1 Chlorobenzene (monochlorobenzene)[a]

5.1.1 General aspects

Chlorobenzene (monochlorobenzene) is widely used as a solvent and as an intermediate in the manufacture of dyestuffs, pesticides, and other chemicals (1). It can also be formed upon chlorination of water.

5.1.2 Routes of exposure

5.1.2.1 Water

Chlorobenzene has been reported in groundwater, surface water, and drinking-water at levels of 0.005–10 μg/litre (1, 2). Assuming a water consumption of 2 litres/day and an absorption efficiency of 100%, the daily intake can vary from 0.01 to 20 μg/day.

5.1.2.2 Air

There are no reports of the compound being detected in ambient air. The only data concerning exposure to chlorobenzene via air are from the industrial working environment. Reported industrial exposures vary from 0.004 to 0.3 mg/litre (2).

---

[a] The references for monochlorobenzene are listed together with those for dichlorobenzene at the end of section 5.2.

5.1.2.3 *Food*

No data are available on the intake of chlorobenzene from food.

5.1.3 Metabolism

Chlorobenzene is metabolized in mammals to diphenolic derivatives, probably via epoxide formation. These chlorophenols are then excreted as conjugates. Chlorobenzene may also be metabolized to *p*-chlorophenylmercapturic acid (*2*).

5.1.4 Health effects

5.1.4.1 *Observations in man*

Chlorobenzene is irritating to the respiratory system and is a central nervous system depressant (*1*).

Several cases of chlorobenzene intoxication due to inhalation have been reported. There are few, if any, usable human exposure data for chlorobenzene alone. Data are available from studies of exposure to chlorobenzenes in combination with other materials.

5.1.4.2 *Observations in other species*

There is enough evidence to suggest that chlorobenzene causes dose-related target organ toxicity, although data are lacking with respect to an acceptable chronic toxicity study.

The no-observed-adverse-effect level in dogs after 3 months was 27.25 mg/kg of body weight per day. Studies in rats showed no observed adverse effects after 3 and 6 months of doses of 12.5 and 14.5 mg/kg of body weight per day, respectively. Another study in rats showed no observed adverse effects after the administration of 0.001 mg/kg of body weight per day for 7 months (*2*).

5.1.4.3 *Mutagenicity*

No data are available.

5.1.4.4 *Teratogenicity*

No data are available.

5.1.4.5 *Carcinogenicity*

There is no information in the literature to indicate that monochlorobenzene is carcinogenic.

5.1.4.6 *Guideline value*

The dose levels producing no detectable adverse effects range from 0.001 mg/kg of body weight per day to 14.5 mg/kg of body weight per day for rats. Choosing the higher no-detectable-adverse-effect dose of 14.5 mg/kg of body weight per day derived from the short-term study of rats and applying a high safety factor of 1000–10 000, a tentative acceptable daily intake of 0.0015–0.015 mg/kg of body weight can be derived. For a 70-kg man, this value would represent an intake of 0.1– 1 mg/day. Allocating 10 % of this dose to water consumption results in a tentative toxicological value in drinking-water of 5–50 µg/litre.

As the odour threshold concentration for monochlorobenzene in water is 30 µg/litre, a value that approaches the calculated limits based on health effect considerations, the recommended guideline value for monochlorobenzene in drinking-water is 10 % of the threshold odour value, i.e., 3 µg/litre.

5.2 Dichlorobenzenes

5.2.1 General aspects

The dichlorobenzenes (DCBs) are a class of three isomeric halogenated aromatic compounds. 1,2-dichlorobenzene (1,2-DCB) and 1,3-dichlorobenzene (1,3-DCB) are liquids at normal environmental temperatures, while 1,4-dichlorobenzene (1,4-DCB) is a solid; all are relatively volatile.

The major uses of 1,2-DCB are as a process solvent in the manufacture of toluene diisocyanate and as an intermediate in the synthesis of dyestuffs, herbicides, and degreasers, but chiefly as an intermediate in the production of pesticides. The primary use of 1,4-DCB is as an air deodorant and insecticide and moth repellant (1). 1,3-DCB may occur as a contaminant of 1,2- or 1,4-DCB, but no information is available concerning its commercial production and use. Both 1,2- and 1,4-DCB are produced almost entirely as by-products during the production of monochlorobenzene.

5.2.2 Routes of exposure

The production, use, transport, and disposal of dichlorobenzenes have resulted in widespread dispersal and in contamination of the environment. Dichlorobenzenes have been detected in rivers, groundwater, municipal and industrial discharges, drinking-water, air, and soil.

5.2.2.1 Water

1,2- and 1,4-dichlorobenzenes are frequently found in potable water sources prior to treatment, at levels of 1–10 µg/litre. In drinking-water these compounds have been found at levels of 0.001–1 µg/litre (5).[a]

Assuming a human water consumption of 2 litres/day and an absorption efficiency of 100 %, the daily intake from water can vary

---

[a] CAMPBELL, I. I. *Maximum acceptable limit in drinking water for dichlorobenzenes.* US Environmental Protection Agency (Unpublished document, prepared for WHO, 1980).

from 6×10^{-3} μg (median level in drinking-water of 3 ng/litre) to 6 μg (maximum reported level of total DCB of 3 μg/litre).

5.2.2.2 Air

Data on air contamination by DCBs are very limited. Values ranging from 0.002 to 50 mg of 1,2-DCB/m³ have been measured in outdoor air in California. 1,4-DCB was not detected. In domestic houses in Tokyo, concentrations of 1,4-DCB are much higher, ranging from 105 μg/m³ (bedroom) to 1700 μg/m³ (wardrobe). Assuming a daily inspired volume of 20 m³ of air (adult male) and an absorption efficiency by inhalation of 50%, the daily intake can vary from 0.02 mg (lowest suburban concentration reported) to 20 mg (reported in wardrobe air as a result of use of 1,4-DCB).

5.2.2.3 Food

Food may also be contaminated by DCB. Pork has been tainted by the presence of 1,4-DCB in the air breathed by the animals; tainting of eggs has been reported when the hens were exposed to 20–38 mg of 1,4-DCB per m³ of air. 1,4-DCB has also been detected in fish in Japanese coastal waters. No data are available from which to estimate specific exposure to DCBs resulting from the consumption of food.

5.2.3 Metabolism

The dichlorobenzenes may be absorbed through the lungs, gastro-intestinal tract, and intact skin. The relatively low water solubility and high lipid solubility of halobenzenes favour their penetration of most membranes by diffusion, including pulmonary and gastrointestinal epithelia, brain, hepatic parenchyma, renal tubules, and placenta.

Studies of 1,2- and 1,4-DCB given in single doses to chinchilla rabbits by stomach tube showed that 1,2-DCB is mainly metabolized by oxidation to 3,4-dichlorophenol and excreted primarily in the urine as conjugates of glucuronic and sulfuric acids. 1,4-DCB was metabolized mainly by oxidation to 2,5-dichlorophenol and excreted only as glucuronides and ethereal sulfates. In humans, also, 2,5-dichlorophenol is indicated as the principal metabolite of 1,4-DCB.

5.2.4 Health effects

5.2.4.1 Observations in man

Most reported cases (16 of 22) of human poisoning by DCBs since 1939 have resulted from long-term exposure, primarily by inhalation of vapours, but some have also resulted from exposure by ingestion (3 of 22) or skin absorption (3 of 22). Most toxic exposures have been occupational in nature but some have been due to the use or misuse of DCB products in the home. Most case reports (15 of 22) have involved exposure to agents containing primarily 1,4-DCB and the remainder primarily 1,2-DCB. In a few cases, mixtures including 1,3-DCB were involved. The target body systems or tissues have been one or more of the following: liver, blood (or reticuloendothelial system, including bone marrow and/or immune components), central nervous system, respiratory tract, and integument. The clinical findings in these reports imply a wide spectrum of target organs for the DCBs. For example, at least 17 of the 22 reported clinical cases have involved general toxic or irritative symptoms (e.g., fatigue or weakness, anorexia, weight loss, nausea, headache, irritation, and malaise) and the same number have displayed symptoms or signs indicating involvement of the circulatory system, including blood and/or bone marrow or other reticuloendothelial components (e.g., anaemia, leukaemia, leukopenia or leukocytosis, polynucleosis, bone marrow hyperplasia, leukoblastosis, haemorrhagic tendency, splenomegaly, and jaundice).

5.2.4.2 *Observations in other species*

Intubation of 10 guinea-pigs with 1,2-DCB (50 % in olive oil) in single oral doses of 800 mg/kg of body weight resulted in loss of body weight, but all the subjects survived. Doses of 2000 mg/kg of body weight were fatal in all cases. In a test of repeated doses of 1,2-DCB in olive oil emulsified with acacia, groups of white rats were dosed by stomach tube five days a week to give a total of 138 doses in 192 days at dose levels of 18.8, 188, and 276 mg/kg of body weight per day. Positive findings in the high-dose animals included increased liver and kidney weights, decreased spleen weight, and slight to moderate cloudy swelling on microscopic examination of the liver. In the intermediate-dose group, liver and kidney weights were slightly increased. No adverse effects were noted at the low dose level. Two drops of undiluted 1,2-DCB in rabbits' eyes caused pain and conjunctival irritation, which cleared completely within one week.

In a chronic toxicity test, rats were given 1,2-DCB at daily doses of 0.001, 0.01 and 0.1 mg/kg of body weight. After 9 months, the high dose disturbed higher cortical function in the central nervous system and caused decreased haemoglobin, thrombocytosis, and neutropenia, and inhibited bone marrow mitotic activity. The dose level of 0.1 mg/kg of body weight was "liminal", and the low dose level (0.001 mg/kg) was "subliminal".

White female rats were fed 1,4-DCB in oil (emulsified with acacia) by stomach tube five days a week to give a total of 138 doses in 192 days. At the high dosage level of 360 mg/kg of body weight per day, increased liver and kidney weights, and hepatic cirrhosis and focal necrosis were observed. At the intermediate dose level of 188 mg/kg of body weight per day, increased liver and kidney weights were observed. No adverse effects were noted at the low dose level (18.8 mg/kg per day).

5.2.4.3 *Mutagenicity*

Insufficient data are available to permit a decision regarding mutagenicity.

5.2.4.4 *Teratogenicity*

No data are available.

5.2.4.5 *Carcinogenicity*

No reports of specific carcinogenicity tests of DCBs in animals or of pertinent epidemiological studies in humans are available.

Although no strong direct evidence of the carcinogenicity of DCBs is available, there are some data suggesting that they should be regarded as suspect carcinogens, pending the availability of better data. Available data in rats do not provide evidence for the carcinogencity of 1,4-DCB. No adequate data on 1,2-DCB are available. Epidemiological studies provide inadequate data for evaluating the carcinogenicity of DCBs in humans (6).

5.2.4.6 *Guideline value*

Published values for non-adverse-effect levels range from 0.001 mg/kg of body weight to 13.4 mg/kg of body weight per day. Very different results can be derived depending on which figures are used.

Using a no-detectable-adverse-effect dose of 13.4 mg/kg of body weight per day, derived from a short-term study with rats, and applying a safety factor of 1000–10 000, a tentative acceptable daily intake of 0.00134–0.0134 mg/kg of body weight can be calculated for both 1,2- and 1,4-DCB. For an adult of 70 kg this represents an intake of 0.1–1 mg/day. Allocating 10 % of this dose to water consumption, and assuming a consumption of 2 litres/day, a tentative toxicological value for drinking-water of 5–50 μg/litre, can be derived.

These values are in excess of the levels quoted as threshold odour values. For the 1,2-isomer the threshold odour value is given as approximately 3 μg/litre; 10 % of this value, i.e., 0.3 μg/litre, is recommended as a reasonable value. Similarly, as the threshold odour value for the 1,4-isomer is only 1 μg/litre, a recommended value is 0.1 μg/litre.

REFERENCES

1. *Toxicological appraisal of halogenated aromatic compounds following groundwater pollution*. Copenhagen, WHO Regional Office for Europe, 1980.
2. *Ambient water quality criteria for chlorinated benezenes*. Washington, DC, US Environmental Protection Agency, 1979 (EPA 440/5-80-028).
3. VAN GEMERT, J. L. & NETTENBREYER, A. H. *Compilation of odour threshold values in air and water*. Leidschendam, The Netherlands. National Institute for Water Supply, 1977 (2260 AD).

4. ZOETEMAN, B. C. J. *Sensory assessment of water quality*. Oxford. Pergamon Press, 1980.
5. *Ambient water quality criteria for dichlorobenzenes*. Washington, DC, US Environmental Protection Agency, 1979 (440/5-80-039).
6. *Some industrial chemicals and dyestuff*. Lyon, International Agency for Research on Cancer, 1982 (IARC Monographs on the evaluation of the carcinogenic risk of chemicals to humans, vol. 29).

6. BENZENE AND LOWER ALKYLBENZENES

Benzene and lower alkylbenzenes, such as toluene and ethylbenzene, are widely used in the chemical industry as intermediates in the production of various chemicals, e.g., phenol and cyclohexane. Lower alkylbenzenes are components of gasoline and are used as solvents for paints and coatings. Because of the suppression of the evaporation process, benzene and lower alkylbenzenes can be present in groundwater at higher levels than those usually found in surface-water. As a result of spills and dumping of chemical wastes, concentrations of the chemicals in groundwater up to levels of several mg per litre have recently been found. Levels in drinking-water do not generally exceed 1 μg/litre. Except for benzene, there seems to be no potential health risk from the levels of alkylbenzenes found in drinking-water.

6.1 General aspects

Benzene and toluene are produced mainly from petroleum or as by-products in the manufacture of gas and coke. They are used in large quantities by the chemical industry, the three major uses being for the production of styrene, cumene (used to manufacture phenol and acetone), and cyclohexane (used in manufacturing nylon). Much of the toluene produced is used in the production of benzene. Relatively minor (but still significant) quantities are used in a variety of industries (e.g., plastics, paints, detergents, and as gasoline additives), either as intermediates for many syntheses, or as solvents. Other alkylbenzenes (ethylbenzene, xylenes) are also used extensively as solvents or as chemical intermediates.

6.2 Routes of exposure

Benzene and the lower alkylbenzenes are volatile and comparatively unreactive in the environment. They become widely dispersed in the environment through movement of air masses and there is continuous recycling between air and water bodies through rain and volatilization from water surfaces. Ultimately, degradation is likely, as a result of biological and microbial oxidation.

6.2.1 Air

Levels of benzene and toluene in urban air are commonly of the order of $100 \,\mu g/m^3$, the major part of which originates from gasoline-related emissions (automotive emissions, gasoline handling). Solvent losses and emissions from industrial activities also contribute. Average levels in the air of a number of cities worldwide have been compiled and found generally to be less than $90 \,\mu g/m^3$ for benzene (1) and $112.5-150 \,\mu g/m^3$ for toluene (2). A mean exposure level to benzene in the USA was calculated to be $9.5 \,\mu g/m^3$ (range, $< 3.5 \,\mu g/m^3$ to $1 \, mg/m^3$) (3).[a]

6.2.2 Food

There are few data on levels of benzene or toluene in foods, although there are indications that benzene occurs naturally in some fruits, fish, vegetables, nuts, dairy products, beverages, and eggs (3), and toluene is found in fish from areas adjacent to petrochemical activities (4).

6.2.3 Water

The major sources of benzene and toluene in water are atmospheric deposition (through rain and snow) and chemical plant effluents (with minor contributions from urban runoff and sewage plants) (5). Benzene levels of up to $179 \,\mu g/litre$ have been reported in chemical plant effluents, although levels in finished drinking-water are generally much lower; levels of $0.1-0.3 \,\mu g/litre$ have been found in four city drinking-water supplies in the USA (6), whilst levels below $0.01 \,\mu g/litre$ have been reported in Canada (1). Levels of up to $19 \,\mu g$ of toluene per litre have been reported (2). Volatile hydrocarbons, such as benzene and toluene, would be expected to evaporate rapidly into the atmosphere from bodies of water (half-lives of 37.3 minutes and 30.6 minutes respectively for benzene and toluene at $25 \,°C$) (2), a fact that may explain why levels in groundwater are occasionally much higher than those in surface-water. Noteworthy is one observation in which benzene in finished water appeared to originate from anthracite filters used in water treatment (7).

6.2.4 Occupational exposure

Levels of benzene in the air of industries in which benzene is produced or used are generally three orders of magnitude higher than those in the general atmospheric environment; levels are commonly in the range 0.3–$9 \, mg/m^3$ (time-weighted average) (1). Individuals exposed to a benzene concentration of $3.0 \, mg/m^3$ (time-weighted average) receive a yearly dose about thirty times higher than non-occupationally exposed individuals (1).

---

[a] In the original publications, the concentrations were given in parts per billion. Conversion factors of 1 ppb = $3.0 \,\mu g/m^3$ for benzene and 1 ppb = $3.75 \,\mu g/m^3$ for toluene have been used.

6.2.5 Exposure estimates for man

The total uptake of benzene by urban dwellers from background (i.e., non-occupational) exposures has been estimated to be about 125 mg/year, 90 mg of which comes from food (1). It should be noted that the information on benzene levels in food is very scanty, so this background level should be considered only as an approximate reference point. Furthermore, this estimate does not take into account possible sporadic uses of benzene-containing products by the consumer. Nevertheless, it does suggest that levels commonly found in drinking-water are minimal (compared with intake from food and air) and are probably only of marginal importance. Similar conclusions apply to toluene.

6.3 Metabolism

The metabolism and excretion of benzene in humans and animals appear to follow similar pathways (8). Regardless of the route of administration, predominantly unchanged benzene is eliminated in the expired air. Conjugated metabolic products, typically large quantities of phenol accompanied by smaller amounts of catechol, hydroquinol, and hydroxyhydroquinol are excreted in the urine. The liver is the major site of both oxidation and conjugation.

Toluene is rapidly and extensively metabolized to hippuric acid, which is excreted in the urine; most of a dose of toluene is almost completely eliminated within 12 hours as unchanged toluene in expired air or as hippuric acid (2).

6.4 Health effects

A number of reviews on benzene toxicity have appeared recently (1, 9–11). Acute exposure to benzene results in central nervous system depression. Although most studies of benzene toxicity have involved exposure by the inhalation route, limited animal studies suggest that exposure by other routes of administration leads to similar sequelae.

Chronic exposure to benzene leads to haemopoietic tissue changes in the form of anaemia and leukopenia. Epidemiological studies and several case reports suggest a relationship between benzene exposure and leukaemia, evidence that has led to its categorization as a human carcinogen by a Working Group convened by the International Agency for Research on Cancer (12).

In animals, benzene exposure has been shown to affect immunological defence mechanisms and in numerous studies it has been shown possible to induce chromosome damage, both in exposed animals and using cell culture techniques.

Acute exposure to toluene by the inhalation route results in central nervous system depression (6). However, toluene appears to produce no irreversible tissue injury and its major metabolite, benzoic acid, is considered relatively nontoxic. Although most toxicity studies have involved inhalation exposure, studies on chronic oral administration have been conducted in rats, the highest dosage group being given toluene at 590 mg/kg of body weight five times weekly for 193 days, after which no adverse effects were observed (5). Toluene does not appear to be teratogenic, mutagenic, or carcinogenic (2), although one study in mice reported teratogenic effects when massive doses were used (13). There are numerous studies involving long-term industrial exposure to toluene without any detectable changes in blood characteristics or liver damage (6). The above data indicate that toluene is of relatively low toxicity and a maximum permissible concentration in water of 14.3 mg/litre has been calculated (2). In view of the fact that this figure is very much higher than the levels found in water it was decided not to recommend a guideline value for toluene. For an excess cancer risk of 1 in 10^5 per lifetime and rounding to the nearest decade, a guideline value of 10 μg/litre is recommended for benzene in drinking-water.

REFERENCES

1. HOLLIDAY, M. ET AL. *Benzene: Human health implications of benzene at levels found in the Canadian environment and workplace.* Ottawa, Health and Welfare Canada, 1978 (Environmental Health Directorate Report No. 79-EHD-40).
2. *Ambient water quality criteria for toluene.* Washington, DC, US Environmental Protection Agency, 1980 (EPA-440/5-80-75).
3. MARA, S. J. & LEE, S. S. *Assessment of human exposure to atmospheric benzene.* Washington, DC, US Environmental Protection Agency, 1978 (EPA-450/3-78-031).
4. OGATA, M. & MIYAKE, Y. Identification of substances in petroleum causing objectionable odour in fish. *Water research,* 7: 1493 (1973).
5. WOLFE, M. A. ET AL. Toxicological studies of certain alkylated benzenes and benzene. *Archives of industrial health,* 14: 387 (1956).
6. NATIONAL RESEARCH COUNCIL. *Drinking water and health.* Washington, DC, National Academy of Sciences, 1977.
7. SMILLIE, R. D. ET AL. Low molecular weight hydrocarbons in drinking water. *Journal of environmental health,* A13: 187 (1978).
8. RUSCH, G. M. ET AL. *Benzene metabolism in benzene toxicity: a critical evaluation.* Washington, DC, American Petroleum Institute, 1977, pp 23–36.
9. LASKIN, A. & GOLDSTEIN, B. D., ed. *Benzene toxicity: a critical evaluation.* Washington, DC, American Petroleum Institute, 1977, 147 pp.
10. NATIONAL RESEARCH COUNCIL *Health effects of benzene: a review.* Washington, DC, National Academy of Sciences, 1976, 23 pp.
11. *Some antithyroid and related substances, nitrofurans and industrial chemicals.* Lyon, International Agency for Research on Cancer, 1974 (IARC Monographs on the evaluation of the carcinogenic risk of chemicals to humans, vol. 7).
12. *Chemicals and industrial processes associated with cancer in humans.* Lyon, International Agency for Research on Cancer, 1979 (IARC Monographs on the evaluation of carcinogenic risk of chemicals to man, Suppl. 1).
13. NAWROT, P. S. & STAPLES, R. E. Embryo-foetal toxicity and teratogenicity of benzene and toluene in the mouse. *Teratology,* 19: 41A (1979).

7. PHENOL AND CHLOROPHENOLS

Chlorophenols are used as biocides and are found as a result of chlorination of water containing phenol. Chlorophenols are well known for their low taste and odour thresholds. Taste thresholds for the most odorous compounds (mono- and dichlorophenols) are as low as 1 μg/litre. For aesthetic reasons, therefore, individual (chloro)phenols should not, as a general rule, be present in drinking-water above the 0.1 μg/litre level; exceptions are phenol and pentachlorophenol, which have taste thresholds of 100 μg/litre. Provided that no chlorination is applied, phenol may therefore be accepted at levels up to 100 μg/litre in drinking-water.

For chlorophenols, in cases where a guideline value of 0.1 μg/litre cannot be achieved, it must be borne in mind that some chlorophenols exert toxic effects at somewhat higher concentrations.

Without extreme concentration procedures, spectrophotometric techniques will only detect chlorophenols at levels above 1 μg/litre, which is higher than the taste threshold concentration of several chlorophenols. If chlorophenols are suspected of being involved in taste problems, direct taste assessment by taste panels, and analytical determination by chromatographic techniques will need to be used.

The best approach to controlling pollution of drinking-water by chlorophenols is to prevent the contamination of the source water by phenol and chlorinated phenolic pesticides. When high phenol levels are present in the raw water these should be reduced, as far as possible, before chlorination is applied. Lower substituted chlorophenols, once present in the water, may be removed by oxidation processes, while the higher substituted ones can be effectively removed only by activated carbon adsorption.

7.1. Chlorinated phenols of toxicological significance

7.1.1 General aspects

Chlorinated phenols are known to be present in drinking-water, resulting either from contamination of raw water sources or from chlorination of water containing phenolic compounds. Phenols may be present in raw water owing to the discharge of wastewaters from coke distillation plants, the petrochemical industry, and numerous other industries where phenols serve as intermediates. They are also present in municipal wastewaters. When water containing phenol itself is

233

chlorinated, the main reaction products are 2- and 4-chlorophenol, 2,4-dichlorophenol, and 2,4,6-trichlorophenol. 2,4-dichlorophenol is produced commercially as an intermediate in the manufacture of the herbicide 2,4-D, related biocides, and pentachlorophenol. This last compound is used as a wood preservative, 2,4,5-trichlorophenol as a fungicide, and 2,4,6-trichlorophenol as an antiseptic. 2,4,6-trichlorophenol is also a major metabolite of the insecticide lindane. The major uses of 2,3,4,6-tetrachlorophenol are as an insecticide and wood preservative.

7.1.2 Occurrence

Contaminated raw water, including groundwater, may contain $1-10 \mu g$ of phenol and mono- and dichlorophenols per litre. Similar levels have been reported in drinking-water. Tri- and tetrachlorophenol have been detected in raw water at levels of $1-10 \mu g$/litre and occasionally higher. The concentrations usually found in drinking-water are, however, 1 or 2 orders of magnitude lower. In view of the apparently low frequency of their occurrence in water, chlorophenols other than those shown in Table 3 can, for the time being, be excluded from further consideration with respect to effects on public health (*1–4*).

7.1.3 Preliminary screening of health effects

From a consideration of the anticipated maximum levels in drinking-water and the existing literature on the toxicity and organoleptic data, it is evident that for several chlorinated phenols, toxicity-based limits will be much higher than taste- and odour-based limits. To be able to select the compounds of primary toxicological concern, a preliminary compilation of the relevant data is presented in Table 3 (*5–7*). As this table shows, the limits for some chlorinated phenols calculated from consideration of their toxicity are much higher than organoleptic considerations allow. 2,4,6-trichlorophenol and pentachlorophenol are exceptions and toxicological limits for these compounds have been calculated.

Limits for the other chlorophenols, as well as phenol itself, are proposed on the basis of organoleptic considerations.

7.1.4 Complementary organoleptic considerations

It is a matter of concern whether the low organoleptic thresholds of mono- and dichlorophenols can in practice guarantee the safety of a tasteless drinking-water, when little is known about the toxicity of several of the compounds. Consideration of the toxicology of those chlorophenols that are well documented indicates that if water is free

Table 3. Criteria for setting limits for chlorinated phenols of toxicological concern

Compound	Indication of maximum levels		Odour threshold concentration[a] (μg/l)	Taste threshold concentration[a] (μg/l)	Typical criterion levels based on	
	Raw water (μg/l)	Drinking water (μg/l)			Toxicity (μg/l)	Carcinogenicity (μg/l)
Phenol	100	1	1000	100	3000	—
2-Chlorophenol	10	1	1	1	—	—
4-Chlorophenol	10	1	1	1	—	—
2,4-Dichlorophenol	10	10	1	1	3000	—
2,6-Dichlorophenol	10	1	10	1	—	—
2,4,5-Trichlorophenol	1	< 0.1	100	1	2600	—[b]
2,4,6-Trichlorophenol	1	1	100	1	—	12
2,3,4,6-Tetrachlorophenol	< 0.1	< 0.1	1000	1	—	—
Pentachlorophenol	10	1	1000	100	21	—

— Insufficient data available.
[a] Thresholds reported in the literature vary considerably. Values higher and lower than those indicated have been reported.
[b] 2,3,7,8-tetrachlorodibenzo-p-dioxin (TCDD) is an impurity of technical 2,4,5-trichlorophenol. In the environment the compounds behave differently and they should therefore be treated separately.

from taste and odour, it is unlikely to present any direct health risks due to chlorinated phenols. However, it must be realized that in general the absence of any adverse taste or smell does not guarantee the safety of drinking-water. Water with a noticeable taste and odour should be investigated for the possible presence of chlorophenolic compounds.

7.2 2,4,6-Trichlorophenol

7.2.1 General aspects

2,4,6-trichlorophenol is a yellow solid with a melting point of 69.5 °C and a boiling point of 246 °C. It is slightly soluble in water (< 0.1 g per 100 ml) (8) and soluble in organic solvents. It is manufactured for use as a wood preservative, bactericide, and fungicide. 2,4,5-trichlorophenol is one of the products formed when drinking-water containing low levels of phenol is disinfected using chlorine.

7.2.2 Routes of exposure

7.2.2.1 Water

2,4,6-trichlorophenol has been detected in river-water at levels up to 1 μg/litre. In 1978, the River Rhine in the Netherlands contained levels ranging from 0.04 to 0.63 μg/litre. Similar concentrations of 2,4,5-trichlorophenol were also found by Wegman.[a] Levels in drinking-water are generally lower but data are scarce. When water containing phenol is chlorinated, concentrations of 2,4,6-trichlorophenol up to a level of a few micrograms per litre may be formed. It is estimated that the maximum exposure of a 70-kg adult from a daily intake of 2 litres of water containing 1 μg/litre of 2,4,6-trichlorophenol would be 0.00003 mg/kg of body weight per day.

7.2.2.2 Food

Another possible route of ingestion is from dairy products since chlorophenolic antiseptics are widely used in the dairy industry (9). 1,3,5-trichlorobenzene has been shown to be metabolized to 2,4,6-trichlorophenol (10). Pentachlorocyclohexane is also converted to 2,4,6-trichlorophenol in corn and pea plants (11). Ingestion can take place via the consumption of fish and shellfish. No quantitative data are available on inhalation studies. In view of the paucity of quantitative data on other routes of exposure, it has been assumed that water is the major route of ingestion.

---

[a] WEGMAN, R. C. C. Unpublished data of the National Institute for Public Health, Bilthoven, The Netherlands, 1980.

7.2.3 Metabolism

2,4,6-trichlorophenol is rapidly cleared from the body, predominantly in the urine (*12*). Very little further is known of the metabolism of this compound.

7.2.4 Health effects

The intraperitoneal LD_{50} for rats, using olive oil as the solvent for 2,4,6-trichlorophenol, is 276 mg/kg of body weight (*13*). As with other chlorinated phenols, 2,4,6-trichlorophenol is capable of increasing body temperature and produces convulsions at high doses.

7.2.4.1 *Mutagenicity*

2,4,6-trichlorophenol increased the mutation rate in a strain of *Saccharomyces cerevisiae* (*14*) but was not found to be mutagenic in the *Salmonella* mammalian microsome Ames test, with or without metabolic activation (*15*).

7.2.4.2 *Carcinogenicity*

In studies with F344 rats and B6C3-F1 mice at the National Cancer Institute in the USA (*16*), the compound was found to increase tumour incidence. In male rats, dose-related increases in lymphoma and leukaemia were observed. Leukocytosis and monocytosis of the peripheral blood and hyperplasia of the bone marrow occurred in both male and female rats. In both the male and female mice, the incidence of hepatocellular carcinomas or adenomas was increased significantly in a dose-related manner.

Epidemiological data are inadequate for evaluating the carcinogenicity of 2,4,5- and 2,4,6-trichlorophenols (*8, 17*).

The guideline value, calculated via the linear multistage extrapolation model, assuming a lifetime cancer risk of 1 per 100 000, is 12 μg/litre whereas the quoted threshold taste level is only 1 μg/litre. To provide a drinking-water of acceptable potability a guideline value of 0.1 μg/litre is recommended.

7.3 Pentachlorophenol

7.3.1 General aspects

Pentachlorophenol is a widely used fungicide and wood preservative with an estimated world production of 40 000 tonnes per annum. It has a melting point of 190 °C, a boiling point of 310 °C, and a water solubility of 14 mg/litre at 20 °C.

7.3.2 Routes of exposure

Contaminated surface-water can contain up to $10 \mu g$ of penta-chlorophenol per litre. In the River Rhine in the Netherlands in 1978, levels ranged from 0.15 to $1.5 \mu g$/litre. Levels in drinking-water are generally well below $1 \mu g$/litre. Levels in worms of 0.2 mg/kg of body weight and in mammals of 0.1–1.3 mg/kg of body weight have been reported.

7.3.3 Metabolism

Pentachlorophenol is well absorbed from the gastrointestinal tract and may also be absorbed through the skin. Most of the absorbed pentachlorophenol in humans is excreted unchanged in the urine. Approximately 20 % of the systemic dose is dechlorinated to form tetrachlorohydroquinone and trichloro-1-hydroquinone in rats and mice.

7.3.4 Health effects

The acute oral LD_{50} of pentachlorophenol for rats is 27 mg/kg of body weight. Clinical signs of poisoning include profuse sweating, thirst, elevated temperature, rapid pulse and respiration, and ultimately cardiac arrest.

Some cases of Hodgkin's disease and leukaemia have been reported in woodworkers using pentachlorophenol (18), but epidemiological studies are inadequate for evaluation (19, 20).

At least 30 fatal cases of pentachlorophenol poisoning have been reported. Pentachlorophenol has been shown to cause chloracne in rabbits. Furthermore, it produces damage to kidneys and liver of experimental animals while similar indications have been obtained for occupationally exposed populations. Pentachlorophenol has been shown to be embryotoxic and fetotoxic in experimental animals. It also increases the frequency of mutation of yeast but showed no mutagenicity in the Ames test. No carcinogenicity has been demonstrated in experimental animals. The available studies on carginogenicity in experimental animals are inadequate (8).

In the USA, the National Research Council (21) has calculated an ADI of $3 \mu g$/kg of body weight per day for pentachlorophenol. Assuming a 70-kg man drinks 2 litres of water per day, and assigning 10 % of the ADI to water, a guideline value of $10 \mu g$/litre may be calculated for drinking-water.

REFERENCES

1. *Ambient water quality criteria for phenol.* Washington, DC, US Environmental Protection Agency, 1980 (EPA 440/5-80-066).
2. *Ambient water quality criteria for chlorinated phenols.* Washington, DC, US Environmental Protection Agency, 1980 (EPA 440/5-80-032).

3. *Ambient water quality criteria for 2,4-dichlorophenol.* Washington, DC, US Environmental Protection Agency, 1980 (EPA 440/5-80-042).
4. *Ambient water quality criteria for pentachlorophenol.* Washington, DC, US Environmental Protection Agency, 1980 (EPA 440/5-80-065).
5. VAN GEMERT, L. J. & NETTENBREIJER, A. H. *Compilation of odour threshold values in air and water.* Leidschendam, The Netherlands, National Institute for Water Supply, 1977.
6. ZOETEMAN, B. C. J. *Sensory assessment of water quality.* Oxford, Pergamon Press, 1980.
7. DIETZ, F. and TRAUD, J. Geruchs- und Geschmacks-Schwellenkonzentrationen von Phenolkorpern. *GWF-Wasser/abwasser,* 119:H6 318 (1978).
8. *Some halogenated hydrocarbons.* Lyon, International Agency for Research on Cancer, 1979 (IARC Monographs on the evaluation of the carcinogenic risk of chemicals to humans. vol. 20).
9. STANNARD, D. J. & SCOTTER, A. The determination of phenol residues in dairy products. *New Zealand journal of dairy science and technology,* 12: 140 (1977).
10. KOHLI, J. ET AL. The metabolism of higher chlorinated benzene isomers. *Canadian journal of biochemistry,* 54: 203 (1976).
11. MOZA, P. ET AL. Beitrage zur okologischen chemie LXXXIX Orientierende versuche zum metabolisms Von-pentachlorocyklohex-1-en in hoheren pflanzen in hydrokultur. *Chemosphere,* 6: 255 (1974).
12. KORTE, F. ET AL. Ecotoxicologic profile analysis, a concept for establishing ecotoxicologic priority list for chemicals. *Chemosphere,* 7: 79 (1978).
13. FARQUHARSON, M. E. ET AL. The biological action of chlorophenols. *British journal of pharmacology,* 13: 20 (1958).
14. FAHRIG, R. ET AL. Genetic activity of chlorophenols and chlorophenol impurities. In: Rao, K. R., ed., *Pentachlorophenol: chemistry, pharmacology and environmental toxicology.* New York, Plenum Press, 1978.
15. RASANEN, L. ET AL. The mutagenicity of MCPA and its soil metabolites, chlorinated phenols, catechols and some widely used slimicides in Finland. *Bulletin of environmental contamination and toxicology,* 18: 565 (1977).
16. NATIONAL CANCER INSTITUTE. *1979 Bioassay of 2,4,6-trichlorophenol for possible carcinogenicity.* Washington, DC, US Department of Health, Education and Welfare, 1979 (Technical Service Report Series, No. 155).
17. THEISS, J. C. ET AL. In: *Long-term hazards of polychlorinated dibenzodioxins and polychlorinated dibenzofurans.* Lyon, International Agency for Research on Cancer 1978 (IARC Internal Technical Report No. 78/001).
18. GREEN, M. H. Familial and sporadic Hodgkin's disease associated with occupational wood exposure. *Lancet,* 2: 626 (1978).
19. HARDELL, L. & SANDSTRÖM, A. Case-control study: Soft tissue sarcomas and exposure to phenoxyacetic acids or chlorophenols. *British journal of cancer,* 39: 711 (1979).
20. ERIKSSON, M. ET AL. Soft-tissue sarcomas and exposure to chemical substances: a case reference study. *British journal of industrial medicine,* 38: 27 (1981).
21. NATIONAL RESEARCH COUNCIL. *Drinking water and health.* Washington, DC, National Academy of Sciences, 1977.

8. TRIHALOMETHANES

Trihalomethanes in drinking-water occur principally as products of the reaction of chemicals used in oxidative treatment reacting with the naturally occurring organic materials present in the water. Their formation is particularly associated with the use of chlorine. The four most frequently occurring trihalomethanes are chloroform, bromo-dichloromethane, dibromochloromethane, and bromoform. The total concentration of these four trihalomethanes in drinking-water may vary up to 1000 μg/litre, but it is frequently less than 100 μg/litre.

Chloroform has been shown to cause cancer in two species of laboratory animal. The three bromine-containing trihalomethanes are only now being subjected to lifetime cancer bioassay tests similar to those in which chloroform was shown to be a carcinogen. These other trihalomethanes are, however, known to be more active than chloroform in the Ames *Salmonella* test for mutagenesis.

It is important to recognize that chlorine is an effective water disinfectant and the hazards of disease arising from microbiological contaminants resulting from incomplete disinfection are substantial. This is particularly true in developing countries where it is estimated that waterborne disease causes thousands of deaths per day. Chlorine is the most convenient and easily controlled disinfectant and is widely used.

8.1 General aspects

Trihalomethanes are halogen-substituted, single-carbon compounds having the general formula CHX_3, where X may be fluorine, chlorine, bromine, or iodine, or combinations thereof. With respect to drinking-water contamination, discussion may appropriately be confined to four members of the group: chloroform ($CHCl_3$), bromodichloromethane ($CHBrCl_2$), dibromochloromethane ($CHBr_2Cl$), and bromoform ($CHBr_3$). Chloroform is the most commonly encountered of these and available information pertains almost exclusively to this substance.

The most important use for chloroform is as the starting material for the manufacture of chlorodifluoromethane, which is used as a refrigerant, aerosol propellant, and in the synthesis of polytetrafluoro-ethene. Chloroform is an important solvent and degreasing agent. It has also been used in small quantities as an anaesthetic, in liniments, permanent wave lotions, dentifrices, and fumigants, and as the active ingredient and preservative in antitussive formulations (*1, 2*). The estimated world production of chloroform in 1973 was 245×10^6 kg (3).

Bromoform is used in industry as a gauge fluid, as a heavy liquid in solids separation, and as an intermediate in the synthesis of other chemicals (1). There are no known commercial applications for the mixed halogen derivatives.

8.2 Routes of exposure

8.2.1 Water

Trihalomethanes occur in water primarily from reactions between chlorine (and adventitiously present bromide ion) and naturally occurring organic compounds. Data gathered in many countries (4–6) have shown that the levels of trihalomethanes in finished water that has been chlorinated are generally very much higher than the levels in raw water, in which they are often undetectable.

Table 4 shows the levels found in a survey of potable waters in the USA. Other surveys have given similar results.

Table 4. Trihalomethane content of finished waters in the USA

Compound	No. of locations	Range of concentrations (mg/litre)	Median concentration (mg/litre)
Chloroform	80	0.0001–0.311	0.021
Bromodichloromethane	78	0.0003–0.116	0.006
Chlorodibromomethane	72	0.0004–0.110	0.0012
Bromoform	26	0.0008–0.092	< 0.005

Studies have indicated that, for a given chlorine dose, the rate (and hence degree) of trihalomethane formation is increased at higher humic acid concentrations, higher temperatures, and higher pH (6).

Trihalomethane formulation has been shown to proceed within the distribution system provided that a free chlorine residual exists (7).

The literature contains no information on the occurrence of bromodichloromethane, chlorodibromomethane, or bromoform.

8.2.2 Air

Chloroform has been detected in rural atmospheres at levels between 100 and 180 ng/m^3 (8). Maximum levels of chloroform found in the atmosphere at any one location range from <0.05 to 73.5 μg/m^3; mean levels are between 0.045 and 5.0 μg/m^3 (9). Atmospheric concentrations of chloroform at ground level at sites in the northern and southern hemispheres were 130 ng/m^3 and <15 ng/m^3, respectively (10).

8.2.3 Food

Two specimens of barley treated with a gaseous fumigant mixture containing chloroform, after airing at 17 °C and 30 °C, respectively, were found initially to contain chloroform residues of 123 and 132 mg/kg. After 60 days it was found that residues had disappeared from the specimen aired at 30 °C but were still present at a level of 16 mg/kg in that aired at 17 °C. Corn and sorghum behaved in a similar manner.

The following concentrations of chloroform were found in foodstuffs in the United Kingdom in 1973: dairy produce, 1.4–33 mg/kg; meat, 1–4 mg/kg; oils and fats, 2–10 mg/kg; beverages, 0.4–18 mg/kg; fruits and vegetables, 2–18 mg/kg (*11*).

8.3 Metabolism

Exposure to air containing chloroform in a concentration of 13.2–31.8 g/m³ for 3–10 minutes resulted in 73 % absorption (*12*).

Inhaled chloroform rapidly enters the bloodstream and is transported to the tissues. In mice, body fat was found to be the important storage site for chloroform; lesser amounts were found in the brain, lungs, kidneys, muscles, and blood. Metabolism of chloroform takes place in the liver, and whole-body autoradiography studies have shown the gradual transfer of radioactivity from the fat storage depots to the liver (*13*). Chloroform appears to be capable of crossing the human placental barrier since concentrations of chloroform in cord blood were found to be higher than those in the maternal blood (*14*).

In humans, up to 50.6 % of an oral dose (7 mg/kg of body weight) was metabolized to CO_2 but there was considerable variation between individuals and up to 68.3 % of the ingested chloroform was expired unchanged (*15*). Much of the absorbed chloroform is eliminated during its first passage through the liver and lungs and only 50–65 % of a 500-mg dose was found to be available for general circulation to the rest of the body (*16*). Of a single 500-mg dose ingested by human volunteers, 18–67 % was exhaled unchanged within 8 hours (*15*). The major metabolites of chloroform are excreted through the lungs (as CO_2) or through the kidney (as inorganic chloride) (*17*).

8.4 Health effects

8.4.1 Toxicity

This discussion of the health effects of trihalomethanes will deal mainly with chloroform, which, because of its former use as an inhalation anaesthetic and because it is the predominant trihalomethane in drinking-water, has received the most intensive study. The toxic effects of the other trihalomethanes are likely to be similar to those of chloroform.

Chloroform is a central nervous system depressant. It also affects liver and kidney function. The immediate effect of chloroform intoxication is loss of consciousness, which may be followed by coma and death (18, 19). Renal damage is noted 24–48 hours after exposure and hepatic injury is seen after 2–5 days; thus, symptoms of poisoning may occur several days after recovery from chloroform anaesthesia (18).

The mean lethal dose in man is considered to be about 44 g or 630 mg/kg of body weight for a 70-kg man; however, ingestion of more than 250 g of chloroform has been survived (20). The lowest published lethal dose of chloroform in man is 210 mg/kg of body weight (21). Ingestion of as little as 440 mg causes gastric irritation and increased peristalsis, as well as some local narcosis in the intestinal tract (19).

Except for studies on the carcinogenicity of chloroform, few investigations of the chronic toxicity of chloroform have been carried out. The long-term oral administration of chloroform at a dose of 0.4 mg/kg of body weight did not produce any changes in the investigated indices in albino rats, and the only effect of this dose on guinea-pigs was an increase of vitamin C in the adrenals (22). The safety of adding chloroform to toothpaste and mouth-rinse was assessed in two long-term studies, involving 229 human subjects (23). Daily consumption of chloroform was estimated to be 0.34–0.96 mg/kg of body weight over a 1–5-year period. Results from this study showed no hepatotoxicity based on liver function tests. Reversible hepatotoxicity was the only effect observed in a 47-year-old man who, each day for 10 years, consumed between 12 and 20 ounces (336–560 g) of a chloroform-containing cough suppressant; his daily dose of chloroform was estimated to be between 23 and 37 mg/kg of body weight (24).

Bromoform is considered to produce toxic symptoms that are similar to those of chloroform. An LD_{50} of 1820 mg/kg of body weight was reported when bromoform was administered to mice by the subcutaneous route (25). There appears to be little information on the toxic effects of the other trihalomethanes. A maximum tolerated dose of bromodichloromethane was determined in strain A/st male mice to be 100 mg/kg of body weight; the dose was injected intraperitoneally 6 times over a 2-week period. The maximum tolerated dose of bromoform, obtained using the same protocol, was also 100 mg/kg of body weight (26).

8.4.2 Carcinogenicity

A study was conducted in the USA by the National Cancer Institute (27). Chloroform dissolved in corn oil was administered by gavage to Osborne-Mendel rats and B6C3-F1 mice at two dose levels five times per week. Dose levels of 90 or 180 mg/kg of body weight were given to the male rats for 78 weeks; the female rats received doses of 125 or 250 mg/kg for the first 22 weeks and the same dose as the males thereafter. After 111 weeks the rats were killed and a statistically significant incidence of kidney epithelial tumours was found in the males

(24% in the high-dose and 8% in the low-dose groups) but not in the females. There was an increase in thyroid tumours in the female rats but it was not considered significant. The male mice first received doses of 100 or 200 mg/kg of body weight, and the females were initially dosed with 200 or 400 mg/kg. After 18 weeks, the doses were changed to 150 and 300 mg/kg for males and to 250 and 500 mg/kg for females. Highly significant increases in hepatocellular carcinoma were found in both sexes: 98% of the males and 95% of the females at the high dose and 36% of the males and 80% of the females at the low dose. Nodular hyperplasia was a frequent finding in low-dose male mice that had not developed hepatocellular carcinoma. It should be emphasized, however, that the doses used in the NCI studies were extremely high and, as a greater than 10% weight loss was observed in the animals, can be considered higher than a true maximum tolerated dose.

8.4.3 Epidemiology

Cantor et al. (28) looked at the association between the rates for 16 different cancers and the levels of trihalomethanes in the drinking-water; the chloroform and non-chloroform components were studied separately. Exposure information came from the National Organics Reconnaissance Survey and the EPA Region V Survey of 1975. Seventy-six counties in which more than 50% of the population was served by the measured water supply were included in the study. The most consistent finding was an association between bladder cancer mortality rates and trihalomethane levels. The association was observed in both sexes and was proportional in strength to the percentage of the population served by the studied water supply. The correlations noted were stronger with the brominated trihalomethanes than with chloroform. However, in its review of 13 epidemiology studies, the National Academy of Sciences Safe Drinking Water Committee concluded that a causal relationship had not been established by the studies that had been reported (29).

Hogan et al. (30) used much the same data base and applied various statistical procedures in order to determine the appropriateness of the statistical model. When a weighted regression analysis was applied, the results were similar to those of previous studies showing positive correlations between rectal-intestinal and bladder cancer mortality rates and chloroform levels in drinking-water.

8.4.4 Guideline value

Chloroform has several adverse effects on the health of humans. Safe levels of consumption for most of the effects observed directly in man would be difficult to estimate because adequate quantitative studies have not been conducted. Of the hazards that could potentially arise at concentrations approximating to those observed in drinking-water, the most serious are the carcinogenic effects observed in experimental animals and the suggestion of similar effects in humans exposed to elevated trihalomethane levels in drinking-water.

Estimation of safe levels of chloroform is made by applying a multistage linear extrapolation model to the data obtained in rats in the NCI bioassay of chloroform. The rat data were used in preference to the data obtained in the B6C3-F1 mouse because there is serious doubt about the mechanism responsible for producing liver tumours in the latter model when exposed to hepatotoxic agents such as chloroform. Data obtained subsequent to the NCI bioassay clearly indicate that this animal is much more sensitive than the rat to liver damage due to chloroform (31). Such damage occurs at doses lower than those used in the NCI study, suggesting an epigenetic mechanism. Extrapolation models for such effects do not exist. Additionally, rates of chloroform metabolism in the rat approximate more closely to those observed in man than do the rates in the mouse.

Given the above considerations, a guideline value of $30 \mu g/litre$ is recommended for chloroform in drinking-water. At an average consumption of 2 litres per day, such a concentration would lead to less than 1 additional case of cancer in a population of 100 000 in a lifetime. Additionally, it should be pointed out that the risk associated with inadequate disinfection would be much higher than that resulting from concentrations of chloroform considerably greater than the recommended value.

REFERENCES

1. HARDIE, D. W. F. *Chloroform.* In: Kirk, R. E. & Othmer, D. T., ed. *Encyclopedia of chemical technology.* 2nd ed. Vol. 5. New York, Interscience Publishers, 1964. p. 119.
2. NATIONAL INSTITUTE FOR OCCUPATIONAL SAFETY AND HEALTH. *Criteria for a recommended standard. Occupational exposure to chloroform.* Washington, DC, US Department of Health, Education and Welfare, 1974, p. 75–114.
3. PEARSON, C. R. & McCONNELL, G. Chlorinated C_1 and C_2 hydrocarbons in the marine environment. *Proceedings of the Royal Society, Series B,* **189**: 305–332 (1975).
4. ROOK, J. J. Formation of haloforms during chlorination of natural waters. *Water treament and examination,* **23**: 234 (1974).
5. ROOK, J. J. Haloforms in drinking water. *Journal of the American Water Works Association,* **68**: 186 (1976).
6. STEVENS, A. A. ET AL. Chlorination of organics in drinking water. In: Jolley, R. L., ed., *Water chlorination. Environmental impact and health effects,* Ann Arbor, MI, Ann Arbor Science, 1975.
7. HEALTH & WELFARE, CANADA, *National survey for halomethanes in drinking water.* Ottawa, Health & Welfare, Canada, 1977 (7-EHD-9).
8. RUSSELL, J. W. & SHADOFF, L. A. The sampling and determination of halocarbons in ambient air using concentration on porous polymer. *Journal of chromatology,* **134**: 375 (1977).
9. LILLIAN, D. ET AL. Atmospheric fates of halogenated compounds. *Environmental science and technology,* **9**: 1042 (1975).
10. COX, R. A. ET AL. Photochemical oxidation of halocarbons in the troposphere. *Atmospheric environment,* **10**: 305 (1976).

11. McCONNELL, G. ET AL. Chlorinated hydrocarbons and the environment. *Endeavour*, **34**: 13 (1975).
12. LEHMANN, K. B. & HASEGAWA, D. Studies of the absorption of chlorinated hydrocarbons in animals and humans. *Archiv für Hygiene und Bakteriologie*, **72**: 327 (1910).
13. COHEN, E. N. & HOOD, N. Application of low-temperature autoradiography to studies of the uptake and metabolism of volatile anesthetics in the mouse. I. Chloroform. *Anesthesiology*. **30**: 306 (1969).
14. DOWTY, B. J. ET AL. The transplacental migration and accumulation in blood of volatile organic constituents. *Pediatric research*, **10**: 696 (1976).
15. FRY, R. J. ET AL. Pulmonary elimination of chloroform and its metabolite in man. *Archives internationales de pharmacodynamie et de thérapie*, **196**: 98 (1972).
16. CHIOU, W. L. Quantitation of hepatic and pulmonary first-pass effects and its implications in pharmacokinetic study. I. Pharmacokinetics of chloroform in man. *Journal of pharmacokinetics and biopharmaceutics*, **3**: 193 (1975).
17. VAN DYKE, R. A. ET AL. A metabolism of volatile anesthestics. I. Conversion *in vivo* of several anesthetics to $^{14}CO_2$ and chloride. *Biochemical pharmacology*, **13**: 1239 (1964).
18. WHIPPLE, G. H. & SPERRY, J. A. Chloroform poisoning—liver necrosis and repair. *Bulletin of the Johns Hopkins University*, **20**: 278 (1909).
19. SECHER, O. Physical and chemical data on anaesthetics. *Acta anaesthesiologica scandinavica*, **42** (Suppl.): 1 (1971).
20. GOSSELIN, R. E. ET AL. *Clinical toxicology of commercial products*. 4th ed. Baltimore, MD, The Williams and Wilkins Co., 1976.
21. DREISBACH, R. H. *Handbook of poisoning*. Los Altos, CA, Lange Medical Publications, 1974. p. 275.
22. MIKLASHEVSKII, V. E. ET AL. Toxicity of chloroform administered perorally. *Gigiena i sanitarija*, **31**: 320 (1966).
23. DE SALVA, S. ET AL. Long-term safety studies of a chloroform containing dentifrice and mouth-rinse in man. *Food and cosmetics toxicology*, **13**: 529 (1975).
24. WALLACE, C. J. Hepatitis and nephrosis due to cough syrup containing chloroform. *California medicine*, **73**: 442 (1950).
25. KUTOB, S. D. & PLAA, G. L. A procedure for estimating the hepatotoxic potential of certain industrial solvents. *Toxicology and applied pharmacology*, **4**: 354 (1962).
26. CARDEIHAC, P. T. & NAIR, K. P. C. Inhibition by castration of aflatoxin induced hepatoma in carbon tetrachloride-treated rats. *Toxicology and applied pharmacology*, **26**: 393 (1973).
27. NATIONAL CANCER INSTITUTE. *Carcinogenesis bioassay of chloroform*. Bethesda, MD, National Cancer Institute, 1976.
28. CANTOR, K. P. Association of cancer mortality rates and trihalomethane level in municipal drinking water supplies. (Abstract) *American journal of epidemiology*, **106**: 230 (1977).
29. US ENVIRONMENTAL PROTECTION AGENCY. *Federal register*, 68698-68703 (1979).
30. HOGAN, M. D. ET AL. Association between chloroform levels in finished drinking water supplies and various site-specific cancer mortality rates. *Journal of environmental pathology and toxicology*, **2**: 873 (1979).
31. BULL, R. J. ET AL. In depth biochemical, pharmacological and metabolic studies of trihalomethanes in water. *Proceedings of the NCI/EPA/NIOSH Workshop on Environmental and Occupational Carcinogenesis, 1980* (in press).

PART V. AESTHETIC CONSTITUENTS AND CHARACTERISTICS

1. ALUMINIUM

1.1 General description

Aluminium compounds are abundant in nature and are often found in water. The salts of aluminium are used extensively in water treatment for the removal of colour and turbidity. Compared with the aluminium intake from food, that from water is small. Ingested aluminium salts do not appear to exert any deleterious effects on man. The incidence of discoloration in drinking-water in distribution systems increases if the aluminium level exceeds 0.1 mg/litre in the final water. A guideline value of 0.2 mg/litre in drinking-water is therefore recommended, based on aesthetic considerations. This value represents a compromise, taking into consideration that, although some discoloration may occur at this level, lower levels may be hard to achieve in certain cases where aluminium compounds are used in water treatment. In those cases, special attention should be paid to the maintenance of the distribution system.

1.2 Occurrence

Aluminium is distributed widely in nature and is a constituent of all soils, plants, and animal tissues (1–4). As a consequence of this wide natural distribution and the activities of man, aluminium is present in air, food, and water, both natural and polluted (2, 5).

1.2.1 Water

Industrial wastes, erosion, leaching of minerals and soils, contamination from atmospheric dust, and precipitation are the main pathways by which aluminium enters the aquatic environment. The level of aluminium in water varies considerably and may exceed 10 mg/litre in the vicinity of aluminium-processing plants (6). The concentration in any particular water is controlled by pH, the type and concentration of complexing agents that may be present, the oxidation state of the mineral components, and the redox potential of the system; many acidic waters contain naturally high levels of aluminium, presumably through the leaching process.

In water-treatment processes, coagulation with aluminium salts, such as alums or sodium aluminate, is used extensively for removal of finely divided mineral and organic material, especially if the water is coloured or turbid. The level of aluminium in drinking-water is, therefore, affected by the use of these salts in water-treatment processes. Most of the aluminium used as coagulant is removed as insoluble aluminium salts, which settle or are removed by filtration. Careful operation of these processes is essential for the successful treatment of water, to ensure both an aesthetically pleasing product and effective disinfection (7).

Although in the treatment of water the aim is to leave as little aluminium as possible in the water, some always remains. Levels in treated water have been reported to be between < 0.01 and 2 mg/litre (2). It is worth pointing out that these values represent an "alum" dose twelve times greater. Levels of aluminium in the final water above about 0.3 mg/litre usually reflect faults in the coagulation, sedimentation, or filtration stages of treatment.

During distribution, a portion of the aluminium may sediment out and a gradual reduction throughout the length of the supply system may be observed (8). This aluminium will accumulate in the system, especially where flows are low, and, together with iron, manganese, silica, organic material, and microorganisms, form sediments that may readily be disturbed by changes in flow and appear at the consumer tap, rendering the water aesthetically unacceptable (9). In the presence of aluminium, levels of iron normally too low to cause problems may produce obvious discoloration of water. It has been shown (10) that the incidence of discoloration in water in distribution systems, and therefore the frequency of consumer complaints, increases if the aluminium level exceeds 0.1 mg/litre in the final water.

1.3 Routes of exposure

Aluminium present in drinking-water contributes only a small proportion of the estimated daily human intake. The bulk of this intake is derived from food and, based on published information (11–14), a total aluminium consumption of 88 mg per person per day has been estimated. A consumption of 2 litres of water daily containing 1.5 mg of aluminium per litre will provide only 3.0 mg of aluminium per person per day, i.e., less than 4% of the normal daily intake.

1.4 Health aspects

Aluminium does not appear to be an essential nutrient in man. Aluminium salts are not normally absorbed from food and water, but are complexed with phosphate and excreted in the faeces (15). The chronic use of large quantities of aluminium hydroxide in the form of "antacids" can lead to excessive loss of phosphate from the system. Ingested aluminium does not accumulate substantially in the tissues, except in bone (16), whereas aluminium compounds inhaled, in the form of dust, accumulate in the lungs (17) and lymph nodes (18).

Administration of aluminium to rats at the rate of 2.5 mg/kg of body weight per day for six months led to minimal systemic toxicity and minimal gonadotoxic effects (19). There was no effect on the median life-span, longevity, incidence of tumours, or clinical biochemistry of rats fed aluminium in drinking-water at a concentration of 5 mg/litre (20).

There was no evidence of carcinogenesis in a series of studies using different mammals and a range of aluminium salts (21).

Aluminium has been associated with certain neurological disorders, such as dialysis dementia and Alzheimer's disease (20, 21). It is not clear, however, whether the presence of aluminium causes such conditions or is simply an indicator of other factors.

REFERENCES

1. COTTON, F. A. & WILKINSON, G. Advanced inorganic chemistry, 3rd ed. New York, Wiley-Interscience, 1972, pp. 261–262.
2. SORENSON, J. R. J. ET AL. Aluminium in the environment and human health. Environmental health perspectives, 8: 3 (1974).
3. UNDERWOOD, E. J. Trace elements in human and animal nutrition, 3rd ed. New York, Academic Press Inc., 1971.
4. SAAKASHVILI, T. G. & KVIRIKAZE, N. A. Content of certain trace elements in human blood. Trudy Instituta Urol., Akademia Nauk Gruzinskoi SSR, 1: 93 (1962); Chemical abstracts, 61: 4784 (1964).
5. MONIER-WILLIAMS, G. W. Aluminium in food. London, Ministry of Health, 1935 (Reports on Public Health and Medical Subjects).
6. SYLVESTER, R. O. ET AL. Factors involved in the location and operation of an aluminium reduction plant. Proceedings of the 22nd Industrial Waste Conference. Lafayette, IN, Purdue University, 1967, pp. 441–454.
7. RIDGWAY, J. ET AL. Water quality changes—chemical and microbiological studies. In: Water distribution systems. Medmenham, England, Water Research Centre, 1979.
8. AINSWORTH, R. G. ET AL. Deposits, corrosion products and corrosion mechanisms in iron mains. In: Water distribution systems. Medmenham, England, Water Research Centre, 1979.
9. AINSWORTH, R. G. ET AL. The introduction of new water into old distribution systems. Medmenham, England, Water Research Centre, 1980 (TR 146).
10. VOZAR, L. Content of aluminium in the diet and its biological action. Voprosy pitanija, 21: 28 (1962).
11. ZOOK, E. G. & LEHMANN, J. Total diet study: content of ten minerals—aluminium, calcium, phosphorus, sodium, potassium, boron, copper, iron, manganese and magnesium. Journal of the Association of Official Agricultural Chemists, 48: 850 (1965).
12. GABOVICH, R. D. Contents of some trace elements in the food in certain cities and towns of the USSR. Gigiena i sanitarija, 31: 41 (1966).
13. JAULNES, P. & HAMELLA, G. Présence et taux des oligo-éléments dans les aliments et les boissons de l'homme. Annales de la nutrition et del' aliment, 25: B133 (1971).
14. HAMILTON, E. I. & MINSKI, M. J. Abundance of the chemical elements in man's diet and possible relations with environmental factors. Science of the total environment, 1: 375 (1973).
15. THIENES, C. H. & HALEY, T. J. Clinical toxicology. Philadelphia, Lea & Febiger, 1972, pp. 169–170.
16. DEICHMANN, W. B. & GERARDE, H. W. Toxicology of drugs and chemicals. New York, Academic Press, 1969, p. 88.
17. HAMILTON, E. I. ET AL. Concentration and distribution of some stable elements in healthy human tissues from the United Kingdom, environmental study. Science and the total environment, 1: 341 (1973).
18. KRASOVSKII, G. N. ET AL. Experimental study of biological effects of lead and

aluminium following oral administration. *Environmental health perspectives*, **30**: 47–51 (1979).

19. MAHURKAR, S. D. ET AL. Electroencephalographic and radionuclide studies in dialysis dementia. *Kidney international*, **13**: 306 (1978).

20. ELLIOTT, H. L. ET AL. Aluminium toxicity during regular haemodialysis. *British medical journal*, **1**: 1101 (1978).

21. CRAPPER, D. R. ET AL. Brain aluminium distribution in Alzheimer's disease and experimental neurofibrillary degeneration. *Science (Washington)*, **180**: 511 (1973).

2. CHLORIDE

2.1 General description

Chloride is widely distributed in nature, generally in the form of sodium (NaCl), potassium (KCl) and calcium ($CaCl_2$) salts. It constitutes approximately 0.05% of the lithosphere (1). By far the greatest amount of chloride in the environment is present in the oceans.

The presence of chloride in natural waters can be attributed to dissolution of salt deposits (2), contamination resulting from salting of roads to control ice and snow (3–7), discharges of effluents from chemical industries (8), oil-well operations (9), sewage discharges (10), irrigation drainage (11), contamination from refuse leachates (12), and seawater intrusion in coastal areas (1). Each of these sources may result in local contamination of both surface-water and groundwater. The chloride ion is highly mobile, however, and is eventually transported into closed basins or to the oceans (1).

2.2 Occurrence

Chloride is generally present at low concentrations in natural surface-water. Levels in unpolluted water are often less than 10 mg/litre and may often be less than 1 mg/litre (11, 13).

In foods of plant and animal origin, chloride occurs naturally, generally at levels less than 0.36 mg/g (14). The addition of salt during processing or cooking, and at the table, can markedly increase the chloride level in food.

2.3 Routes of exposure

Estimation of the daily intake of chloride in food is complicated by the widespread use of table-salt as a condiment. Approximately 600 mg of chloride per day are ingested in a salt-free diet (15, 16). Because of the addition of salt to food, however, the daily intake of chloride averages 6 g and may range as high as 12 g (17, 18). An average intake of chloride from drinking-water is approximately 100 mg per day (10, 17, 18). The intake from air is negligible.

2.4 Health aspects

Chloride is the most abundant anion in the human body and contributes significantly, along with its associated cations, to the osmotic activity of the extracellular fluid; 88 % of the chloride in the body is extracellular (19). A normal 70-kg human body contains approximately 81.7 g of chloride (19) and 45 litres of water.

Water and electrolyte balance in the body is maintained by adjusting total dietary intake and excretion via the kidneys and intestinal tract. Absorption of chloride is almost complete in normal individuals. Most fluid and electrolyte absorption takes place in the proximal half of the small intestine (20). Normal daily loss of fluid is the equivalent of about 1.5–2 litres of water, together with about 4 g of chloride; 90–95 % of the chloride loss occurs in the urine, 4–8 % in the faeces, and about 2 % in sweat. The total obligatory loss of chloride per day amounts to approximately 530 mg (19). On the basis of this estimate of obligatory loss, a daily dietary intake for adults of 9 mg of chloride per kg of body weight (630 mg for a 70-kg man) is essential (equivalent to slightly more than 1 g of table-salt per person per day). For children up to 18 years of age, a daily dietary intake of 45 mg of chloride per kg of body weight should be sufficient (19).

The taste threshold for chloride in drinking-water is dependent upon the associated cation, but is usually within the range 200–300 mg of chloride per litre. Taste threshold levels for sodium chloride, potassium chloride and calcium chloride in water are 210, 310, and 222 mg/litre, respectively (21, 22). The taste of coffee particularly is affected if it is brewed with water having a chloride concentration of 400 mg/litre as sodium chloride or 530 mg/litre as calcium chloride (23).

Conventional water-treatment processes do not remove the chloride ion from water and although the amount of chloride ingested daily from drinking-water is but a very small proportion of the total daily intake, a guideline value of 250 mg of chloride per litre is recommended, based on organoleptic considerations.

REFERENCES

1. NATIONAL RESEARCH COUNCIL OF CANADA ASSOCIATE COMMITTEE ON SCIENTIFIC CRITERIA FOR ENVIRONMENTAL QUALITY. Effects of alkali halides in the Canadian environment. Ottawa, National Research Council, 1977.
2. NATIONAL RESEARCH COUNCIL. Nutrient and toxic substances in water for livestock and poultry. Washington, DC, National Academy of Sciences, 1974.
3. MURRAY, D. M. & ENNST, V. F. W. An economic analysis of the environmental impact of highway de-icing salts. National technical information service publication, 253: 268 (1976).
4. POLLOCK, J. J. & TOLER, L. G. Effects of highway de-icing salts on groundwater and water supplies in Massachusetts. Washington, DC, Department of the Interior, 1972 (US Geological Survey).
5. TERRY, R. C. Road salt, drinking water and safety. Cambridge, MA, Ballinger, 1974.
6. HUTCHINSON, F. E. Effects of highway salting on the concentration of sodium chloride in private water supplies. Research in life sciences, 15 (1969).
7. RALSTON, J. G. De-icing salts as a source of water pollution, Toronto, Ministry of the Environment, 1971.

8. LITTLE, A. D. *Inorganic chemical pollution of freshwater.* Washington, DC, US Environmental Protection Agency, 1971.
9. PETTYJOHN, W. A. Water pollution by oil-field brines and related industrial wastes in Ohio. *Ohio journal of science,* **71**: 257 (1971).
10. PETTYJOHN, W. A. *Water quality in a stressed environment,* Minnesota, Burgess Publishing Co., 1972.
11. BOND, R. G. & STRAUB, C. P. *Handbook of environmental control,* Vol. 3, Cleveland, Chemical Rubber Co., 1973.
12. SCHNEIDER, W. J. *Hydrologic implications of solid-waste disposal.* Washinghton, DC, Department of the Interior, 1970, pp. F1-F10 (US Geological Survey Circular 601-F).
13. *National water quality data bank.* Ottawa, Inland Waters Directorate, Water Quality Branch, 1976.
14. LONG, C. ET AL. *Biochemists' handbook.* London, E. and F. N. Spon Ltd, 1961.
15. DAHL, L. K. Salt and hypertension. *American journal of clinical nutrition,* **25**: 231 (1972).
16. MENEELY, G. R. A review of sources of and the toxic effects of excess sodium chloride and the protective effect of extra potassium in the diet. *Plant foods for human nutrition,* **23**: 3 (1973).
17. ZOETEMAN, B. C. J. & BRINKMAN, F. J. J. Human intake of minerals from drinking water in the European Communities. In: *Hardness and drinking water and public health.* Proceedings of the European Scientific Colloquium, Luxembourg. Oxford, Pergamon Press, 1976, p. 175.
18. *Sodium, chlorides and conductivity in drinking water. Report on a WHO Working Group.* Cophenhagen, WHO Regional Office for Europe, 1979 (EURO Reports and Studies, No. 2).
19. *Dietary standard for Canada.* Ottawa, Health Protection Branch, Department of National Health and Welfare, 1975.
20. SLADEN, C. E. Absorption of fluid and electrolytes in health and disease. In: McColl, I. & Sladen, G. E., ed. *Intestine absorption in man.* London, Academic Press, 1975, p. 51.
21. WHIPPLE, G. C. *The value of pure water.* New York, John Wiley and Sons, 1907.
22. RICHTER, C. P. & MACLEAN, A. Salt taste threshold of humans. *American journal of physiology,* **126**: 1 (1939).
23. LOCKHARD, E. E. ET AL. The effect of water impurities on the flavor of brewed coffee. *Food research,* **20**: 598 (1955).

3. COLOUR

3.1. General description

Colour in drinking-water may be due to the presence of: coloured organic substances, usually humics; metals such as iron and manganese; or highly coloured industrial wastes, of which pulp and paper and textile wastes are the most common. The primary importance of colour in drinking-water is aesthetic but the sensory effects may be regarded as a health effect.

Experience has shown that consumers whose drinking-water contains aesthetically displeasing levels of colour may seek alternative, possibly unsafe, sources.

Most people can detect levels of colour above 15 TCU (true colour units) in a glass of water. The removal of excess colour, prior to chlorination, will reduce the production of trihalomethanes. Taste due to chlorinated organics is also mitigated. Limiting the colour in potable water also limits the concentration of undesirable substances that are complexed with or adsorbed on to humic material.

The guideline value recommended for colour in drinking-water is less than 15 TCU.

3.1.1 Source

The appearance of colour in drinking-water is caused by the absorption of certain wavelengths of normal "white" light, by the presence of coloured substances, and by light scattering caused by suspended particles (1, 2). Colour measured in water that contains suspended matter is defined as "apparent colour"; "true colour" is measured on water samples from which particulate matter has been removed by centrifugation or filtration, colour then being due to humic substances in true solution (3, 4). In general, the true colour of a given water sample is substantially less than its apparent colour (4).

3.1.2 Measurement

The colour of a water sample may be measured by visual comparison with a series of standard solutions containing known amounts of potassium chloroplatinate and added cobalt(II) chloride. Since the platinum-cobalt standard method was designed to analyse naturally coloured water, difficulty in comparing the colour of a water supply with standard colour solutions may be indicative of pollution.

One TCU corresponds to the amount of colour exhibited under the specified test conditions by a solution containing 1.0 mg of platinum per litre in the form of chloroplatinate ion (4). A colour of 15 TCU can be detected in a glass of water by most consumers and 5 TCU will be apparent in large volumes of water, such as in a white bathtub; few people can detect a colour level of 3 TCU (5).

The colour of natural surface-water generally increases with increasing pH. This is commonly referred to as the "indicator effect" (1). Therefore, it is widely recommended that the pH of the sample be recorded together with the colour measurement to allow for this effect (4).

3.2 Occurrence

Complaints of coloured water generally approach in number those collectively concerned with taste and odour. Colour in natural waters is due mainly to organic matter, primarily humic substances, originating from the decay and aqueous extraction of vegetation into surface-water. Iron and manganese may often be present in groundwater as well as in some surface-waters and impart a colour. Another important source of iron in drinking-water is dissolution of iron pipes conveying the water. Iron and manganese can give rise to red and black water respectively. Copper solubilized from copper pipes may give rise to blue-green discoloration of sanitary ware in addition to a faint blue colour to the water in extreme cases. Highly coloured wastewaters, in particular wastes from the pulp, paper and textile industries, may create coloured water problems.

A colour problem of microbiological origin is the production of "red water", a phenomenon caused by the oxidation of iron (II) to iron (III), as a result of which, the iron precipitates from solution as the hydroxide and imparts a characteristic reddish colour to the water. In severe cases, distribution lines have been blocked by the action of these "iron bacteria". Similarly, a black discoloration may be imparted to drinking-water by the action of bacteria capable of oxidizing dissolved manganese to its insoluble oxides. This colour problem occurs more frequently in groundwater than in surface-water supplies.

3.2.1 Removal of colour

Removal of colour from water may sometimes best be effected in practice by chemical oxidation to supplement the coagulation and filtration. Data recorded in 1976 for about 200 plants in Sweden showed that colour levels ranging from < 5 to 150 TCU for untreated water were reduced to < 5 to 25 TCU for treated water (6).

3.3 Health aspects

Limits for colour in potable water have traditionally been based on aesthetic considerations. It has been noted, however, that supply to consumers of visibly coloured water may lead them to seek a colourless, but possibly unsafe, alternative source of drinking-water (7). Other health-related criteria include the association between colour and production of some chlorinated organic compounds, interference with the water treatment, and increased chlorine consumption.

Few toxicological studies of natural organic colour agents have been undertaken. Drinking-water containing a low-ash preparation of soil fulvic acid in concentrations of 10, 100, and 1000 mg/litre was supplied to male rats for periods of up to 90 days; no significant changes in body weight, food and water intake, organ/body weight ratios, or tissue histology were observed (G. C. Becking & A. P. Yagminas, unpublished observations, 1978). The same fulvic acid preparation was also given daily (for 14 days) to rats by gavage at a dosage of 1000 mg/kg of body weight; no mortality occurred at this dose level. The rate of weight gain, however, was decreased compared with that of control animals and slight changes were noted in some of the kidney enzyme concentrations. In the most relevant study to date, humic material was fed to rats in their drinking-water at two dose levels for periods of 19–35 weeks. The authors concluded that, applying a safety factor of approximately 100, drinking-water containing 2.5 mg of "humic acid" per litre would be safe for human consumption (8).

Very few studies have been devoted to the comparative toxicities of trace metals and their humate complexes with reference to human health (9). It has been shown that the acute mammalian toxicities of iron, lead, barium, silver, copper, and zinc are substantially enhanced on intravenous injection as their humate complexes, but that orally ingested lead humate is at least 60% less toxic than lead acetate (10). Increases of 50 to 100% were found in the amounts of ionic material (calcium, magnesium, iron, manganese, zinc, and sulfate) that permeated the intestine in the presence of humic acid (11). Unfortunately, the substances studied did not include toxic trace metals. No information has been published to date on the bioavailability to mammals of the humate complexes of toxic organic substances.

3.4 Other considerations

The adsorption by humic substances isolated from soil (12, 13) of organic compounds in amounts that can exceed their aqueous solubilities (14, 15) is a matter of potential importance meriting further investigation. Furthermore, owing to their polyanionic electrolytic properties, humic substances play an essential role in the dissolution, transport, and deposition of positively charged inorganic ions, such as the heavy metals. Most metals will complex on contact with humic substances in water. Complex formation can dramatically increase the solubility of the metal; for example, naturally occurring humic substances in water may render iron up to 10^9 times more soluble (*16*).

Some metals under certain circumstances form insoluble complexes with humic substances; this is the basis for the use of iron and aluminium salts in the production of drinking-water.

The fate of complexed toxic metals in water undergoing treatment for production of drinking-water is a particularly relevant question. The most prevalent view is that, although toxic metals associated with the suspended solids in water may be at least partially removed, dissolved trace metals are probably removed only to a negligible extent during conventional treatment (*17*).

Claims that dissolved humic substances cause a taste in drinking-water (*18*) cannot be confirmed as no recent research appears to have been done on the topic. Highly coloured, polluted water will frequently have an associated objectionable taste, but the degree to which this association is causative is unknown. It is known that the organic colouring material in water stimulates the growth of many aquatic microorganisms (*19*), some of which are directly responsible for the production of odour in water.

The relationship between corrosion and incrustation and the humic content of water is both complex and important. Small amounts of humic substances (1–2 mg/litre) assist in the deposition of a protective layer of calcium carbonate in distribution systems (*20*). Where lime has been added as a post-treatment corrective step for corrosive waters (*21*), larger amounts of humic acid may be responsible for the deposition of flow-restrictive "humus mud" in distribution systems. Water containing very little dissolved humic material can be more metal-corrosive than water containing larger amounts (*22*, *23*).

Since humic acid and certain of its metal complexes are poorly soluble at the pH of potable water, they may be partly responsible for turbidity in a water sample. Furthermore, since "dissolved" humic substances in water exist predominantly as colloidal dispersions, and since optical measurements of turbidity are influenced by particles in the colloidal size range (*2*), such colour in water will affect turbidity values.

Difficulty in maintaining a free available chlorine residual in distribution systems may be due to the presence of organic colour in treated water. Although this fact has been known since 1949 (*18*), it was not until the discovery in 1974 (*24*, *25*) of the presence of relatively large amounts of chloroform and other trihalomethanes in chlorinated water that the reaction of chlorine with dissolved humic substances was subjected to careful study. Trihalomethanes are reaction products of chlorine (and adventitiously present bromine and iodine) with humic substances; fortunately, however, methods used for coagulation remove most of the organic precursors from the raw water (26).

Colour can interfere with the chemical analysis of many constituents of water. In *Standard methods for the examination of water and wastewater*, for example, it is stated that colour must be compensated for or removed in colorimetric analytical methods (4).

The metal-complexing properties of humic substances can interfere in non-colorimetric methods of analysis. Humic substances interfere with trace metal analysis when a complexing agent is employed and the sample is concentrated by extraction with an organic solvent (27).

REFERENCES

1. BLACK, A. P. & CHRISTMAN, R. F. Characteristics of colored surface waters. *Journal of the American Water Works Association*, **55**: 753 (1963).
2. BLACK, A. P. & HANNAH, S. A. Measurement of low turbidities. *Journal of the American Water Works Association*, **57**: 901 (1965).
3. SAWYER, C. N. & McCARTY, P. L. *Chemistry for sanitary engineers*, 2nd ed. Toronto, McGraw-Hill Book Company, 1967, p. 299.
4. *Standard methods for the examination of water and wastewater*, 14th ed. Washington, DC, APHA, AWWA, WPCF, 1976.
5. BEAN, E. L. Progress report on water quality criteria. *Journal of the American Water Works Association*, **54**: 1313 (1962).
6. ANON. VAV AD 76 *Vattenbeskaffenhet 1976 [Water quality 1976]*. Stockholm, Swedish Water Works Association, 1977.
7. *Public health service drinking water standards*. Rockville, MD, US Department of Health, Education and Welfare, 1962, p. 21 (Public Health Service Publication No. 956).
8. JANECEK, J. & CHALUPA, J. Biological effects of peat water humic acids on warm-blooded organisms. *Archiv für hydrobiologie*, **65**: 515 (1969).
9. BROWN, V. M. ET AL. Aspects of water quality and the toxicity of copper to rainbow trout. *Water research*, **8**: 797 (1974).
10. KLOCKING, R. Influence of humic acids on the toxicity of lead. *Proceedings of the European society of toxicology*, **16**: 258 (1975).
11. VISSER, S. A. Some biological effects of humic acid in the rat. *Acta biologica et medica Germanica*, **31**: 569 (1973).
12. FLAIG, W. ET AL. In: Gieseking, J. E., ed. *Soil components*. Vol. 1, New York, Springer-Verlag, 1975.
13. FELBECK, G. T. JR. In: McLaren A. D. & Skujins, J., ed. *Soil biochemistry*. Vol. 2. New York, Marcel Dekker, 1971, pp. 54–56.
14. KHAN, S. U. & SCHNITZER, M. The retention of hydrophobic organic compounds by humic acid. *Geochimica cosmochimica acta*, **36**: 745 (1972).
15. HAGUE, R. & SCHMEDDING, D. Studies on the adsorption of selected polychlorinated biphenyl isomers on several surfaces. *Journal of environmental science and health*, **B11**: 129 (1976).
16. SHAPIRO, J. Effect of yellow organic acids on iron and other metals in water. *Journal of the American Water Works Association*, **56**: 1062 (1964).
17. COMMITTEE ON WATER QUALITY CRITERIA, NATIONAL ACADEMY OF SCIENCES. *Water quality criteria 1972*. Washington, DC, US Government Printing Office, 1973, p. 51 (EPA-R-73-033).
18. McKEE, J. E. & WOLF, H. W., ed. *Water quality criteria*, 2nd ed. Sacramento, CA, California State Water Quality Control Board, 1963, p. 198 (Publication No. 3-A).
19. PRAKASH, A. ET AL. Humic substances and aquatic productivity. In: Povoledo, D. &

Golterman, H. L., ed. *Humic substances 1972*. Wageningen, The Netherlands, Pudoc, 1975, pp. 259–268.

20. AMERICAN WATER WORKS ASSOCIATION. *Water quality and treatment*, 3rd ed. Toronto, McGraw-Hill Book Company, 1971, p. 311.

21. GJESSING, E. T. *Physical and chemical characteristics of aquatic humus*. Ann Arbor, MI, Ann Arbor Science, 1976.

22. VAN BENEDEN, G. & LECLERC, E. Les matières humiques. Leur comportement dans les eaux ou géneral, leur role dans la corrosion des metaux. *Technol. water (Czech)*, **8**: 225 (1964).

23. MOORE, M. R. Plumbosolvency of waters. *Nature*, **243**: 223 (1973).

24. ROOK, J. J. Formation of haloforms during chlorination of natural waters. *Journal of water treatment and examination*, **23**: 234 (1974).

25. BELLAR, T. A. ET AL. The occurrence of organohalides in chlorinated drinking waters. *Journal of the American Water Works Association*, **66**: 703 (1974).

26. STEVENS, A. A. & SYMONS, J. M. Measurement of trihalomethane and precursor concentration changes. *Journal of the American Water Works Association*, **69**: 546 (1977).

27. PAKALNS, P. & FARRAR, Y. J. The effect of surfactants on the extraction-atomic absorption, spectrophotometric determination of copper, iron, manganese, lead, nickel, zinc, cadmium and cobalt. *Water research*, **11**: 145 (1977).

4. COPPER

4.1 General description

Copper and its compounds are ubiquitous in the environment and are thus frequently found in surface-water. The nature of the copper in water depends on the pH and carbonate concentration in the water and the other anions in solution (1).

Water treatment processes usually result in the removal of trace metals from water but the copper concentration in drinking-water at the consumer's tap can be higher than in either the source water or the treated water entering the supply. Various chemical and physical characteristics of the distributed water influence the leaching of copper from the distribution system and household plumbing. Water stored in copper vessels tends to maintain the bacterial quality without deterioration (2). Copper in solution imparts a colour and an undesirable taste to drinking-water (3).

4.2 Occurrence

The copper content of soils depends on such factors as geographical location, proximity to industry, and use of fertilizers. Copper concentrations in inorganic-based fertilizers were found to range from 0.01 to 0.05 mg/g (4). The amount of copper present in food will vary with the copper content of the soil from which it derives. Foods such as vegetables, flour, and dairy and meat products normally have a copper content less than 0.01 mg/g.[a] Copper levels in drinking-water vary normally from 0.01 to 0.5 mg/litre (5).

4.3 Health aspects

Copper is an essential element in human metabolism, having roles in erythrocyte formation, release of tissue iron, and the development of bone, the central nervous system, and connective tissue. Copper is usually combined with proteins: haemocuprein in the erythrocytes and ceruloplasmin in the blood plasma contain copper as an integral part of their structure; metallothionein is a copper storage protein. A number of copper-containing enzymes have been isolated, notably cytochrome oxidase, ascorbic acid oxidase, and uricase.

---

[a] Department of National Health and Welfare. Food Monitoring Survey, Project FMO1, Ottawa, Canada. August 1971-January 1976 (unpublished data).

As copper is widely distributed in foods, it is unlikely that human beings, with the exception, perhaps, of infants on an exclusive milk diet, ever develop a dietary deficiency of copper. It is a beneficial adjunct to iron therapy in the treatment of nutritional anaemia in infants. In animals, however, a wide variety of clinical disorders have been associated with copper deficiency (6).

Intake of excessively large doses by man leads to severe mucosal irritation and corrosion, widespread capillary damage, hepatic and renal damage, and central nervous system irritation followed by depression. Severe gastrointestinal irritation and possible necrotic changes in the liver and kidneys could occur. However, copper poisoning is rare in man and higher mammals owing to the powerful emetic action of copper. Application of copper salts to the skin is corrosive and may lead to papulovesicular eczema. Local action on the eye produces serious inflammation. Copper in water has an unpleasant, astringent taste. The taste threshold is above 5.0 mg/litre, although taste is detectable in distilled water at 2.6 mg/litre (7).

4.4 Other aspects

The presence of copper in the water supply, although not constituting a hazard to health, may interfere with the intended domestic uses of water.

Copper in public water supplies enhances corrosion of aluminium and zinc utensils and fittings. Staining of laundry and plumbing fixtures occurs when copper eoncentrations in water exceed 1.0 mg/litre, and this value is recommended as a guideline value.

REFERENCES

1. McKee, J. E. & Wolf, H. W. *Water quality criteria.* Sacramento, CA, California State Water Quality Control Board, 1963.
2. Dhabadgaonkar, S. M. Metallic copper for disinfection of water in rural areas. *Journal of Indian Water Works Association,* 12: 43 (1980).
3. Page, G. G. Contamination of drinking water by corrosion of copper tubes. *New Zealand journal of sciences,* 16: 349 (1973).
4. Van Loon, J. C. & Lichwa, J. A study of the atomic absorption determination of some important heavy metals in fertilizers and domestic plant sludges. *Environment letters,* 4: 1 (1973).
5. Zoeteman, B. C. J. & Brinkman, F. J. J. In: *Hardness of drinking water and public health.* Proceedings of the European Scientific Colloquium, Luxembourg, 1975. Oxford, Pergamon Press, 1976, p. 173.
6. WHO Technical Report Series, No. 532, 1973 (*Trace elements in human nutrition*: report of a WHO Expert Committee).
7. Cohen, J. M. et al. Taste threshold concentrations of metals in drinking water. *Journal of the American Water Works Association,* 52: 660 (1960).

5. HARDNESS[a]

5.1 General description

Water hardness is the traditional measure of the capacity of water to react with soap, hard water requiring a considerable amount of soap to produce a lather. Scaling of hot-water pipes, boilers, and other household appliances is due to hard water. Water hardness is caused by dissolved polyvalent metallic ions. In freshwater, the principal hardness-causing ions are calcium and magnesium; the ions strontium, iron, barium, and manganese also contribute (1). Hardness is usually measured by the reaction of the polyvalent metallic ions present in a water sample with a chelating agent such as EDTA and is expressed as an equivalent concentration of calcium carbonate (1, 2). Hardness may also be estimated by determination of the individual concentrations of the components of hardness, their sum being expressed in terms of an equivalent quantity of calcium carbonate. The degree of hardness of drinking-water has been classified in terms of its equivalent $CaCO_3$ concentration as follows:

soft	0–60 mg/litre
medium hard	60–120 mg/litre
hard	120–180 mg/litre
very hard	180 mg/litre and above.

Hardness has also been classified in terms of equivalent concentration of CaO or $Ca(OH)_2$. In the SI system, it is recommended that hardness be expressed as moles of Ca^{2+} per cubic metre (3).

Although hardness is caused by cations, it may also be discussed in terms of carbonate (temporary) and non-carbonate (permanent) hardness (4). Carbonate hardness refers to the amount of carbonates and bicarbonates in solution that can be removed or precipitated by boiling. This type of hardness is responsible for the deposition of scale in hot-water pipes and kettles. Non-carbonate hardness is caused by the association of the hardness-causing cations with sulfate, chloride, or nitrate and is referred to as "permanent hardness" because it cannot be removed by boiling.

Alkalinity, an index of the buffering capacity of water, is closely linked to hardness. For the most part, alkalinity is produced by anions or molecular species of weak acids, mainly hydroxide, bicarbonate, and

---

[a] The problems associated with hardness in drinking-water have also been discussed in Part III, section 9, p. 106.

carbonate; other species, such as borates, phosphates, silicates, and organic acids also contribute to a small degree if present. Whichever solute species contributes to the alkalinity of water, it is always expressed in terms of an equivalent quantity of calcium carbonate. When the alkalinity of a surface-water is due to the presence of carbonates and/or bicarbonates, the alkalinity value is usually close to the hardness value (5).

5.2 Occurrence

The principal natural sources of hardness in water are sedimentary rocks, seepage, and runoff from soils. Hard water normally originates in areas with thick topsoil and limestone formations (4). Groundwater is generally harder than surface-water. Groundwater rich in carbonic acid and dissolved oxygen usually possesses a high solubilizing potential towards soil or rocks that contain appreciable amounts of the minerals calcite, gypsum, and dolomite, and consequently hardness levels up to several thousand milligrams per litre can result (4, 6).

The two main industrial sources of hardness are the inorganic chemical and the mining industries (4, 7). In the building industry calcium oxide is used in mortar, stucco, and plaster. It also finds use in pulp and paper production, sugar refining, petroleum refining, tanning, and as a water and wastewater treatment chemical (8). Magnesium is also used in various processes in the textile, tanning, and paper industries. Alloys of magnesium find extensive use in moulds and die castings, portable tools, luggage, and general household goods. The salts of magnesium are used in the production of magnesium metal, fertilizers, ceramics, explosives, and medicines (9).

5.3 Health aspects

As outlined in section 5.1, the major contributing factors to hardness of water are calcium and magnesium ions. There is no evidence of adverse health effects specifically attributable to high levels of calcium or magnesium in drinking-water.

Apart from the domestic disadvantage resulting from the use of water possessing a high degree of hardness, another possible disadvantage may arise from the association of magnesium with the sulfate ion resulting in a water possessing laxative properties.

The taste threshold for the calcium ion in drinking-water varies from 100 to 300 mg/litre, depending upon the anions present; for the magnesium ion the taste threshold is less than this value (10). Part III, which deals with the health aspects of inorganic constituents of drinking-water, should be consulted for further details on the relationship between water hardness and cardiovascular disease (see p. 109). Guideline values are not proposed for calcium or magnesium in water as a guideline value is proposed for total hardness based on aesthetic considerations.

5.4 Other aspects

Soft water has a greater tendency to cause corrosion of pipes, and consequently, certain heavy metals such as copper, zinc, lead, and cadmium may be present in the distributed drinking-water (*11–14*). The degree to which this corrosion and solubilization of metals occurs is also a function of pH, alkalinity, and dissolved oxygen concentration. In some communities, corrosion is so severe that special precautions must be adopted with the supply (*15*).

In areas with very hard water, household pipes can become choked with deposited scale (*16*); hard water also deposits incrustations on kitchen utensils as well as increasing soap consumption. Hard water can thus be both a nuisance and an economic burden to the consumer. Public acceptance of water hardness varies among communities; it is often related to the hardness to which the consumer has, over the years, become accustomed, and in many communities a water hardness greater than 500 mg/litre is tolerated. A hardness level of about 100 mg of $CaCO_3$ per litre provides an acceptable balance between corrosion and the problems of incrustation, although, from aesthetic considerations, 500 mg/litre is recommended as a guideline value (*17*).

REFERENCES

1. *Quality criteria for water*, Washington, DC, US Environmental Protection Agency, 1976 (EPA-440/9-76-023).
2. SEKERKA, I. & LECHNER, J. F. Simultaneous determination of total, non-carbonate and carbonate water hardnesses by direct potentiometry. *Talanta*, **22**: 459 (1975).
3. GLOHMANN, A. Harte des Wassers. In: Amavis, R. et al., ed. *Hardness of drinking water and public health*. Proceedings of the European Scientific Colloquium, Luxembourg, 1975. Oxford, Pergamon Press, 1976, p. 129.
4. SAWYER, C. N. & McCARTY, P. L. *Chemistry for sanitary engineers*, 2nd ed. New York, McGraw-Hill, 1967 (Series in sanitary science and water resources engineering).
5. THOMAS, J. F. J. *Industrial water resources of Canada*. Water Survey Report No. 1. Scope, Procedure and Interpretation of Survey Studies, Ottawa, Queen's Printer, 1953.
6. DE FULVIO, S. & OLORI, L. Definitions and classification of naturally soft and naturally hard waters. Chemical and physical characteristics of the water in some member states of the European community. In: Amavis, R. et al., ed. *Hardness of drinking water and public health*. Proceedings of the European Scientific Colloquium, Luxembourg, 1975. Oxford, Pergamon Press, 1976, p. 95.
7. BIESECKER, J. E. & GEORGE, J. R. Stream quality in Appalachia as related to coal-mine drainage, 1965. In: Pettyjohn, W. A., ed. *Water quality in a stressed environment*. Minnesota, Burgess Publishing Company, 1972.
8. McQUARRIE, M. C. Lime. In: *McGraw-Hill encyclopedia of science and technology*. New York, McGraw-Hill, 1966.
9. BECH, A. V., ed. The technology of magnesium and its alloys. In: *McGraw-Hill encyclopedia of science and technology*, New York, McGraw-Hill, 1966.
10. ZOETEMAN, B. C. J. *Sensory assessment of water quality*. Oxford, Pergamon Press, 1980.

11. NERI, L. C. Some data from Canada. In: Amavis R., et al., ed. *Hardness of drinking water and public health.* Proceedings of the European Scientific Colloquium, Luxembourg 1975. Oxford, Pergamon Press, 1976, p. 343.
12. SHARRETT, A. R. & FEINLEIB, M. Water constituents and trace elements in relation to cardiovascular diseases. *Preventive medicine,* **4**: 20 (1975).
13. CRAUN, G. F. & McCABE, L. J. Problems associated with metals in drinking water. *Journal of the American Water Works Association,* **67**: 593 (1975).
14. NERI, L. C. & HEWITT, D. Review and implications of ongoing and projected research outside the European communities. In: Amavis, R. et al., ed. *Hardness of drinking water and public health.* Proceedings of the European Scientific Colloquium, Luxembourg, 1975. Oxford, Pergamon Press, 1976, p. 443.
15. MULLEN, E. D. & RITTER, J. A. Potable-water corrosion control. *Journal of the American Water Works Association,* **66**: 473 (1974).
16. COLEMAN, R. L. Potential public health aspects of trace elements and drinking water quality. *Annals of the Oklahoma Academy of Science,* **5**: 57 (1976).
17. BEAN, E. L. Quality goals for potable water. *Journal of the American Water Works Association,* **60**: 1317 (1968).

6. HYDROGEN SULFIDE

6.1 General description

Hydrogen sulfide is a flammable, poisonous gas with a characteristic odour of rotten eggs (*1*). Hydrogen sulfide and sulfides of the alkali and alkaline earth metals are soluble in water (*2*). Soluble sulfide salts dissociate in water into sulfide ions, which react with the hydrogen ions to form hydrosulfide ion (HS^-) or hydrogen sulfide (H_2S). The relative concentrations of these species are a function of the pH of the water, hydrogen sulfide concentration increasing with decreasing pH (*1, 3*). The sulfide ion is present in appreciable concentrations above pH 10.

In well-aerated water, hydrogen sulfide is oxidized to sulfate. Biological oxidation to elemental sulfur (*3*) occurs and sulfides form an indispensable link in nature in the "sulfur cycle" (*4*).

6.2 Occurrence

Sulfide occurs naturally in mineral ores, oil, and coal deposits (*5*). Copper, lead, zinc, nickel, and other mined base metals may occur as simple or complex sulfides. Iron sulfides are often associated with these ores (*1*).

Sulfides are also present in industrial wastes from petroleum and petrochemical plants, chemical plants, gas works, paper mills, heavy water plants, and tanneries (*1, 3, 6, 7*). Sulfides are generated by sulfate-reducing bacteria (*1, 3, 8–10*). The growth of sulfate-reducing bacteria in distribution systems can be a major cause of taste and odour problems in drinking-water.

Sulfide levels in Mississippi river water were in the region of 0.092 mg/litre; at St Paul, Minnesota, pond and well-water (*11*) were found to contain 0.16 and 0.19 mg/litre, respectively.

Levels of atmospheric hydrogen sulfide naturally range from 1.5×10^{-4} to 4.6×10^{-4} mg/m^3 (*6*). In industrialized areas, levels are significantly higher.

Sulfides are present in a number of raw and cooked foods. Chives and garlic contain both dimethyl sulfide and dimethyl trisulfide (*12*). The *S*-methylmethionine level in onion bulbs is less than 0.001 mg/g; in tomatoes it averages 0.003 mg/g, and in cabbage the concentration reaches 0.05 mg/g (*13*).

Dimethyl sulfide is an important flavouring compound in British ales (0.0002–0.0037 mg/litre) and European lager beers (0.003–0.114

mg/litre) (*14*). It has also been found in eggs, milk, and dairy products (*13*).

The hydrogen sulfide content of cooked meat ranges in concentration from 0.276 mg/kg for minced beef to 0.394 mg/kg for minced lamb (*15*). The hydrogen sulfide concentration of heated dairy products ranges from 0.80 mg/litre in skimmed milk (0.1 % fat) to 1.84 mg/litre in cream (30.5 % fat) (*16*).

Hydrogen sulfide and other soluble sulfides are utilized in pigment and dye manufacturing, tanning, pulp, and chemical processing. They are also employed in the production of cosmetics (*17*). Spring-waters that contain elevated concentrations of hydrogen sulfide are used for therapeutic baths and have been consumed for medicinal purposes.

6.3 Routes of exposure

Data on the sulfide content of foodstuffs are incomplete and thus the daily dietary intake of sulfides has not been estimated. Exposure may result from the consumption of beer and ale, seafoods, cooked meat, warm milk, and asparagus and other vegetables. In England, the "maximum likely daily intake" of dimethyl sulfide from artificially flavoured sweets, soft drinks, creams and jellies was estimated to be 1.7 mg (*17*).

The inhalation of 20 m^3 of air containing hydrogen sulfide in natural concentrations would result in a daily intake of 0.003–0.01 mg.

The intake of sulfide from drinking-water cannot be estimated owing to the absence of data on sulfide levels in drinking-water.

6.4 Health aspects

The alkali sulfides are absorbed rapidly from the intestine (*18*). Hydrogen sulfide is transformed to alkali sulfide in the blood and tissues. Excretion of sulfide is via the kidneys and lungs, whereas metallic sulfides are excreted by the bowels (*18*). Sulfide may also be oxidized to sulfate and thiosulfate and excreted by the kidneys (*18*).

Hydrogen sulfide blocks the action of certain enzyme systems, some of which are involved directly in cellular oxidative processes. Such inhibitive effects have been reported for dehydrogenase (succinic), phosphatase (ATPase), oxidase (dopa oxidase), carbonic anhydrase, dipeptidase, benzamidase, and some iron-containing enzymes (*6*). The mechanism of enzyme inhibition is not completely understood, but is considered to be via the formation of metal sulfides, which decrease the availability of cations to the enzymes.

Alkaline sulfides irritate the epithelium of the mucous membranes (*18*). Oral ingestion of alkali sulfides produces nausea, vomiting, and epigastric pain (*1, 18*). An oral dose of 10–15 g of sodium sulfide is fatal (*18*). Daily oral ingestion of dimethyl sulfide at a dose of 250 mg/kg of body weight over a period of 14 weeks was found to produce no ill effects in rats; this dose is equivalent to a daily intake of 17.5 g by a 70-kg adult human (*17*).

Inhaled hydrogen sulfide gas causes death by paralysing the respiratory centre (*19*); the threshold for acute inhalation poisoning is between 700 and 1000 mg/m^3 (length of exposure was not specified). Exposure levels of 1400–2100 mg/m^3 stimulate the nervous system and are fatal within 30 minutes; levels greater than 2800 mg/m^3 result in instantaneous death by means of nervous system paralysis. The progression of symptoms includes sudden fatigue, dizziness, intense anxiety, loss of olfactory function, collapse, respiratory arrest, and death. Olfactory paralysis prevents detection of hydrogen sulfide beyond concentrations of 225 mg/m^3 (*6*). Hydrogen sulfide at levels of 0.12 mg/m^3 in air causes mental depression; levels of 1.5–43 mg/m^3 result in conjunctivitis and visual disorders (*6*). Levels of 70–700 mg/m^3 produce chronic intoxication resulting in psychic changes, dizziness, confused somnolence, tachycardia, coughing, and vomiting (*6, 19*).

Hydrogen sulfide in water has a disagreeable taste and odour and its presence is an important cause of consumer complaints. The taste and odour thresholds for hydrogen sulfide in solution are estimated to be in the range 0.05–0.10 mg/litre (*1, 9*). For sulfides, the taste and odour threshold is about 0.2 mg/litre (*1*). It is unlikely that a person would consume a harmful dose of sulfide because of the unpleasant taste and odour at concentrations much lower than the toxic levels; consequently, the guideline value has been chosen as not detectable by consumers.

6.5 Other aspects

Hydrogen sulfide in association with soluble iron causes black deposits in piping and on fixtures and produces black stains on laundered items.

REFERENCES

1. McKee, J. E, & Wolf, H. W. *Water quality criteria*, 2nd ed. Sacramento, CA, California State Water Quality Control Board, 1963. pp. 156–7, 271–2, 277, 335–336.
2. Senko, M. J. & Plane, R. A. *Chemical principles and properties*, 2nd ed. New York, McGraw-Hill Publishing Co., 1974, pp. 639–640.
3. *Quality criteria for water*. Washington, DC, US Environmental Protection Agency, 1976, pp. 410–416.
4. Smith, R. L. *Ecology and field biology*, 2nd ed. New York, Harper & Row Publishers, 1974, pp. 88–89, 123–128.
5. National Research Council of Canada. *Sulfur and its inorganic derivatives in the Canadian environment*. Associate Committee on Scientific Criteria for Environmental Quality. Ottawa, National Research Council, 1977.
6. Booras, S. G. *Hydrogen sulfide: health effects and recommended air quality standard*. Illinois Institute for Environmental Quality, 1974 (NTIS PB-233 843).
7. Colby, P. J. & Smith, L. L., Jr. Survival of Walleye eggs and fry on paper fibre sludge deposits in Rainy River, Minnesota. *Transactions of the American Fisheries Society*, **96**: 278–296 (1967).
8. Adelman, I. R. & Smith, L. L., Jr. Toxicity of hydrogen sulfide to goldfish

(Carassius Auratus) as influenced by temperature, oxygen, and bioassay techniques. *Journal of the Fisheries Research Board of Canada*, **29**: 1309 (1972).

9. ACREE, T. E. & SPLITTSTOESSER, D. F. Prevention of H_2S in wine fermentation. *New York food and life sciences bulletin*, **5**: 19 (1972).

10. KADOTA, H. & ISHIDA, Y. Production of volatile sulfur compounds in microorganisms. *Annual review of microbiology*, **26**: 127 (1972).

11. BRODERIUS, S. J. & SMITH, L. L., JR. Direct determination and calculation of aqueous hydrogen sulfide. *Analytical chemistry*, **49**: 424 (1977).

12. CREWE, R. M. & ROSS, F. P. Biosynthesis of alkyl sulphides by an ant. *Nature*, **254**: 448 (1975).

13. HATTULA, T. & GRANROTH, B. Formation of dimethyl sulphide from s-methyl-methionine in onion seedlings (Allium Cepa). *Journal of the science of food and agriculture*, **25**: 1517 (1974).

14. NIEFIND, H. J. & SPAETH, G. Some aspects of the formation of dimethyl sulfide through brewer's yeast and beer spoilage microorganisms. *Proceedings of the Annual Meeting of the American Society of Brewing Chemists*, **33**: 54 (1975).

15. KUNSMAN, J. E. & RILEY, M. L. A comparison of hydrogen sulfide evolution from cooked lamb and other meats. *Journal of food science*, **40**: 506 (1975).

16. THOMAS, E. L. ET AL. Determination of hydrogen sulphide in heated milks by gas liquid chromatographic head space analysis. *Journal of dairy science*, **59**: 1865 (1975).

17. BUTTERWORTH, K. R. ET AL. Short-term toxicity of dimethyl sulphide in the rat. *Food and cosmetics toxicology*, **13**: 15 (1975).

18. THIENES, C. H. & HALEY, T. J. *Clinical toxicology*, 5th ed. Philadelphia, Lea & Febiger, 1972, pp. 59–60.

19. HENKIN, R. I. Effects of vapor phase pollutants on nervous system and sensory function. In: Finkel, A. J. & Duel, W. C., ed. *Clinical implications of air pollution research*. Acton, MA. Public Science Group Inc., American Medical Association., 1974, pp. 193–212.

7 IRON

7.1 General description

Iron is the fourth most abundant element by weight in the earth's crust. In water it occurs mainly in the divalent and trivalent (ferrous and ferric) states (*1*). Both cast iron and steel pipes are employed for drinking-water distribution to the consumer. In the production of potable water, various salts of iron are used as coagulating agents (*2*).

7.2 Occurrence

Iron in surface-water is generally present in the ferric (Fe III) state. The concentration of iron in well aerated water is seldom high, but under reducing conditions, which may exist in some groundwater, lakes or reservoirs, and in the absence of sulfide and carbonate, high concentrations of soluble ferrous iron may be found (*3*). Concentrations of iron greater than 1 mg/litre have been reported to occur in groundwater (*4*). The presence of iron in natural waters can be attributed to the dissolution of rocks and minerals, acid mine drainage, landfill leachates, sewage, or iron-related industries.

Iron is generally present at low concentrations in the atmosphere as a result of emissions from the iron and steel industry, thermal power plants, and incineration, but few data are available on levels of iron in the atmosphere.

The iron content of foods varies considerably. Cereals (mean 0.0295 mg/g) and meat (0.0262 mg/g) appear to be the main dietary sources of this element (*5*), the iron concentration of most other natural foods being less than 0.020 mg/g (*6, 7*). Levels may be somewhat higher in foods fortified with iron or in food cooked in iron utensils (*8*). Evidence suggests that the iron content of foodstuffs decreases during boiling (*9*).

7.3 Routes of exposure

The daily intake of iron from typical diets in developed countries has been estimated to be in the range of 15 to 22 mg (*5, 9, 10*).

Concentrations of iron in drinking-water are normally less than 0.3 mg/litre, and the intake from food is substantially higher than that from drinking-water.

The contribution made by airborne iron to human exposure is negligible.

7.4 Health aspects

Iron is an essential element in human nutrition (*7, 11*). It is contained in a number of biologically significant proteins, for example, haemoglobin and cytochromes, and also in many oxidation-reduction enzymes. Estimates of the minimum daily requirement for iron vary from 7 to 14 mg, depending upon age and sex; pregnant women may require in excess of 15 mg per day (*12*). The average daily requirement is considered to be 10 mg.

Individual iron requirements (which depend on age, sex, and physiological state) regulate the amount of iron absorbed from the diet, the amount varying from 1 to 20 % (*12, 13*). In most individuals, about 10 % of ingested iron is absorbed (*7, 12*). Obligatory losses (in faeces, urine, and perspiration) amount to 1 mg per day (*12*). Between 60 and 70 % of absorbed iron is used in the production of haemoglobin; 5 % is utilized in myoglobin production, the excess being stored primarily in the liver, bone marrow, and spleen.

Iron ingestion in large quantities results in a condition known as haemochromatosis (normal regulatory mechanisms do not operate effectively) wherein tissue damage results from iron accumulation. This condition rarely develops from simple dietary overloading (*7, 13, 14*). However, the condition has resulted from prolonged consumption of acid foodstuffs cooked in iron kitchenware (*14*). Haemochromatosis has not been demonstrated in test animals receiving extremely high doses of iron (*15*). Small children have been poisoned following the ingestion of large quantities of iron tablets (*16*).

7.5 Other aspects

The presence of iron in drinking-water supplies is objectionable for a number of reasons unrelated to health (*4, 9, 17*). Under the pH conditions existing in drinking-water supplies, ferrous salts are unstable and precipitate as insoluble ferric hydroxide, which settles out as a rust-coloured silt. Such water often tastes unpalatable and stains laundry and plumbing fixtures. The iron that settles out in the distribution system gradually reduces the flow of water. Iron also promotes the growth of "iron bacteria". These microorganisms derive their energy from the oxidation of ferrous iron to ferric iron, and in the process deposit a slimy coating on the piping.

The above problems usually arise, especially in distribution systems, when the iron concentration approaches 0.3 mg/litre, and consequently, whenever possible, it would be prudent to maintain levels below this value.

REFERENCES

1. *Quality criteria for water*. Washington, DC, US Environmental Protection Agency, 1976.

2. Cox, C. R. *Operation and control of water treatment processes.* Geneva, World Health Organization, 1964 (WHO Monograph Series No. 49).

3. Hem, J. D. *Chemical factors that influence the availability of iron and manganese in aqueous systems.* Washington, DC, The Geological Society of America Inc., 1972 (Special Paper 140) p. 17.

4. Dart, F. J. *The hazard of iron.* Ottawa, Water and Pollution Control Canada, 1974.

5. Méranger, J. C. & Smith, D. C. The heavy metal content of a typical Canadian diet. *Canadian journal of public health,* **63**: 53 (1972).

6. Gormican, A. Inorganic elements in foods used in hospital menus. *Journal of the American Dietetic Association,* **56**: 397 (1976).

7. Watt, B. K. & Merrill, A. L. *Composition of foods—raw, processed, prepared.* Washington, DC, Revised USDA Agriculture Handbook, Vol. 8 (1963).

8. Bowering, J. & MacPherson Sanchez, A. A conspectus of research on iron requirements of man. *Journal of nutrition,* **106**: 985 (1976).

9. Zoeteman, B. C. J. & Brinkmann, F. J. J. Intake of minerals by man. In: *Hardness of drinking water and public health.* Proceedings of the European Scientific Colloquium, Luxembourg, 1975. Oxford, Pergamon Press, 1976.

10. Kirkpatrick, D. C. & Coffin, D. E. The trace metal content of representative Canadian diets in 1970 and 1971. *Canadian Institute of Food Science and Technology journal,* **7**: 56 (1974).

11. Moore, C. V. Iron. In: Goodheart, R. S. & Shils, M. E., ed., *Modern nutrition in health and disease.* Philadelphia, Lea & Febiger, 1973, p. 297.

12. *Dietary standard for Canada.* Ottawa, Department of National Health and Welfare, 1975.

13. Hopps, H. C. *Ecology of disease in relation to environmental trace elements–particularly iron.* Washington, DC, The Geological Society of America Inc., 1972 (Special Paper 140, p. 1).

14. Jacobs, A. Iron overload—clinical and pathological aspects. *Seminars in hematology,* **14**: 89 (1977).

15. Brown, E. B. et al. Studies in iron transportation and metabolism. *Journal of laboratory and clinical medicine,* **12**: 862 (1957).

16. Stein, M. et al. Acute iron poisoning in children. *Western journal of medicine,* **125**: 289 (1976).

17. Iron. In: McKee, J. E. & Wolf, H. W., ed., *Water quality criteria.* Sacramento, CA, California State Water Quality Control Board, 1971 (Publication 3-A, p. 202).

8. MANGANESE

8.1. General description

Airborne manganese sources are widespread and its distribution over large distances is favoured by the fact that emitted manganese is predominantly associated with smaller dust particles (*1*).

When present in natural surface-water, manganese occurs in both the dissolved and suspended forms. Anaerobic groundwater often contains elevated levels of dissolved manganese.

8.2 Occurrence

Freshwater may contain from one to several thousand micrograms of manganese per litre depending on the location. Manganese concentrations in various lakes and rivers in Canada, the Federal Republic of Germany, the United Kingdom, the USA, and the USSR were found to range from 1 μg/litre to around 600 μg/litre (*1, 2*). Higher levels of manganese sometimes found in freely flowing river-water are usually associated with industrial pollution. The reducing conditions that may exist in underground water and in some lakes and reservoirs are conducive to very high levels of manganese.

Concentrations in the ambient air of non-industrialized areas average 0.05 μg/m^3, whereas in industrialized areas concentrations up to 0.3 μg/m^3 have been found.

The manganese content of foodstuffs varies considerably. Low concentrations are found in dairy products (0.0–1.9 mg/kg), meats (0.0–0.8 mg/kg), and fish (0.0–0.1 mg/kg); higher concentrations occur in grains and cereals (1.2–30.8 mg/kg), nuts (0.4–35.1 mg/kg), and vegetables (0.2–12.7 mg/kg) (*1*). Extremely high manganese concentrations have been found in the leaves of tea and a cup of tea can contain from 1.4 to 3.6 μg.

8.3 Routes of exposure

Uptake of manganese occurs via inhalation and ingestion from both food and drinking-water. The greatest exposure to manganese is from foods, and adults have been estimated to consume between 2.0 and 8.8 mg/day. Infants, during the first six months of life, have been shown to consume from 2.5 to 25 μg/kg of body weight per day (*1*). The average daily intake of manganese in North America lies in the range 3.0–4.1 mg (average 3.6 mg) (3–5). The average daily manganese requirement for normal physiological function is estimated to be 3–5 mg (6).

Manganese intake through drinking-water can vary considerably, normally being substantially lower than intake from food. Available data indicate that exposure via this source would normally be less than 0.1 mg/day, but can be an order of magnitude higher (7–9).

The daily atmospheric exposure to manganese of a non-occupationally exposed population living in a non-industrialized area has been estimated to be in the range of 2–10 μg (9).

A daily intake of 3–7 mg will give a body burden of 12–20 mg in a 70-kg man (11, 14).

8.4. Health aspects

Manganese is an essential element in animals and man but only about 3% of ingested manganese is absorbed (10). It is required as a cofactor in a number of enzyme systems; it plays a role in the proper functioning of flavoproteins and in the synthesis of sulfated mucopolysaccharides, cholesterol, and haemoglobin, and in many other important metabolic processes (11). It has recently been suggested that its presence in drinking-water is inversely related to cardiovascular mortality (12). Absorbed manganese leaves the bloodstream quickly and concentrates in the liver. Manganese concentrated in the liver is conjugated with the bile salts. Manganese has a rather short biological half-life (10). The storage capacity of the liver for manganese is limited to about 1–1.3 mg/kg (wet weight). In animals, experimentally induced or naturally occurring manganese deficiency has resulted in a variety of symptoms (1). Although no specific syndrome due to manganese deficiency has been described in man, it has been suggested that there may be an association between manganese deficiency and disorders such as anaemia, bone changes in children (13), and lupus erythematosus.

The main routes of absorption of manganese are the respiratory and gastrointestinal tracts (15). Owing to the low solubility of manganese in the gastric juice, only 3–4% of that orally administered is absorbed by the gastrointestinal tract (14). Intimately linked with the absorption of manganese is the absorption of iron (16). Anaemia leads to an increased absorption of both iron and manganese, and in anaemic individuals absorption of manganese is increased more than twofold. Absorption is inversely related to the level of calcium in the diet (5) and directly related to the level of potassium (17).

Manganese in the body is regulated primarily by excretion rather than by both absorption and excretion. Some of the metal is excreted through the pancreatic secretions and some directly through the wall of the gut. Very little, 0.1–2%, is eliminated in the urine. Manganese is regarded as one of the least toxic elements. Chronic ingestion experiments in rabbits, pigs, and cattle at 1–2 mg/kg of body weight dose levels have shown no effects other than a change in appetite and reduction in metabolism of iron to form haemoglobin (11).

Except for one isolated incident, manganese intoxication from drinking-water has not been documented. In 1941, the cause of an encephalitis-like disease in Japan was attributed to contaminated well-water, which had a manganese concentration of 14 mg/litre. However, the concentrations of other metals, especially zinc, were also excessive, and whether the high concentration of manganese alone was responsible for the disease was never unequivocally established (18). In another area in Japan, a manganese concentration of 0.75 mg/litre in a drinking-water supply had no apparent adverse effect on the health of its consumers (19).

In rabbits, chronic parenteral administration of manganese produced marked degenerative changes in the seminiferous tubules, resulting in infertility (20). In contrast to these findings is the observation in mice that the administration of small amounts of manganese prevents the necrotic effect of cadmium on the testes (18). In other studies, it was noted that copper acts synergistically with manganese.

There is no evidence that manganese is carcinogenic. On the contrary, several studies suggest that manganese may have an anticarcinogenic effect. A preliminary report cited by Spivey Fox (21) indicated that a higher incidence of human cancer in Finland was geographically related to soils low in easily soluble manganese. Manganese has been reported to inhibit aminoazo dye metabolism during carcinogenesis (18).

No adverse health effects in humans were noted with daily manganese intake levels as follows (5):

	Average (mg)	Range (mg)
Food	3.000	2.0–7.0
Water	0.005	0.0–1.0
Air	0.002	0.0–0.029

8.5 Other aspects

The presence of manganese in drinking-water supplies may be objectionable for a number of reasons unrelated to health. At concentrations exceeding 0.15 mg/litre manganese imparts an undesirable taste to beverages and stains plumbing fixtures and laundry (22). When manganous compounds in solution undergo oxidation, manganese is precipitated resulting in problems of incrustation. Even at concentrations of approximately 0.02 mg/litre, manganese will form coatings on piping which may slough off as a black precipitate (23). The growth of certain nuisance organisms is also supported by manganese (22, 24); these organisms concentrate manganese and give rise to taste, odour, and turbidity problems in the distributed water.

The guideline value for drinking-water is recommended as 0.1 mg/litre, based on considerations of the staining properties of manganese.

This is a compromise value and obviously, to avoid staining problems, concentrations of this metal in solution should be kept as low as possible.

REFERENCES

1. *Manganese*, Geneva, World Health Organization, 1981 (Environmental Health Criteria, No. 17)
2. *National Water Quality Data Bank*. Ottawa, Inland Waters Directorate, Water Quality Branch. Environment, Canada, 1976.
3. KIRKPATRICK, D. C. & COFFIN, D. E. The trace metal content of representative Canadian diets in 1970 and 1971. *Canadian Institute of Food Science and Technology journal*, **7**: 56 (1974).
4. MÉRANGER, J. C. & SMITH, D. C. The heavy metal content of a typical Canadian diet. *Canadian journal of public health*, **63**: 53 (1972).
5. SCHROEDER, H. A. ET AL. Essential trace metals in man: manganese. A study in homeostasis. *Journal of chronic diseases*, **19**: 545 (1966).
6. KAY, H. O. Micro-nutrient elements—a recapitulation. *Journal of food technology*, **2**: 99 (1967).
7. US ENVIRONMENTAL PROTECTION AGENCY. *Scientific and technical assessment report on manganese*. Washington, DC, Office of Research and Development 1975 (Report No. EPA-600/6–75–002).
8. CRAUN, G. F. & McCABE, L. J. Problems associated with metals in drinking water. *Journal of the American Water Works Association*, **67**: 593 (1975).
9. *Guidelines for Canadian drinking-water quality*. Quebec, Supply and Services, 1980 (supporting documentation).
10. WHO Technical Report Series, No. 647, 1980 (*Recommended health-based limits on occupational exposure to heavy metals*: report of a WHO Study Group).
11. NATIONAL RESEARCH COUNCIL (Committee on Medical and Biological Effects of Environmental Pollutants). *Manganese*. Washington, DC, National Academy of Sciences, 1973.
12. MASIRONI, R. International studies on trace elements in the etiology of cardiovascular diseases. *Nutrition reports international*, **7**: 51 (1973).
13. PIER, S. M. The role of heavy metals in human health. *Texas reports on biology and medicine*, **33**: 85 (1975).
14. MENA, J. The role of manganese in human disease. *Annals of clinical and laboratory science*, **4**: 487 (1974).
15. RODIER, J. Manganese poisoning in Moroccan miners. *British journal of industrial medicine*, **12**: 21 (1955).
16. MENA, J. ET AL. Chronic manganese poisoning. *Neurology*, **19**: 1000 (1969).
17. UNDERWOOD, E. J. *Trace elements in human and animal nutrition*, 3rd ed. New York, Academic Press, 1971.
18. *Environmental health criteria programme for manganese and its compounds* (Japanese Report). Geneva, World Health Organization. 1974.
19. SUZUKI, Y. [Environmental contamination by manganese.] *Japanese journal of industrial health*, **12**: 529–533 (1970).
20. CHANDRA, S. & TANDON, S. K. Enhanced manganese toxicity in iron-deficient rats. *Environmental physiology and biochemistry*, **3**: 230 (1973).
21. SPIVEY FOX, M. R. In: Lee, D. H. K., ed. *Metallic contaminants and human health*. New York, Academic Press, 1972.
22. GRIFFIN, A. E. Significance and removal of manganese in water supplies. *Journal of the American Water Works Association*, **52**: 1326 (1960).
23. BEAN, E. L. Potable water—quality goals. *Journal of the American Water Works Association*, **66**: 221 (1974).
24. WOLFE, R. S. Microbial concentration of iron and manganese in water with low concentrations of these elements. *Journal of the American Water Works Association*, **52**: 1335 (1960).

9. OXYGEN, DISSOLVED

9.1 General description

The primary effect of dissolved oxygen in water is on oxidation–reduction reactions, involving iron, manganese, copper, and compounds that contain nitrogen and sulfur. In certain distribution systems, there may be a tendency for the level of dissolved oxygen to fall with residence time. Although such changes are normally indicative of corrosion processes, it is also possible that microbial respiration of organic material, especially in sediments and deposits, within pipes may be responsible. Thus dissolved oxygen may decrease without a marked increase in the concentration of iron in the water (1). Conversely, water containing high levels of iron as a result of corrosion may show little depletion of dissolved oxygen content.

9.2 Relationship to other water-quality parameters

In the corrosion of iron, proportionately little oxygen is required, thus

$$2Fe = 2Fe^{2+} + 4e$$

$$4e + 2H_2O + O_2 = 4OH^-$$

1 mg of oxygen per litre will produce 3.5 mg of ferrous iron per litre so that a large amount of iron corrosion may occur with little perceptible change in dissolved oxygen.

When the available oxygen in water has been depleted, anaerobic corrosion processes proceed involving the activity of the sulfate-reducing bacteria that may be present, thus reducing sulfate to sulfide.

$$8H^+ + 8e + SO_4^{2-} = S^{2-} + 4H_2O$$

Frequently, depletion of the level of dissolved oxygen below about 80% saturation leads to an increased incidence of consumer complaints, especially regarding taste, odour, and discoloured water (1).

Depletion of oxygen in drinking-water is often associated with other problems. Under anaerobic conditions microbial reduction of nitrate to nitrite (1) and also sulfate to sulfide occurs, often giving rise to odour problems.

Anaerobic waters may not be exceptionally corrosive but the products of corrosion are often less adhesive to the pipe walls and hence more likely to cause consumer complaints of discoloration. Under anaerobic conditions, the concentration of ferrous iron in solution may be increased throughout the distribution system. Increasing the oxygen concentration in water as, for example, following contact with air in storage tanks and water towers, will cause insoluble ferric iron to be deposited resulting in discoloration of the water distant from the source of the problem (2).

There are many disadvantages in distributing a water low in dissolved oxygen. It is recommended that water in a distribution system should always contain adequate dissolved oxygen. It is difficult to recommend a guideline value, however, as other constituents in the water influence the acceptable level.

REFERENCES

1. Ridgway, J. et al. Water quality changes–chemical and microbiological studies. In: Water distribution systems: maintenance of water quality and pipeline integrity. Medmenham, England, Water Research Centre, 1979.
2. Hall, E. S. & Smith, I. G. Rusty water cured by oxygen injection. Water services, 78: 941 (1974).

10. pH LEVEL

10.1 General description

The pH of a solution is the negative common logarithm of the hydrogen ion activity, a_{H^+}:

$$pH = -\log_{10}(a_{H^+}).$$

In a dilute solution, the hydrogen ion activity is approximately equal to the concentration of hydrogen ion.

The pH of an aqueous sample is usually measured electrometrically with a glass electrode (*1, 2*). Temperature exerts significant effects on pH measurement (*1, 2*).

10.2 Factors influencing pH range

The pH of an aqueous system is a measure of the acid-base equilibrium achieved by various dissolved compounds and, in most natural waters, is controlled by the carbon dioxide–bicarbonate–carbonate equilibrium system (*3*). This system involves various constituent equilibria, all of which are affected by temperature. In pure water, a decrease in pH of about 0.45 occurs as the temperature is raised by 25 °C (*4*). In water with a buffering capacity imparted by bicarbonate, carbonate, and hydroxide ions, this temperature effect is modified (*4*).

The pH of most raw water sources lies within the range 6.5–8.5 (*5*).

Hydrogen ion concentration may be significantly altered during water treatment. Chlorination tends to lower the pH, whereas water softening using the excess lime/soda ash process raises the pH level.

10.3 Relationship to corrosion, incrustation, and other water-quality parameters

Corrosion in water mains and water treatment plants can be a large economic burden (*6*). In addition to the corrosion problem there is a loss of distribution capacity and the concomitant increase in pumping costs that results in cases of calcium carbonate deposition (*7*).

Metals used in distribution systems, such as cast iron, steel, and copper, because of their thermodynamic instability, tend to corrode in contact with water. The deterioration of concrete, asbestos-cement, and cement-lined cast-iron pipe, all of which are commonly used in

281

distribution systems, may also occur. Natural waters contain gases, colloidal matter, and a variety of electrolyte and non-electrolyte material; these, together with pH, determine the extent of corrosion in a system (8) and define the "aggressivity" of the water. The presence of anions that form soluble compounds with a metal increase the "corrosiveness" of the water with respect to that metal, whereas anions that form insoluble compounds may increase the metal's passivity.

The role of pH in the corrosion of metals used in water distribution has been summarized by Drane (8).

Calcium carbonate deposition may control corrosion. Factors affecting this process are temperature, pH, total dissolved solids, hardness, carbon dioxide, and alkalinity. Under practical conditions it is extremely difficult and may be impossible to control the calcium carbonate/bicarbonate equilibrium. Accordingly, a number of semi-empirical and empirical relationships, using easily measured parameters, have been developed. The most widely used relationship is that developed by Langelier (4, 9).

The effectiveness of corrosion protection, by altering pH and alkalinity, depends upon a judicious balance of the carbonate/bicarbonate equilibrium system. Water that is exactly in equilibrium, i.e., just stabilized, with respect to calcium carbonate, will normally be corrosive to both iron and steel owing to failure to deposit calcium carbonate. Supersaturated water, on the other hand, will form substantial scale unless suitably treated. This scale, depending upon its porosity and its power of adhesion to the metal, may or may not inhibit corrosion (7).

Waters that are excessively hard do not usually create severe corrosion problems, but they are prone to excessive incrustation. Hard water softened by the lime/soda ash treatment will have a pH of the order of 10.9 and will possess scale-forming tendencies (10). Stabilization can be achieved by recarbonation, addition of carbon dioxide to a pH of 9.7–10, or by the addition of 0.25–0.5 mg of sodium polyphosphate per litre (10). Recarbonation to pH 8.6, to stabilize the water against subsequent excessive calcium carbonate deposition in the distribution system, has been recommended (11).

Biological slime on distribution pipe surfaces may prevent the removal of oxidation products from and the penetration of oxygen to the pipe walls, thus inhibiting corrosion. Alternatively, excessive growths generating carbon dioxide could create regions of locally low pH at the pipe surface. This could lead to localized corrosion even though the bulk water might possess favourable stability and agressivity indices (12). The growth of iron bacteria is very pH-dependent, occurring over the range 5.5–8.2, with an optimum pH of about 6.5 (13). Red-water complaints are often the result of sudden growth of iron bacteria, which produce ferric hydroxide as the metabolic end product. Under favourable conditions iron bacteria may develop so rapidly that severe blocking of water pipes can occur in a matter of weeks.

pH is related, in several different ways, to almost every other water quality parameter as aqueous chemical equilibria invariably involve hydrogen (and hydroxyl) ions.

The formation of gaseous hydrogen sulfide, yielding "bad-egg" odours in waters prone to sulfur contamination, is thermodynamically favoured at pH values less than about 7.0 (*14*). In the chlorination process, the objectionable pungent odour (*15*) of nitrogen trichloride tends to be formed in greater concentrations at pH values less than 7.0 (*16*). It is claimed that at high pH levels drinking-water acquires a bitter taste (*17*).

Colour intensity in a given water sample is increased by raising the pH (*18*). This effect, known as the "indicator effect", has led to the suggestion that all colour measurements for water quality control be carried out at the standard pH of 8.3 (*19*).

The efficiency of the coagulation and flocculation process is markedly dependent on pH, and it is standard practice in water treatment to adjust pH so that optimum floc formation is achieved (*20, 21*). Filtration efficiency in certain instances is also sensitive to pH (*22*).

Most microorganisms usually tolerate the pH range commonly found in water sources (*13, 23, 24*). The microbiological integrity of water is dependent upon the pH level, which influences the effectiveness of chlorine disinfection. The germicidal efficiency of chlorine in water is lower at high pH values; this has been attributed to the reduction in hypochlorous acid concentration with increasing pH (*25–27*). Within the pH range encountered in drinking-water, the effectiveness of both ozone and chlorine dioxide as alternative disinfectants is unchanged (*16*).

Corrosion in the water supply system is a major source of metal contamination in drinking-water (*28*). Two of the potentially most troublesome metals are lead and cadmium. At all pH levels above 6 in pure water, lead is immune to corrosion. In the presence of carbonates and bicarbonates, lead is passive between about pH 4 and pH 12, but is subject to corrosion above pH 12 (*29*). The supply of drinking-water with a low alkalinity and a fairly low pH to households that had lead plumbing resulted in high levels of lead in the drinking-water (*30*). In pure water, cadmium appears passive between about pH 9 and pH 13.5, whereas, according to experimental data, corrosion is only significant below pH 6 (*29*).

10.4 Health aspects

A direct relationship between human health and the pH of drinking-water is impossible to ascertain because pH is so closely associated with other aspects of water quality.

In an epidemiological study carried out on drinking-water supplies in which pH was one of the parameters considered, Taylor and co-workers (*31*) were unable to establish any significant correlation between the incidence of viral hepatitis A and finished water pH; however, it must be appreciated that this study was primarily directed towards the effects of pH on disinfection efficiency.

In so far as pH affects the various processes in water treatment that contribute to the removal of viruses, bacteria, and other harmful organisms, it could be claimed that pH has an indirect effect on health.

The recommended guideline value for pH is 6.5–8.5, although it is recognized that some problems could arise within a distribution system with pH levels below 7.

REFERENCES

1. pH value. In: *Standard methods for the examination of water and waste water*, 14th ed. Washington, DC, APHA, AWWA, WPCF, 1976, p. 460.
2. Standard method of test for pH of water and waste water. In: *Annual book of ASTM standards, Part 31*. Philadelphia, American Society for Testing and Materials, 1976, p. 178.
3. GOLDMAN, J. C. ET AL. *Water research*, 6: 637 (1972).
4. LANGELIER, W. F. Effect of temperature on the pH of natural waters. *Journal of the American Water Works Association*, 38: 179 (1946).
5. WEBBER, W. J., JR. & STUMM, W. Mechanism of hydrogen ion buffering in natural waters. *Journal of the American Water Works Association*, 55: 1553 (1963).
6. HUDSON, H. E., JR. & GILCREAS, F. W. Health and economic aspects of water hardness and corrosiveness. *Journal of the American Water Works Association*, 68: 201 (1976).
7. MCCLANAHAN, M. A. & MANCY, K. H. Effect of pH on quality of calcium carbonate film deposited from moderately hard and hard water. *Journal of the American Water Works Association*, 66: 49 (1974).
8. DRANE, C. W. Natural waters. In: Shreir, L. L., ed. *Corrosion*, 2nd ed. London, Newnes-Butterworths, Chapter 2.
9. LANGELIER, W. F. Chemical equilibria in water treatment. *Journal of the American Water Works Association*, 38: 169 (1946).
10. DYE, J. F. & TUEPKER, J. L. Chemistry of the lime-soda process. In: *Water quality and treatment*, 3rd ed. Toronto, McGraw-Hill, 1971, p. 313.
11. SAWYER, C. N. & MCCARTY, P. L. Residual chlorine and chlorine demand. In: *Chemistry for sanitary engineers*, 2nd ed. Toronto, McGraw-Hill, 1967, p. 363.
12. O'CONNOR, J. T. ET AL. Deterioration of water quality in distribution systems. *Journal of the American Water Works Association*, 67: 113 (1975).
13. SHAIR, S. Iron bacteria and red water. *Industrial water engineering*, March-April: 16 (1975).
14. POURBAIX, M. *Atlas of electrochemical equilibria in aqueous solutions*, 2nd ed. Houston, National Association of Corrosion Engineers, 1974, p. 545.
15. AMERICAN WATER WORKS ASSOCIATION RESEARCH FOUNDATION. *Handbook of taste and odor control experiences in the US and Canada*. Denver, CO, AWWA 1976.
16. MORRIS, J. C. Chlorination and disinfection—state of the art. *Journal of the American Water Works Association*, 63: 769 (1971).
17. *Statement of basis and purpose for the national secondary drinking water regulations*. Washington, DC, US Environmental Protection Agency, 1977.
18. BLACK, A. P. & CHRISTMAN, R. F. Characteristics of coloured surface waters. *Journal of the American Water Works Association*, 55: 753 (1963).
19. SINGLEY, J. E. ET AL. Correction of color measurements to standard conditions. *Journal of the American Water Works Association*, 58: 455 (1966).
20. SAWYER, C. N. & MCCARTY, P. L. Chemical coagulation of water. In: *Chemistry for sanitary engineers*, 2nd ed. Toronto, McGraw-Hill, 1967, p. 341.
21. MAUDLING, J. S. & HARRIS, R. H. Effect of ionic environment and temperature on the coagulation of color-causing organic compounds with ferric sulfate. *Journal of the American Water Works Association*, 60: 460 (1968).
22. Committee Report. Coagulation-filtration practice as related to research. *Journal of the American Water Works Association*, 66: 502 (1974).

23. DAVIS, B. D. ET AL. *Microbiology*, 2nd ed. New York, Harper and Row, 1973, pp. 92–93.
24. RUDOLFS, W. ET AL. Literature review on the occurrence and survival of enteric, pathogenic, and relative organisms in soil, water, sewage, and sludges, and on vegetation. *Sewage and industrial wastes*, **22**: 1261 (1950).
25. BUTTERFIELD, C. T. ET AL. Influence of pH and temperature on the survival of coliforms and enteric pathogens when exposed to free chlorine. *Public health reports*, **58**: 1837 (1943).
26. SMITH, W. W. & BODKIN, R. E. The influence of hydrogen ion concentration on the bactericidal action of ozone and chlorine. *Journal of bacteriology*, **47**: (A17) 445 (1944).
27. SCARPINO, P. V. ET AL. A comparative study of the inactivation of viruses in water by chlorine. *Water research*, **6**: 959 (1972).
28. CRAUN, G. E. & MCCABE, L. J. Problems associated with metals in drinking water. *Journal of the American Water Works Association*, **67**: 593 (1975).
29. POURBAIX, M. *Atlas of electrochemical equilibria in aqueous solutions*, 2nd ed. Houston, National Association of Corrosion Engineers, 1974, pp. 488–491.
30. MCFARREN, E. F. ET AL. *Water quality deterioration in the distribution system.* Kansas City, MO, Water Quality Technology Conference, 1977.
31. TAYLOR, F. B. ET AL. The case for water-borne infectious hepatitis. *American journal of public health*, **56**: 2093 (1966).

11. SODIUM

11.1 General description

The increased pollution of surface and groundwater during the past decades has resulted in a substantial increase in the sodium content of drinking-water in different regions of the world. Waterworks treatment processes and the practice of domestic water softening may also contribute to an increase in sodium levels in drinking-water.

Sodium sulfate is employed in the manufacture of pigments and colours, the pulp and paper industries, and many other present-day manufacturing activities whose effluents contain enhanced sodium concentrations (1).

Snow and ice control on roads accounts for the largest single use of sodium chloride in many countries and the quantity used for this purpose is steadily increasing. Sodium chloride is also used in the production of caustic soda, chlorine, and many industrial chemicals. Significant quantities are used in the food processing, slaughtering and meat packing, dairy, fishing, grain, and brewing industries (2).

11.2 Occurrence

Sodium is the most abundant of the alkali elements, compounds of sodium being widely distributed in nature, and constitutes 26 g/kg of the earth's crust.

Soils contain sodium within the range 1–10 g/kg, it being mainly present as silicate minerals, such as amphiboles and feldspars.

Some groundwaters contain high concentrations of sodium and, under certain circumstances, this can lead to increased salinity in rivers and streams.

The control of ice and snow, by salting of highways, also results in an increased sodium burden in soils. Runoff may be sufficiently high to lead to contamination of public water supplies. It is estimated that between 25% and 50% of the salt used on a road infiltrates the groundwater (3).

Sewage, industrial effluents, seawater intrusion in coastal areas, and the use of sodium compounds for corrosion control and water-softening processes all contribute to sodium concentration in water because of the high solubility of sodium salts and minerals. Sodium concentrations vary considerably, depending on regional and local hydrological and geological conditions, the time of year, and salt utilization patterns. Sodium levels in groundwater vary widely but normally range between 6 and 130 mg/litre (*4*). Higher levels may be associated with saline soils. In surface-water the sodium concentration may be less than 1 mg/litre or exceed 300 mg/litre, depending upon the geographical area (*5–7*).

Reported sodium concentrations in public water supplies range from less than 1 mg/litre to over 1000 mg/litre (*7*). Water softening can dramatically increase the sodium concentration in a supply (*8*).

Sodium occurs naturally in all foods, varying considerably for different types, and food processing can have a marked effect on these levels. Fresh peas contain about 9 mg of sodium per kg whereas the level is 2.3 g/kg in drained canned peas and 1 g/kg in frozen peas (*9*). Fresh fruits and vegetables contain from less than 0.01 g/kg to about 1 g/kg, whereas cereals and cheeses may contain between 10 and 20 g/kg (*10*).

11.3 Routes of exposure

Daily sodium intake by individuals varies considerably owing to variation in the sodium content of foods and personal variation in the use of salt as a food-seasoning agent. It was found that in a sample of 3833 people, 45% of the males and 30% of the females routinely added salt to their food (*11, 12*). The average daily intake of sodium for a Canadian male, aged between 20 and 64 years, has been estimated to be 3600 mg (*10*). Drinking-water contributes only a very small percentage of the daily intake of salt compared with the intake from food.

11.4 Health aspects[a]

Sodium is the most abundant extracellular cation and, together with its associated anions, contributes significantly to the osmotic activity of the extracellular fluid.

Water and electrolyte balances are maintained by dietary intake in food and water and loss in urine, faeces, perspiration, and expired air. A normal 70-kg man contains approximately 69 g of metabolically active sodium and 45 litres of water.

The control of water and sodium balance is achieved through a complex series of interrelated processes, involving both nervous and hormonal systems. The balance is maintained by a sodium loss rather than by control of absorption through the gut. The most important factor controlling loss is the mineralocorticoid hormone, aldosterone.

Because the body has very effective methods of controlling sodium levels, sodium is not an acutely toxic metal. The deaths of 6 out of 14 infants were reported after they had mistakenly been given sodium in a concentration of 21 140 mg/litre in their feed (*13*). Toxic symptoms of sodium poisoning include general involvement of the central nervous system with increase in sensitivity. There are some studies that show a

---

[a] For further discussion of the health aspects of the sodium content of drinking-water, see Part III, section 16, p. 145.

positive correlation between sodium intake and hypertension in man (*14*) and others that do not (*15–17*).

A daily sodium intake in the range 1600–9600 mg is generally considered to have no adverse effect on the health of normal individuals (*18*). To ensure a total dietary intake level of 500 mg/day would require limiting the concentration of sodium in drinking-water to about 20 mg/litre which would incur considerable extra water treatment expense using the technologies at present available (*19*).

To ensure that drinking-water is tasteless to the majority of the consumers, the salt composition in the water should approximate to the salt content of saliva. The average sodium content of saliva is 300 mg/litre but may well exceed this value by a factor of two.

The taste threshold of sodium in water depends upon the associated anion and the temperature of the solution. Sodium carbonate has the lowest taste threshold and the bicarbonate salt the highest. At room temperature the various threshold values for sodium were found to be about 20 mg/litre for Na_2CO_3, 150 mg/litre for NaCl, 190 mg/litre for $NaNO_3$, 220 mg/litre for Na_2SO_4, and 420 mg/litre for $NaHCO_3$.

The recommended guideline value is 200 mg/litre, which is based on the above taste thresholds and not on health considerations.

REFERENCES

1. KILLIN, A. F. Sodium sulphate. In: *Canadian minerals yearbook*. Ottawa, Department of Energy, Mines and Resources, 1974.
2. KILLIN, A. F. Salt. In: *Canadian minerals yearbook*. Ottawa, Department of Energy, Mines and Resources, 1974.
3. McCONNELL, H. H. & LEWIS, J. Add salt to taste. *Environment*, **14**: 38 (1972).
4. BOND, R. G. & STRAUB, C. P. Genetic types of subterranean waters in relation to their salinity. In: *Handbook of environmental control. Vol. 3. Water supply and treatment*, 1st ed. Cleveland, OH, Chemical Rubber Co., 1973, p. 85.
5. WEILER, R. R. & CHAWLA, V. K. *Dissolved mineral quality of Great Lakes waters*. Proceedings of the 12th Conference on Great Lakes Research, Ann Arbor, MI, 1969, p. 801.
6. DOBSON, H. H. *Principal ions and dissolved oxygen in Lake Ontario*. Proceedings of the 10th Conference on Great Lakes Research, 1967, p. 337.
7. *Sodium, chlorides and conductivity in drinking water*. Report on a WHO working group. Copenhagen, WHO Regional Office for Europe, 1979 (EURO Reports and Studies, No. 2).
8. ELLIOTT, G. B. & ALEXANDER, E. A. Sodium from drinking water as an unsuspected cause of cardiac decompensation. *Circulation*, **23**: 562 (1961).
9. MENEELY, G. R. A review of sources and the toxic effects of excess sodium chloride and the protective effect of extra potassium in the diet. *Qualitas plantarum. Plant foods for human nutrition*, **23**: 3 (1973).
10. GORMICAN, A. Inorganic elements in foods used in hospital menus. *Journal of the American Dietetic Association*, **56**: 397 (1970).
11. *Statement of basis and purpose for the national interim primary drinking water regulation*, Washington, DC, US Environmental Protection Agency, 1975.
12. NATIONAL HEART AND LUNG INSTITUTE. *The public and high blood pressure*. Washington, DC, US Department of Health, Education and Welfare, 1973 (Publication No. (NIH) 74-356).

13. FINBERG, L. ET AL. Mass accidental poisoning in infancy. *Journal of the American Medical Association*, **184**: 187 (1963).
14. DAHL, L. K. Salt and hypertension. *American journal of clinical nutrition*, **25**: 231 (1972).
15. KERKENDALL, W. M. The effects of dietary sodium on the blood pressure of normotensive man. In: Genest, J. & Koiw, E., ed. *Hypertension*, Heidelberg, Springer-Verlag, 1972, p. 360.
16. EVANS, J. G. & ROSE, G. Hypertension. *British medical bulletin*, **27**: 37 (1971).
17. DAUBER, T. R. ET AL. In: Stamber, J. et al., *Environmental factors in hypertension*. New York, Grune and Stratton Inc., 1967.
18. DAHL, L. K. Possible role of salt intake in the development of essential hypertension. In: Cottier, P. & Bock, K. D., ed. *Essential hypertension: an international symposium*. Heidelberg, Springer-Verlag, 1960, p. 53.
19. NATIONAL ACADEMY OF SCIENCES AND NATIONAL ACADEMY OF ENGINEERING. *Water quality criteria, 1972*. Washington, DC, US Government Printing Office, 1974.

12. SULFATE

12.1 General description

The majority of sulfates are soluble in water, the exceptions being the sulfates of lead, barium, and strontium (*1*). Dissolved sulfate is considered to be a permanent solute of water. It may, however, be reduced to sulfide, volatilized to the air as H_2S, precipitated as an insoluble salt, or incorporated in living organisms (*2*).

Sulfates are discharged into the aquatic environment in the wastes from many different industries (*1, 3*). Atmospheric sulfur dioxide (SO_2), formed by the combustion of fossil fuels and emitted by the metallurgical roasting processes, may also contribute to the sulfate content of surface-water. Sulfur trioxide (SO_3), produced by the photolytic or catalytic oxidation of sulfur dioxide, combines with water vapour to form sulfuric acid, which is precipitated as "acid rain" or snow (*3*).

12.2 Occurrence

The concentration of sulfate in most freshwaters is very low, although levels of 20–50 mg/litre are common in the eastern USA, Canada, and most of Europe (*4, 5*).

The average sulfate concentration in the public water supplies in 23 of the larger cities in European communities is reported to be 64 mg/litre (range, 9–125 mg/litre) (*6*). According to data collected over a 5-year period (1969–1973), at some 600 water sources supplying approximately 60% of the total population of England, Scotland and Wales, United Kingdom water supplies contain sulfate levels ranging from 4 to 303 mg/litre.[a] Aluminium sulfate, which is extensively used as a flocculant for water treatment, may add 20–50 mg of sulfate per litre to the final water. Sulfate is not removed from water by conventional water treatment methods. Sulfate concentrations of bottled mineral water marketed in European communities average 223 mg/litre (range, 0–1182 mg/litre) (*6*).

12.3 Routes of exposure

Data concerning the daily dietary intake of sulfates are scarce. Sulfate compounds used in the USA as additives in foods are estimated to

---

[a] Figures quoted by P. Powell (Water quality and health division, Water Research Centre, Medmenham Laboratory, Medmenham, Marlow, Buckinghamshire, England).

contribute an average of 453 mg to the daily sulfate intake of Americans (7, 8).

Daily intake of sulfate from drinking-water, particularly if bottled mineral water is used, is extremely variable.

12.4 Health aspects

Sulfate is poorly absorbed from the human intestine (8); it slowly penetrates the cellular membranes of mammals and is rapidly eliminated through the kidneys (9).

The reported minimum lethal dose of magnesium sulfate in mammals is 200 mg/kg of body weight (10). Sulfate doses of 1.0–2.0 g have a cathartic effect on humans, resulting in the purgation of the alimentary canal (1). Infants ingesting sulfate equivalent to 21 mg/kg of body weight per day may also suffer from this effect. Magnesium sulfate at concentrations above 1000 mg/litre acts as a purgative in normal humans, but concentrations below this are apparently physiologically harmless (1, 10). Sensitive people are responsive to magnesium sulfate levels as low as 400 mg/litre and new users or those imbibing occasionally may be affected by concentrations in excess of 700 mg/litre. The human system adapts in the course of time to higher concentrations of sulfate in drinking-water (11).

Taste threshold concentrations for the most prevalent sulfate salts are: 200–500 mg/litre for sodium sulfate; 250–900 mg/litre for calcium sulfate; and 400–600 mg/litre for magnesium sulfate (1, 11).

Essentially on the basis of the above values, which are also allied to the cathartic effect of sulfate, a guideline value of 400 mg/litre is proposed.

12.5 Other aspects

High sulfate concentrations in water may contribute to the corrosion of metals in the distribution system, particularly in waters having low alkalinity.

REFERENCES

1. McKEE, J. E. & WOLF, H. W. *Water quality criteria*, 2nd ed. Sacramento, CA, California State Water Quality Control Board, 1963, pp. 136, 213, 247, 270, 275–277.
2. NATIONAL RESEARCH COUNCIL, *Drinking water and health, part 1.* Washington, DC, National Academy of Sciences, 1977.
3. DELISLE, C. E. & SCHMIDT, J. W. The effects of sulphur on water and aquatic life in Canada. In: *Sulphur and its inorganic derivatives in the Canadian environment.* Ottawa, National Research Council of Canada, 1977, pp. 227–284.
4. HITCHCOCK, D. R. Biogenic contributions to atmospheric sulfate levels. *Proceedings of the Second National Conference on Complete Water Re-use.* Chicago, American Institute of Chemical Engineers, 1975. pp. 291–310.
5. KATZ, M. The Canadian sulphur problem. In: *Sulphur and its inorganic derivatives in the Canadian environment.* Ottawa, National Research Council of Canada, 1977, pp. 21–67.

6. AMAVIS, R., ET AL., ed. *Hardness of drinking water and public health.* Oxford, Pergamon Press, 1976, pp. 176–199.
7. Subcommittee on Research of GRAS (Generally Recognized as Safe) List (Phase II). *Food ingredients,* Washington, DC, National Academy of Sciences, 1972 (DHEW No. FDA 70–22).
8. NOVIKOV, YU. V. & ERISMAN, F. F. The 'potable water' standard (GOST 2874–73). A new stage in development of water hygiene. *Vestnik Akademii Meditsinskikh Nauk SSR,* No. 3, 59 (1975). English translation No. 3, 76 (1975).
9. SENNING, A. *Sulfur in organic and inorganic chemistry.* Vol. 2. New York, Marcel Dekker Inc., 1972, p. 160.
10. ARTHUR D. LITTLE, INC. *Inorganic chemical pollution of freshwater.* Washington, DC, US Environmental Protection Agency, 1971 (Water Pollution Control Research Series No. DPV 18010).
11. ZOETEMAN, B. C. J. *Sensory assessment of water quality.* Oxford, Pergamon Press, 1980.

13. TASTE AND ODOUR

13.1   Taste

13.1.1 General description

In the strict meaning of the word, the taste of water is the sensation that results from the interaction between the saliva and substances dissolved in the water, as perceived by receptors located in the taste buds. There are about 3000–10 000 taste buds in the mouth, most of which are located on the upper surface of the tongue, at its tip, sides, and rear surfaces.

When "tasting" water, the senses of both gustation and olfaction are activated and it is extremely difficult to differentiate between them. Consequently, the combined effect of taste and odour is frequently classified as "taste".

The taste perception is much less sensitive than that of smell (1, 2). However, water apparently free from odour may produce an offensive "taste" when taken in the mouth. In such cases, the higher oral temperature is conducive to the release into the nasal cavity of dissolved organic substances from the water. In this concentrated form, the sense of smell perceives the presence of the solutes and thus the "tasting" of water, as opposed to just smelling, is often a more sensitive sensory assessment of quality (3).

In the assessment of drinking-water quality, the sensations of taste and odour are complementary. In general, the sense of taste is most useful in detecting inorganic constituents of drinking-water, while the sense of smell is more useful in detecting organic constituents.

Taste tests in general have received considerable criticism (4–6). In taste threshold tests, the use of distilled water as the standard for "tasteless" water and as a mouth rinse introduces a bias into the results. Water containing salts in concentrations higher or lower than those in saliva will be perceived as different from the saliva by the sense of taste. Therefore, minimum concentrations of ions, such as sodium, chloride, calcium, and bicarbonate, are essential to make a water tasteless. Taste tests that involve large laboratory panels of judges can be time-consuming and difficult for small treatment facilities to perform (4), and the use of consumer panels has been advocated as an alternative (3). Taste tests performed in water treatment plants may underestimate the taste of the water as delivered to the consumer because, at the plant, objectionable tastes can be masked by residual chlorine. The masking effect diminishes as the chlorine residual decreases in the distribution system (*4*). Chemical dechlorination of the water prior to assessment may augment the taste (*4*). Panels of consumers assessing water taste in their homes can overcome these problems (*3*). Additional difficulty arises from differences in test procedures, such as quality of dilution water, number of and motivation of panel members, and variations in the statistical treatment and interpretation of taste-test data.

Suitable methods for determining the taste intensity of drinking-water are similar to those described for odour.

13.1.2 Occurrence

Taste and odour problems in drinking-water supplies account for the largest single class of consumer complaints. They may occur in any type of water and at any time of year. Some are due to natural causes, others to man's industrial activity. They may be associated primarily with the raw water, the treatment method, the distribution system, or with a combination of all three.

Taste and odour survey results in water treatment plants in Canada, the Netherlands, and the USA have been published (*3, 7*). Groundwater sources for supply normally have the fewest taste problems. The majority of the surface-water supplies are subject to seasonal variation in taste and odour problems, which suggests that such problems may be of biological origin.

High concentrations of colour and turbidity in water are often associated with nonspecific taste (and odour) problems (*8*).

Taste acuity is reported to depend on temperature (*9, 10*), and the degree to which taste is influenced by temperature is a function of the specific taste-causing substance (*10*). The growth rate of microorganisms, some of which may produce bad-tasting metabolites, is enhanced by higher temperatures, as also is the rate of formation of offensive-tasting corrosion products.

Where pH controls the equilibrium concentration of the neutral and ionized forms of a substance in solution it can notably influence its taste (and odour).

Several studies of the organoleptic properties of residual chlorine have been performed over the years, but further work will be required before the key questions can be answered. Under ideal conditions, the amount of free available chlorine at the consumer's tap should be sufficiently high to ensure the microbial safety of the water and sufficiently low to avoid objectionable taste and odour problems. Taste and odour thresholds of residual chlorine are thus of considerable interest and the most recent investigation into this subject found that the average taste threshold concentration of free residual chlorine increased from 0.075 mg/litre to 0.450 mg/litre as the pH increased from 5.0 to 9.0 (*11*).

The average threshold was 0.156 mg/litre, with a range of 0.02–0.29 mg/litre at pH 7.0.

Although much more work is clearly called for, it is probable that most treatment plant managers are aware of the appropriate balance between the applied chlorine residual and consumer complaints. In mineralized water (12) and in coffee (13), taste thresholds for chlorine indicate that other constituents causing taste in water can influence the magnitude of the chlorine threshold concentration. Thus, the nature of the raw-water supply will be one major factor in the detectable taste threshold concentration for residual chlorine. In certain countries, it is known that consumers are assured of the safety of their water supply by the presence of a slight taste of chlorine.

Many of the inorganic substances occurring in water exert an unpleasant taste at concentrations much lower than those required for acute toxic effects. Limits for such substances are, therefore, set at concentrations that reflect levels found to be objectionable to consumers. These substances are discussed briefly below.

Taste thresholds in distilled water for the major cations of drinking-water, i.e., calcium, magnesium, sodium, and potassium, have been reported to be approximately 100, 30, 100 and 300 mg/litre, respectively (3, 14). The uncertainty associated with these evaluations is largely due to the influence on taste of their associated anions.

Taste threshold tests for iron, as Fe(II), have shown that the most sensitive 5% of the members of a test panel can detect concentrations of 0.04 mg/litre in distilled water but in a mineralized spring water having a total dissolved solids content of 500 mg/litre the threshold value was 0.12 mg/litre (15). Zinc could be detected at a concentration of 4.3 mg/litre in distilled water, but only at 6.8 mg/litre in mineralized spring water (15).

Reliable data on the taste and odour thresholds for sulfide in water are sparse, and the situation is complicated somewhat by the influence of pH on the sulfide-bisulfide-hydrogen sulfide equilibrium.

13.1.3 Health aspects

The presence of objectionable tastes in a public water supply may cause consumers to seek alternative sources of potable water, which may or may not be subject to the same degree of microbial protection afforded by the rejected supply. This has been exemplified by a survey, conducted by the California State Department of Public Health, which found that dissatisfied consumers were large-scale purchasers of bottled water. In the Netherlands, it was shown that water possessing an offensive taste resulted in a lower consumption of tap-water (3). The taste of water unfortunately provides no assurance that such water is free of pathogens or toxic inorganic chemicals. Fortunately, median taste thresholds of inorganic substances are generally much lower than the concentrations that cause adverse health effects.

In a public water supply, short-term changes in the normal taste may signal changes in the quality of the raw-water source, deficiencies in the treatment process, or chemical corrosion and biological growths in the distribution system.

The objective is, therefore, to provide water that is free of objectionable taste for the majority (90%) of consumers. The most direct way to verify this objective is to seek periodically the views of a selected consumer group. Laboratory panels can be applied to assess water taste using a category scale (e.g., good—not observable—weak—objectionable—bad) or by assessing the taste number by the forced choice method (see section 13.2.1., p. 297). In the latter case, it is recommended that the taste number of the drinking-water be kept below 1. Although consumer panels are most suited for assessing the taste of tap-water, laboratory panels are valuable for assessing the taste of the water during treatment.

Special local circumstances may occur resulting in an unavoidable perceptible taste in the water. In such cases, local health authorities should give priority to disinfection to ensure the control of disease-causing contaminants, such as pathogenic bacteria.

REFERENCES

1. ROSEN, A. A. & BOOTH, R. L. Taste and odour control. In: *Water quality and treatment*, 3rd ed. Toronto, McGraw-Hill, 1971, p. 225.
2. SUFFETT, I. H. & SEGALL, S. Detecting taste and odour in drinking water. *Journal of the American Water Works Association*, **63**: 605 (1971).
3. ZOETEMAN, B. C. J. *Sensory assessment of water quality*. Oxford, Pergamon Press, 1980.
4. BAKER, R. A. Dechlorination and sensory control. *Journal of the American Water Works Association*, **56**: 1578 (1964).
5. BRUVOLD, W. H. Human perception and evaluation of water quality. *CRC critical reviews in environmental control*, **5**: 153 (1975).
6. SWETS, J. A. Is there a sensory threshold? *Science*, **134**: 168 (1961).
7. *Handbook of taste and odour control experiences in the US and Canada*. Denver, CO, American Water Works Association, 1976.
8. RIDDICK, T. M. Zeta potential and polymers. *Journal of the American Water Works Association*, **58**: 719 (1966).
9. *Standard methods for the examination of water and waste water*, 14th ed. Washington, DC, American Public Health Association, 1976, p. 121.
10. PANGBORN, R. M. & BERTOLERO, L. L. Influence of temperature on taste intensity and degree of linking of drinking water. *Journal of the American Water Works Association*, **64**: 511 (1972).
11. BRYAN, P. E. ET AL. Taste thresholds of halogens in water. *Journal of the American Water Works Association*, **65**: 363 (1973).
12. PANGBORN, R. M. ET AL. Sensory examination of mineralized, chlorinated waters. *Journal of the American Water Works Association*, **62**: 572 (1970).
13. CAMPBELL, C. L. ET AL. Effects of certain chemicals in water on the flavour of brewed coffee. *Food research*, **23**: 575 (1958).
14. NATIONAL ACADEMY OF SCIENCES. *Water quality criteria 1972*. Washington, DC, US Government Printing Office, 1973 (EPA-R-73-033).
15. COHEN, J. M. ET AL. Taste threshold concentrations of metals in drinking water. *Journal of the American Water Works Association*, **52**: 660 (1960).

13.2. Odour

13.2.1 General description

The odour of drinking-water may be defined as the sensation that is due to the presence of substances having an appreciable vapour pressure and that stimulate the human sensory organs in the nasal and sinus cavities. The sense of smell will generally respond to much lower concentrations (a few micrograms per litre or less) of a substance than will the sense of taste (a few milligrams per litre or more).

The odour intensity of water is usually measured in terms of its threshold odour number (TON), which is defined as the geometric mean of the dilution ratios with odour-free water, the odour of which is just detectable by a panel of judges under very carefully controlled test conditions (*1*). As in the case of taste threshold measurements, laboratory panel quality ratings or mean threshold values are only estimates of these values for the entire consuming population (*1*). Water with a TON of 2 may, depending upon its nature, stimulate more consumer complaints than other water having a TON of 4 (*2*).

An alternative to the TON determination for odour intensity is the odour number determination by the forced choice method, which has certain advantages (*3*). Each member of a panel of judges is presented with a series of paired samples, one of which is a dilution of the sample being tested, the other being an odour-free control. Each panel member has to judge, for each pair, which of the two has the stronger odour. A choice must be made even when no difference is perceived. The percentage of right responses is calculated for each dilution and corrected for the 50% probability that the right flask has been indicated by chance. The odour number is taken as the dilution for which the corrected right response rate is 50%, calculated from the graph of dilution against corrected response rate. Further details of this method are given by Zoeteman (*3*). Good results have also been obtained by using the less time-consuming "interval scaling method", which is most suitable for drinking-water with very low odour intensity.

Odour intensity measurements are usually nonspecific. However, intensity measurements for specific substances in water are normally reported in terms of their odour threshold concentrations (*4*). This is the concentration of the substance in water whose odour can be detected by 50% of the panel members. The wide variation in individual odour detection has been markedly illustrated by the use of odour threshold concentrations. In a large population, the most sensitive 5% is able to detect odour reliably at one-hundredth of the average odour threshold concentration (*5*). To obtain reliable and responsible data, therefore, odour assessment of water quality should be carried out by large panels. Smaller panels may be employed, if it is more convenient to do so, but it must be appreciated that the accuracy and reliability of the determination will thereby be reduced.

It is important to specify the temperature at which odour intensity measurements are made, since odour intensity is related to the vapour pressure of any odour-causing substance and, hence, will be directly related to the water temperature.

13.2.2 Occurrence

Water odour is predominantly due to the presence of organic substances in water. Very many odour-producing compounds have been reported in water (6, 7).

Objectionable odours in drinking-water may be of either biological or industrial origin, and some odours of natural origin may be due indirectly to human activities; for example, the dumping of raw sewage into the aquatic environment enhances biological growth, which may produce odorous products.

Natural odours tend to be described as earthy, musty, or sour, on the one hand, or as fishy, grassy, or cucumber-like, on the other, involving compounds such as geosmin and decanal (8–11). Industrially derived odours often smell like such substances as petroleum or creosote or have a medicinal odour. Typical examples in this category are naphthalene and the chlorinated benzenes and phenols (5). Groundwater tends to have fewer odour problems, although odours are not restricted to any single type of water nor to any particular season of the year. Odours may also be produced under stagnant water conditions in low-flow sections of distribution systems or in raw- and finished-water reservoirs. Water-purification processes may convert substances with weak odours (such as amines and phenols) into substances possessing very intense odours (such as chloramines and chlorophenols) (12). The proliferation of nuisance organisms, such as iron and sulfur bacteria, in distribution systems may also be a source of odour.

The nonspecific fishy, grassy, and musty odours normally associated with biological growth tend to occur most frequently in warm surface-water in the warmer months of the year (10, 11).

Surveys of taste and odour problems in Canada and the USA identified 50 so-called "nuisance" organisms responsible for odours in drinking-water. The very intense musty odours of substances produced by the actinomycetes group of organisms can be a major source of odour contamination in public water supplies. It has been recommended that raw-water supplies be monitored for actinomycetes (1, 10). Although such monitoring in some localities has positively correlated the presence of actinomycetes with odour problems, other treatment plants have found the converse to be true.

13.2.3 Health aspects

Odour in potable water is almost invariably indicative of some form of pollution of the water source or of malfunction during water treatment or distribution. Odours of biological origin are indicative of increased biological activity, which may include an increased loading of dangerous pathogens on the system. Odours of industrial origin are associated with pollution of the source-water with commercial waste products, some of which may be toxic. Sanitary surveys should include investigations for potential or existing sources of odour, and attempts should always be made to identify the source of an existing odour problem.

Some chemical contaminants, of concern because of their toxic properties, may also cause odour problems. The threshold odour for hydrogen cyanide in water, for example, has been reported to be 0.001 mg/litre (6). A value for cyanide in drinking-water based on this information would be one-hundredth the value recommended elsewhere in these guidelines (see Vol. 1, p. 55). In this and other examples, the sense of smell is more sensitive than the best available analytical instrumentation. Except possibly for chlordane, the odours of pesticides in water are too weak to permit their detection at or below their recommended guideline values (3).

Ideally, drinking-water should have no observable odour to any consumer. However, owing to the large differences in individual odour sensitivity within a population, a more realistic objective is to provide a water free of objectionable odour for the large majority of the population (e.g., 90 %). The most direct way of achieving this objective is through the cooperation of a large consumer panel (e.g., 100 consumers) in a supply area who are asked to make periodic assessments of water odour and taste in their homes. Participants should indicate their observations on a category scale (e.g., good—not perceptible—mildly objectionable—bad). Laboratory panels tend to be more critical in assessing water odour and taste. However, laboratory panels of 10–20 trained persons can also indicate if a water is aesthetically acceptable to the majority of the consumers (3). If the odour number is measured at room temperature by the forced choice method using a selected laboratory panel, it is recommended that the objective value be less than 1, unless local circumstances demand a disinfection practice requiring perceptible free chlorine residuals.

REFERENCES

1. *Standard methods for the examination of water and wastewater*, 14th ed. Washington, DC, APHA, AWWA, WPCF, 1976, p. 75.
2. BAKER, R. A. Dechlorination and sensory control. *Journal of the American Water Works Association*, **56**: 1578 (1964).
3. ZOETEMAN, B. C. J. *Sensory assessments of water quality*. Oxford, Pergamon Press, 1980.
4. BAKER, R. A. Threshold odors of organic chemicals. *Journal of the American Water Works Association*, **55**: 913 (1963).
5. ZOETEMAN, B. C. J. & PIET, G. J. Cause and identification of taste and odour compounds in water. *Science of the total environment*, **3**: 103 (1974).

6. VAN GEMERT, L. J. & NETTENBREIJER, A. H., ed. *Compilation of odour threshold values in air and water.* Voorburg, National Institute for Water Supply; Zeist, Netherlands, Central Institute for Nutrition and Food Research, TNO, 1977.
7. STAHL, W. H., ed. *Compilation of odor and taste threshold values data.* Philadelphia, American Society for Testing and Materials, 1973 (ASTM Data Series Publication No. DS 48).
8. *Handbook of taste and odour control experiences in the US and Canada.* Denver, CO, American Water Works Association, 1976, p. XIV-1.
9. ZOETEMAN, B. C. J. & PIET, G. J. On the nature of odours in drinking water resources of the Netherlands. *Science of the total environment,* **1**: 399 (1972/73).
10. MORRIS, R. L. ET EL. Chemical aspects of Actinomycetes metabolites as contributors of taste and odour. *Journal of the American Water Works Association,* **55**: 1380 (1963).
11. MCKEE, J. E. & WOLF, H. W., ed. *Water quality criteria.* 2nd ed. Sacramento, CA, California State Water Quality Control Board, 1963 (Publication No. 3-A).
12. BURTTSCHELL, R. H. ET AL. Chlorine derivatives of phenol causing taste and odour. *Journal of the American Water Works Association,* **51**: 205 (1959).

14. TEMPERATURE

14.1 General description

In general, the rates of chemical reactions decrease with decreasing temperature. The relative concentrations of reactants and products in chemical equilibria can also change with temperature. Temperature can, therefore, affect every aspect of the treatment and delivery of potable water.

14.2 Physical aspects

Cool drinking-water is preferable to warm. The intensity of taste is greatest for water at room temperature and is significantly reduced by chilling or heating. Increasing the temperature will also increase the vapour pressure of trace volatile compounds in drinking-water and may lead to increased odour.

Turbidity and colour are indirectly related to temperature as the efficiency of coagulation is strongly temperature-dependent. The optimum pH for coagulation decreases as temperature increases (1). In order to achieve the most economical use of a coagulant, therefore, jar tests should be carried out at the temperature of the treated water and not at room temperature (2).

As temperature decreases, the viscosity of water increases, and the rates of sedimentation and filtration decrease. The efficiency of colour and turbidity removal by coagulation, sedimentation, and filtration may be less under winter temperature conditions than in summer. The decreased efficiency of turbidity removal by filtration at lower temperatures is possibly due to a reduction in floc-strength or average particle size (3). Filtration through activated carbon is also affected by temperature; the adsorptivity of activated carbon increases as the temperature drops (4).

14.3 Microbiological aspects

The microbiological characteristics of drinking-water are related to temperature through its effect on water-treatment processes, especially disinfection, and its effect on both growth and survival of microorganisms.

In general, disinfection is aided by increased temperature. Working with *Escherichia coli*, Butterfield and co-workers observed a 5-fold

301

increase in the bactericidal effectiveness of chlorine between 20 and 25 °C compared with that between 2 and 5 °C (5). In a US Army study, Ames & Smith found a 9-fold increase in effectiveness between 8 °C and 40 °C (6). Chambers reported that the influence of temperature on the effectiveness of chlorine disinfection was insignificant at pH values between 7.0 and 8.5, but at higher pH values over the temperature range 4 °C to 22 °C, a 4- to 8-fold increase in effectiveness was observed (7). Similar results have been obtained with viruses (8). It has been reported that the inactivation of *Mycobacterium fortuitum* by ozone increases with temperature (9).

Coagulation and sedimentation of water reduce the number of suspended microorganisms and, as discussed earlier, temperature affects these processes.

At a given pH, higher temperature leads to greater dissociation of hypochlorous acid. The magnitude of this effect on the germicidal efficiency of chlorinated water is, however, of secondary importance to the larger, and opposite, effect of increased germicidal action at higher temperatures.

Published information is somewhat equivocal about the effect of temperature on bacterial survival in water (10). Seasonal variations in coliform counts in raw-water sources have been observed (11). However, temperature would be only one of a number of factors leading to this variation.

At low temperatures, viruses can survive considerably longer than bacteria and survival times of up to 6 months have been reported for poliovirus in tap-water at low temperature (12). However, an epidemiological study of viral hepatitis A in 13 cities in the USA showed no correlation between infection rate and raw water temperature (13).

The survival time in water of the cysts and ova of parasitic worms is shortened by higher temperatures. For example, *Schistosoma* ova die in 9 days at 29–32 °C, in 3 weeks at 15–24 °C, and in 3 months at 7 °C (14).

The growth of nuisance organisms is enhanced by warm water conditions and could lead to the development of unpleasant tastes and odours.

14.4 Chemical aspects

The rate of formation of trihalomethanes in chlorinated drinking-water increases with temperature (15) and is perhaps the single most important factor influencing seasonal variation in trihalomethane concentrations (16).

The effect of temperature on corrosion in water treatment systems demonstrated that corrosion increased as a function of temperature (17). Sodium hydroxide adjustment of the pH halved this increase over the same temperature range. At temperatures below 10 °C, however, water containing sodium hydroxide showed a higher corrosion rate than the untreated water. The corrosion rate is also a function of the dissolved oxygen concentration in the water. Dissolved oxygen variation with temperature is small compared with the much larger (and opposite) change in corrosion rates cited above. The dissolved oxygen content, however, plays an unimportant role in the temperature dependence of corrosion.

The solubility product of calcium carbonate decreases with temperature. At low alkalinities, however (50 mg/litre as calcium carbonate), the decrease in pH with increased temperature actually increases the solubility of calcium carbonate. This effect on the saturation index tends to reduce incrustation by carbonate and at the same time increases the agressivity of the water, leading to increased corrosion in hot-water systems (3).

REFERENCES

1. MAUDLING, J. S. & HARRIS, R. H. Effect of ionic environment and temperature on the coagulation of color-causing organic compounds with ferric sulfate. *Journal of the American Water Works Association*, **60**: 460 (1968).
2. CAMP, T. R. ET AL. Effects of temperature on rate of floc formation *Journal of the American Water Works Association*, **32**: 1913 (1940).
3. AMERICAN WATER WORKS ASSOCIATION. *Water quality and treatment*, 3rd ed. Toronto, McGraw-Hill, 1971, pp. 89, 305.
4. WEBER, W. J. & MORRIS, J. C. Equilibria and capacities for adsorption on carbon. *Journal of the Sanitary Engineering Division, Proceedings of the American Society of Civil Engineers*, **90**: (5A3) 79 (1964).
5. BUTTERFIELD, C. T. ET AL. Influence of pH and temperature on the survival of coliforms and enteric pathogens when exposed to free chlorine. *Public health reports* **58**: 1837 (1943).
6. AMES, M. & WHITNEY-SMITH, W. *Journal of bacteriology*, **47**: 445 (1944).
7. CHAMBERS, C. W. An overview of the problems of disinfection. *Symposium on wastewater treatment in cold climates*. Saskatoon, Canada, University of Saskatchewan, 1974, p. 423 (EPS 3-WP-74-3).
8. WHITE, G. C. Disinfection: The last line of defense for potable water. *Journal of the American Water Works Association*, **67**: 410 (1975).
9. FAROOQ, S. ET AL. Influence of temperature and UV light on disinfection with ozone. *Water research* **11**: 737 (1977).
10. RUDOLFS, W. ET AL. Literature review on the occurrence and survival of enteric, pathogenic, and relative organisms in soil, water, sewage, and sludges, and on vegetation. *Sewage and industrial wates*, **22**: 1261 (1950).
11. RAO. S. S. & HENDERSON, J. *Summary report of microbiological baseline data on Lake Superior 1973*. Ottawa, Environment Canada, Inland Waters Directorate, 1974, (Scientific Series No. 45).
12. HEALTH AND WELFARE CANADA. *Microbiological quality of drinking water*. Ottawa, Health and Welfare Canada, 1977, (77-EHD-2).
13. TAYLOR, F. B. ET AL. The case for water-borne infectious hepatitis. *American Journal of public health*, **56**: 2093 (1966).
14. Temperature. In: *Water quality criteria*, 2nd ed. Sacramento, CA, California State Water Quality Control Board, 1963, p. 283.
15. STEVENS. A. A. ET AL. Chlorination of organics in drinking water. *Journal of the American Water Works Association*, **68**: 615 (1976).
16. SMILLIE, R. D. ET AL. *Organics in Ontario drinking water*, Part II. Toronto, Ontario Ministry of the Environment, 1977.
17. MULLEN. E. D. & RITTER, J. A. Potable water corrosion control. *Journal of the American Water Works Association*, **66**: 473 (1974).

15. TOTAL DISSOLVED SOLIDS

15.1 General description

The total dissolved solids (TDS) in water comprise inorganic salts and small amounts of organic matter. The principal ions contributing to TDS are carbonate, bicarbonate, chloride, sulfate, nitrate, sodium, potassium, calcium, and magnesium (*1*). Total dissolved solids influence other qualities of drinking-water, such as taste, hardness, corrosion properties, and tendency to incrustation.

15.2 Occurrence

Total dissolved solids in water may originate from natural sources, sewage effluent discharges, urban runoff, or industrial waste discharges.

Waters in contact with granite, siliceous sand, well-leached soil, or other relatively insoluble material have TDS levels of less than 30 mg/litre (*2*). Waters in precambrian shield areas generally have TDS levels less than 65 mg/litre (*3*). Waters in areas of palaeozoic and mesozoic sedimentary rock have higher TDS levels, ranging from as little as 195 to 1100 mg/litre (*3*); carbonates, chlorides, calcium, magnesium, and sulfates are the principal ions present (*2, 4*). In addition to these natural leaching processes, sewage and industrial wastes may lead to further increases.

Under arid conditions, the TDS of the smaller streams may increase to levels of 15 g/litre (*4*). Elsewhere, levels in excess of 35 g of TDS per litre have been recorded in briny waters (*3*).

The use of salt for snow and ice control on roads during winter weather contaminates both surface and groundwater sources, increasing the TDS of waters noticeably in some countries. In the winter of 1969–70, Denmark applied 203 000 tonnes of salt to its road system; 2.5 million tonnes were used in Canada in 1974, and in the USA, salt consumption for road de-icing in 1970 was calculated as 9 million tonnes—all contributing to the TDS levels of streams, waterways, and groundwater supplies (*5*).

15.3 Health aspects

There is no evidence of deleterious physiological reactions occurring in persons consuming drinking-water supplies that have TDS levels in excess of 1000 mg/litre (*2, 4, 6*). The results of certain epidemiological studies would appear to suggest that TDS in drinking-water may even have beneficial health effects.

The common dissolved mineral salts are claimed to affect the taste of water (*7–11*). The effects that many of these minerals have on taste have been discussed in the separate reviews of these constituents and in the section dealing with taste (see p. 293). Bruvold et al. (*8*) have rated the palatability of drinking-water according to the TDS level thus:

Excellent:	less than 300 mg/litre
Good:	between 300 and 600 mg/litre
Fair:	between 600 and 900 mg/litre
Poor:	between 900 and 1200 mg/litre
Unacceptable:	greater than 1200 mg/litre

Water with extremely low TDS levels may also be unacceptable because of its flat, insipid taste.

15.4 Other aspects

Certain components of TDS, such as chlorides, sulfates, magnesium, calcium, and carbonates, affect corrosion or incrustation in water distribution systems (*2*).

Total dissolved solids are not generally removed in conventional water-treatment plants.

Although no deleterious physiological effect has been recorded with total dissolved solids in water above 1000 mg/litre, it was considered that it would, as a rule, be unacceptable to exceed this level, which is recommended as a guideline value.

REFERENCES

1. *Quality criteria for water*. Washington, DC, US Environmental Protection Agency, 1976 (EPA-440/9-76-023).
2. RAINWATER, F. H. & THATCHER, L. L. *Methods for collection and analysis of water samples*. Geological Survey Water-Supply Paper, Washington, DC, US Government Printing Office, 1960.
3. GARRISON INVESTIGATIVE BOARD. *Water quality report* (Appendix A). Garrison Diversion Study, Report to the International Joint Commission: US-Canada, Windsor, Ontario, 1977.
4. DURFOR, C. J. & BECKER, E. Constituents and properties of water. In: Pettyjohn, W. A., ed. *Water quality in a stressed environment*. Minnesota, Burgess Publishing Company, 1972.
5. *Sodium, chlorides and conductivity in drinking-water supplies*. Copenhagen, WHO Regional Office for Europe, 1979 (EURO Reports and Studies, No. 2).
6. ONGERTH, H. J. ET AL. The taste of water. *Public health reports*, **79**: 351 (1964).
7. BRUVOLD, W. H. & PANGBORN, R. M. Rated acceptability of mineral taste in water. *Journal of applied psychology*, **50**: 22 (1966).

8. BRUVOLD, W. H. ET AL. Consumer attitudes toward mineral taste in domestic water. *Journal of the American Water Works Association*, **59**: 547 (1967).
9. BRUVOLD, W. H. Scales for rating the taste of water. *Journal of applied psychology*, **52**: 245 (1968).
10. BRUVOLD, W. H. Mineral taste and the potability of domestic water. *Water research*, **4**: 331 (1970).
11. BRUVOLD, W. H. & ONGERTH, H. J. Taste quality of mineralized water. *Journal of the American Water Works Association*, **61**: 170 (1969).

16. TURBIDITY

16.1 General description

Turbidity in water is caused by the presence of suspended matter, such as clay, silt, colloidal organic particles, plankton, and other microscopic organisms. Turbidity is an expression of certain light-scattering and light-absorbing properties of the water sample. It is a parameter whose significance is to a large extent dependent on the measurement technique. The total intensity and angular distribution of light scattered from turbid water represent the overall effects of intraparticle and interparticle interactions. They depend, in a complex manner, on such factors as the number, size, shape, and refractive index of the foreign particles, and on the wavelength of the incident light. Complex though the factors are, a number of generalizations can be made (1, 2).

Five methods may be used in the measurement of water turbidity, but only two of these, nephelometry and turbidimetry, form the basis of present standard methods (3–6).

Historically, turbidity measurements in the waste- and drinking-water fields have been based on the Jackson candle turbidimeter (7). Jackson candle turbidity is an empirical quantity based on the measurement in a special graduated vessel of that depth of sample that is just sufficient to extinguish the image of a burning standard candle observed vertically through the sample. The Jackson turbidity unit (JTU) is defined in terms of that depth; a depth of 21.5 cm corresponds to 100 JTU (5, 6). The Jackson candle turbidimeter is applicable only to turbidities greater than 25 JTU and as such has limited applicability to the monitoring of drinking-water. Improved instruments, such as the Patterson turbidimeter (3), using electrical light sources and mirror optics, can measure lower values. As an alternative to JTU, turbidimeters can be calibrated in terms of the concentration of suspended solids (mg/litre) that gives rise to a certain turbidity (a gravimetric definition). Diatomaceous earths are commonly used to form the standard suspensions. This type of definition (sometimes called the Fuller's earth scale) is arbitrary and is specific for the type and particle size of the particular clay used (8).

The nephelometric method is the current method of choice for turbidity measurement (6, 9–11). Nephelometric turbidimeters measure the intensity of light scattered at 90° to the path of the incident light. Differences in the physical design of such turbidimeters will cause differences in measured turbidity values.

In an attempt to minimize such differences, the light source equipment and detector geometry and the calibration method are specifically defined (6). Suspensions of formazin polymer as turbidity reference standards have been almost universally adopted by the water industry (7). A suspension of formazin, formed by the interaction under specified conditions of hydrazine sulfate (50 mg/litre) with hexamethylamine-tetramine (500 mg/litre), has a defined turbidity of 40 nephelometric turbidity units (NTU) (6, 10). These are also known as formazin turbidity units (FTU). When measured on a candle turbidimeter this standard suspension has a turbidity of about 40 JTU (6).

As defined by the above methods, turbidity represents a nonspecific measurement of suspended solids concentration. Electronic particle counters have recently become available and these are capable of accurately counting and recording the number of suspended particles as a function of size. Generally there is a relationship between turbidity (in the range 0.2–1 NTU) and particle counts; good point-for-point agreement between the two methods does not exist (11).

16.2 Occurrence

The particles that cause turbidity in water range in size from colloidal dimensions (approximately 10 nm) to diameters of the order of 0.1 mm. They may be divided into three general classes: clays; organic particles resulting from decomposition of plant and animal debris; and fibrous particles, e.g., asbestos minerals (12). Clay particles generally have an upper particle-size limit of about 0.002 mm diameter.

Soil particles derived from the land surface by erosion constitute the major part of suspended material in most natural waters. The coarser sand and silt fractions are wholly or partially coated with organic material. Phyllosilicate clay particles as well as non-clay material, such as iron and aluminium oxides and hydroxides, quartz, amorphous silica, carbonates, and feldspar, constitute the clay fraction (12). Clays and organic particles are also often found together as a "clay-organic" complex (12). Humic substances have a much higher ionic exchange capacity than inorganic clays (13), and in many instances the effect of humic components predominates.

Organic turbidity resulting from the accumulation of higher micro-organisms may occur in such large amounts that waters become unsightly and turbid. Examples of turbidity due to microorganisms are the summer blooms of blue-green algae in surface-water, algal debris, and the detritus from iron bacteria in distribution systems (red-water is such a manifestation) (14).

Raw-water turbidity can vary from less than 1 NTU to greater than 1000 NTU. Removal of turbidity may be achieved by simple filtration, or more effectively, by a combination of coagulation, sedimentation, and filtration.

Filtration through sand beds or other single medium filters can consistently produce a water with a turbidity of 1 NTU or less. Continuous monitoring of turbidity throughout the treatment stages is a valuable aid in attaining such a performance.

16.3 Relationship with other water-quality parameters

The turbidity of water is related to or affects many other indicators of drinking-water quality. The particulate matter may also be a source of nutrients and protection for some microorganisms.

There is evidence that a large part of the colour in water arises from colloidal particles, 50% of such colour being due to a "colloidal fraction" of humic substances (15). True colour is, therefore, defined as the colour of water from which the turbidity has been removed (16).

The relationship between high turbidity, in both raw and filtered water, and taste and odour has also long been recognized, and suspended particulate matter in a potable water supply renders the water unattractive to the consumer (17).

The presence of turbidity can have a significant effect on the microbiological quality of drinking-water. The detection of bacteria and viruses in drinking-water may be complicated by the presence of turbidity. In water, microbial growth is most extensive on the surfaces of particles and inside loose, naturally occurring, floc and floc formed during coagulation treatment (see Part I, "Microbiological aspects"). This growth is facilitated because nutrients are adsorbed on to surfaces and attached bacteria are thus able to grow more efficiently compared with those in free suspension (18, 19). Similarly it has been demonstrated that river silt readily adsorbs viruses (20). In the water-treatment process of coagulation, bacteria and viruses become trapped in the floc formed and are removed along with turbidity (21, 22). Breakthrough of filter beds by floc is also accompanied by an increase in virus penetration, even though the turbidity of the finished water remains below 0.5 JTU (23).

Particulate matter, whether organic, inorganic or due to higher microorganisms, can protect bacteria and viruses from the action of disinfectants. Sanderson & Kelly reported the presence of coliform organisms in water with turbidities ranging between 3.8 and 84 NTU, even after treatment with chlorine producing free chlorine residuals between 0.1 and 0.5 mg/litre and a minimum contact time of 30 minutes.[a] Neefe and co-workers showed that chlorination of drinking-water deliberately contaminated with faecal matter was, by itself, insufficient protection against viral hepatitis A (24). Only by coagulation and filtration prior to chlorination could the water be rendered safe to drink. In laboratory tests, the presence in water of various clays and humic acid was shown to protect "*Klebsiella aerogenes*" (invalid) from disinfection by ultraviolet light (25).

---

[a] Comments following a paper by Clark, N. A. et al., reference 21.

Consumption of highly turbid, chlorinated water may be a dangerous health risk (26–29).

The adsorptive capacity of some suspended particulates can lead to the entrapment of undesirable inorganic and organic compounds present in the water and in this way, turbidity can bear an indirect relationship to the health aspects of water quality. Most important in this respect is the organic or humic component of turbidity (*30–33*).

The strength of some metal-humate complexes in the turbidity fraction may complicate the analytical measurement of trace metals in natural waters resulting in an underestimation of the metal (*34*).

Organic molecules are also adsorbed by natural organic matter. Herbicides such as 2,4–D, Paraquat, and Diquat can be adsorbed on to clay-humic acid particulates, the adsorption being greatly influenced by metal cations present in the humic material (*35*). The presence of turbidity, therefore, may also interfere with the detection of biocides in water samples.

In so far as turbidity is used as a measure of the efficiency of the removal of particulate matter throughout the purification process of water, low turbidity in the finished product is an indication of the effectiveness of the coagulation, sedimentation, and filtration processes.

16.4 Health aspects

A turbidity in excess of the guideline value of 5 NTU is generally objectionable to consumers (*36*). The perception of higher turbidity in water at the consumer's tap than in that entering the distribution system may indicate post-treatment contamination, corrosion, or other distribution problems. Consequently, as excessive turbidity can protect microorganisms from the effects of disinfection, stimulate the growth of bacteria in the water, and itself exert a significant chlorine demand, it is vitally important in producing safe drinking-water, using chlorine as disinfectant, that turbidity should be kept low, preferably below 1 NTU.

REFERENCES

1. BLACK, A. P. & HANNAH, S. A. Measurement of low turbidities. *Journal of the American Water Works Association*, **57**: 901 (1965).
2. McCLUNEY, W. R. Radiometry of water turbidity measurements. *Journal of the Water Pollution Control Federation*, **47**: 252 (1975).
3. EDEN, G. E. The measurement of turbidity in water. *Procedings of the Society of Water Treatment and Examination*, **14**: 27 (1965).
4. AWWA TASK GROUP. Progress toward a filtrability index test. *Journal of the American Water Works Association*, **51**: 1539 (1959).
5. Standard methods of test for turbidity of water, D1889–71 (1977). In: *1980 Annual book of ASTM standards, Part 31*. Philadelphia, American Society for Testing and Materials, 1980, p. 260.

6. AMERICAN PUBLIC HEALTH ASSOCIATION. *Standard methods for the examination of water and wastewater*, 14th ed. Washington, DC, American Water Works Association, 1976, p. 131.

7. HACH, C. C. Understanding turbidity measurement. *Industrial water engineering*, **9**: 18 (1972).

8. PACKHAM, R. F. The preparation of turbidity standard. *Proccedings of the Society of Water Treatment and Examination*, **11**: 64 (1962).

9. *National interim primary drinking water regulations*. Washington, DC, US Environmental Protection Agency, 1976, p. 12 (EPA-570/9-76-003).

10. Turbidity. In: *Methods for chemical analysis of water and wastes*. Washington, DC, US Environmental Protection Agency. 1976, p. 295 (EPA-625-6-74-003a).

11. Beard, J. D. & Tanaka, T. S. A comparison of particle counting and nephelometry. *Journal of the American Water Works Association*, **59**: 533 (1977).

12. NATIONAL RESEARCH COUNCIL. *Drinking water and health*. Washington, DC, National Academy of Sciences, 1977, Chapter IV.

13. NARKIS, N. & REBHUN, M. The mechanism of flocculation processes in the presence of humic substances. *Journal of the American Water Works Association*, **67**: 101 (1975).

14. MACKENTHUN, K. M. & KEUP, L. E. Biological problems encountered in water supplies. *Journal of the American Water Works Association*, **62**: 520 (1970).

15. PEMMANEN, V. Humus fractions and their distribution in some lakes in Finland. In: Povoledo, D. & Golterman, H. L., ed., *Humic substances, their structure and function in the biosphere*. Wageningen, The Netherlands, Pudoc, 1975, p. 207.

16. AMERICAN PUBLIC HEALTH ASSOCIATION. *Standard methods for the examination of water and wastewater*, 14th ed. Washington, DC, American Water Works Association, 1976, p. 64.

17. ATKINS, P. F. & TOMLINSON, H. D. Evaluation of daily carbon chloroform extracts with CAM. *Water sewage works*, **110**: 281 (1963).

18. BROCK, T. D. *Principles of microbial ecology*. New Jersey, Prentice-Hall Inc., 1966, pp. 72–74.

19. STOTZKY, G. Influence of clay minerals on microorganisms. III. Effect of particle size, cation exchange capacity, and surface area on bacteria. *Canadian journal of microbiology*, **12**: 1235 (1966).

20. BERG, G. Removal of viruses from sewage, effluents, and waters. 2. Present and future trends. *Bulletin of the World Health Organization*, **49**: 461 (1973).

21. CLARKE, N. A. ET AL. Human enteric viruses in water: source, survival and removability. In: *Proceedings of the International Conference in Water Pollution Research, London, 1962. Advances in water pollution research*, **2**: 523 (1964).

22. FOLIGUET, J. M. & DONCOEUR, F. Elimination des enterovirus au cours du traitement des eaux d'alimentation par coagulation-floculation-filtration. *Water research*, **9**: 953 (1975).

23. ROBECK, G. G. ET AL. Effectiveness of water treatment processes in virus removal. *Journal of the American Water Works Association*, **54**: 1275 (1962).

24. NEEFE, J. R. ET AL. Inactivation of the virus of infectious hepatitis in drinking water. *American journal of public health*, **37**: 365 (1947).

25. BITTON, G. ET AL. Effect of several clay minerals and humic acid on the survival of *Klebsiella aerogenes* exposed to ultraviolet irradiation. *Applied microbiology*, **23**: 870 (1972).

26. DENNIS, J. M. 1955–56 Infectious hepatitis epidemic in Delhi, India. *Journal of the American Water Works Association*, **51**: 1288 (1959).

27. SYMONS, J. M. & HOFF, J. C. Rationale for turbidity maximum contaminant level. Presented at *3rd Water Quality Technology Conference, Atlanta*. Washington, DC, American Water Works Association, 1975.

28. HUDSON, H. E. High-quality water production and viral disease. *Journal of the American Water Works Association*, **54**: 1265 (1962).

29. TAYLOR, F. B. ET AL. The case for water-borne infectious hepatitis. *American journal of public health*, **56**: 2093 (1966).

30. SCHNITZER, M. & KAHN, S. U. *Humic substances in the environment.* New York, Marcel Dekker Inc., 1972, pp. 204–251.
31. CHAU, Y. K. & LUM-SHUE-CHAN, K. Measurement of complexing capacity of lake waters. In: Povoledo, D. & Golterman, H. L., ed., *Humic substances, their structure and function in the biosphere.* Wageningen, The Netherlands, Pudoc, 1975, p. 11.
32. OLIVER, B. G. Heavy metal levels of Ottawa and Rideau River sediments. *Environmental science and technology*, 7: 135 (1973).
33. RAMAMOORTHY, S. & RUST, R. R. Mercury sorption and desorption characteristics of some Ottawa River sediments. *Canadian journal of earth sciences*, 13: 530 (1976).
34. GARDINER, J. The chemistry of cadmium in natural water-1. A study of cadmium complex formation using the cadmium specific-ion electrode. *Water research*, 8: 23 (1974).
35. KAHN, S. U. Adsorption of 2,4-D from aqueous solution by Fulvic acid-clay complex. *Environmental science and technology*, 4: 236 (1974).
36. *Public health service drinking water standards*, Rockville, MD, US Department of Health, Education and Welfare, 1962, p. 21.

17. ZINC

17.1 General description

Zinc is an abundant element and constitutes approximately 0.04 g/kg of the earth's crust (*1*). The most common zinc mineral is sphalerite (ZnS), which is often associated with the sulfides of other metallic elements, for example, lead, copper, cadmium, and iron (*2*). The natural zinc content of soils is estimated to be between 1 and 300 mg/kg (*3*).

The atmospheric concentrations of zinc vary considerably depending on such factors as proximity to point sources. In rural locations, zinc concentrations are typically between 10 and 100 ng/m³, while levels in urban areas most commonly fall within the range of 100 to 500 ng/m³ (*4*).

The carbonates, oxides, and sulfides of zinc are sparingly soluble in water, while the highly soluble chloride and sulfate salts tend to hydrolyse to form zinc hydroxide and zinc carbonate. As a result the concentration of zinc in natural water is generally low. Adsorption on to sediments further depletes the levels of dissolved zinc (*5*).

17.2 Occurrence

The concentration of zinc in tap-water can be considerably higher than that in surface-water owing to the leaching of zinc from galvanized pipes, brass, and zinc-containing fittings. Zinc concentrations in tap-water generally vary between 0.01 and 1 mg/litre (*6*).

17.3 Routes of exposure

Zinc is important nutritionally and in the USA comprehensive tables of the zinc content of foods are available (*7, 8*). Meats and dairy products are richest in zinc, while cereals and nuts are also important sources (*8*). The zinc content of some important food groups is (*9*):

beef, pork, and lamb	20–60 mg/kg
milk	3–5 mg/kg
fish and seafood	in excess of 15 mg/kg
legumes and wheat	15–50 mg/kg
leafy vegetables and fruits	less than 2 mg/kg (fresh weight)

Food is by far the largest source of zinc for humans. The average daily intake for a "normal" man is reported to be 12 mg (*10*). The

313

average daily intake of zinc from drinking-water probably does not exceed 400 μg. Air is a negligible source of zinc for man.

17.4 Health aspects

Zinc is an essential element for both animals and man and is necessary for the functioning of various enzyme systems, including alkaline phosphatase, carbonic anhydrase, and alcohol dehydrogenase (11). More than 70 zinc metallo-enzymes are known (12).

The recommended dietary intake of zinc, depending upon age and sex, is between 4 and 15 mg/day. Pregnant women and mothers of newborn babies require up to 16 mg/day (13).

In Egypt and the Islamic Republic of Iran, an endemic zinc deficiency syndrome (among young men) has been reported (14, 15). This syndrome, characterized by retarded growth and other signs of immaturity, including anaemia, is probably caused by low intestinal absorption of zinc. A complete cure is effected by oral administration of large daily doses of zinc as the sulfate (16).

In man and animals, zinc absorption is affected by many factors, for example, intake of protein, vitamins, and metals (17). Low zinc intake and low body weight increase zinc absorption, whereas high oral doses of zinc, calcium, and phytate reduce uptake; the fraction of ingested zinc that is absorbed is difficult to determine because zinc also appears to be excreted into the gut (18, 19). However, zinc does not accumulate in tissues and the proportion absorbed is thought to be inversely related to the amount ingested (22).

In human serum and plasma, zinc levels are about 1 mg/litre, whereas in whole blood the concentration is about five times higher owing to a high concentration (10 mg/litre) in the red blood cells (20). The highest body concentration of zinc occurs in the prostate (100 mg/kg wet weight), but high levels also occur in bone, muscle, liver, and pancreas (21).

Zinc may be considered nontoxic. The low toxicity of zinc and efficient homoeostatic control mechanisms make chronic zinc toxicity from drinking-water and dietary sources an unlikely hazard in man. Symptoms of zinc toxicity in humans include vomiting, dehydration, electrolyte imbalance, abdominal pain, nausea, lethargy, dizziness, and lack of muscular coordination (17). Acute renal failure caused by zinc chloride has been reported (23).

Daily doses of 150 mg of zinc interfere with copper and iron metabolism because zinc is a metabolic antagonist of both these metals. However, where dietary intake of copper and iron are adequate there is little problem even with high zinc doses. Zinc is also a metabolic antagonist of cadmium. High zinc intakes may, therefore, be expected to afford some protection against the toxic effects of cadmium exposure from the environment (24).

Taste threshold tests indicate that 5% of a population distinguished between zinc-free water and that containing zinc at a level of 4.3 mg/litre (as zinc sulfate) (*25*). The detection levels with other zinc salts were somewhat higher.

17.5 Other aspects

Zinc imparts to water an undesirable astringent taste; in addition, water containing zinc at concentrations in excess of 5.0 mg/litre may appear opalescent and develop a greasy film on boiling. This value is recommended as a guideline value. To avoid any of the problems mentioned, however, the level of zinc in water should be kept well below this value.

REFERENCES

1. BROWING, E. *Toxicity of industrial metals.* 2nd ed. London, Butterworths, 1969, p. 348.
2. *Quality criteria for water.* Washington, DC, US Environmental Protection Agency, 1976, p. 481.
3. LEVINSON, A. A. *Introduction to exploration geochemistry.* Calgary, Applied Publishing Co., 1974, p. 44.
4. NRIAGU, J. O., ED. *Zinc in the environment. Part I: ecological cycling.* New York, John Wiley & Sons, 1980.
5. HEM, J. D. Zinc. In: *Study and interpretation of the chemical characteristics of natural water.* Washington, DC, US Geological Survey, 1970, p. 125 (Water-Supply Paper 1473).
6. ZOETEMAN, B. C. J. & BRINKMAN F. J. J. Human intake of minerals from drinking-water in the European communities. In *Hardness of drinking water and public health.* Proceedings of the European Scientific Colloquium, Luxembourg, 1975. Oxford, Pergamon Press, 1976, p. 173.
7. MURPHY, E. W. ET AL. Provisional tables on the zinc content of foods. *Journal of the American Dietetic Association,* **66**: 345 (1975).
8. FREELAND, J. H. & COUSINS, R. J. Zinc content of selected foods. *Journal of the American Dietetic Association,* **68**: 526 (1976).
9. WHO Technical Report Series, No. 532, 1973 (*Trace elements in human nutrition*: report of a WHO Expert Committee).
10. WARREN, H. V. Some trace element concentrations in various environments. In: Howe, G. M. & Loraine, J. A., ed. *Environmental medicine.* London, William Heinemann Medical Books Ltd., 1973, p. 9.
11. PARISIC, A. F. & VALLEE, B. L. Zinc metalloenzymes: characteristics and significance in biology and medicine. *American journal of clinical nutrition,* **22**: 1222 (1969).
12. Symposium on trace elements. *Medical clinics of North America,* **60**: 4 (1976).
13. FOOD AND NUTRITION BOARD. *Recommended dietary allowances.* 8th revised version. Washington, DC, National Academy of Sciences, 1974, pp. 99–101.
14. PRASAD, A. S. ET AL. Syndrome of iron, anemia, hepatosplenomegaly, hypogonadism, dwarfism and geophagia. *American journal of medicine,* **31**: 532 (1961).
15. HALSTED, J. A. ET AL. Zinc deficiency in man. The Shiraz experiment. *American journal of medicine,* **53**: 277 (1972).
16. MICHAELSON, G. Zinc therapy in acrodermatitis enteropathica. *Acta dermatologica,* **54**: 377 (1974).
17. PRASAD, A. S. & OBERLEAS, D., ed. *Trace elements in human health and disease. Vol. 1: zinc and copper.* New York, Academic Press, 1976, p. 470.
18. BECKER, W. M. & HOEKSTRA, W. G. The intestinal absorption of zinc. In: Skorylla, S. C. & Valdron-Edwards, D., ed. *Intestinal absorption of metal ions, trace elements and radionuclides.* New York, Pergamon Press, p. 229.

19. HONSTEAD, J. F. & BRADY, D. N. The uptake and retention of 32P and 65Zn from the consumption of Columbia River fish. *Health physics*, **13**: 455 (1967).
20. SUNDERMAN, F. W. In: Goyer R. A. & Mehlman, M. A., ed. *Advances in modern toxicology*. Washington, DC, Hemisphere Publishing Corporation, 1976.
21. HALSTED, J. A. ET AL. *Journal of nutrition*, **104**: 345 (1974).
22. HETH, D. A. ET AL. Effect of calcium, phosphorus and zinc on zinc-65 absorption and turnover in rats fed semipurified diets. *Journal of nutrition*, **88**: 331 (1966).
23. CSATA, S. ET AL. Akute Niereninsuffizienz als folge einer. Zinkchloridvergiftung. *Zeitschrift für Urologie*, **61**: 327 (1968).
24. UNDERWOOD, E. J. *Trace elements in human and animal nutrition*, 4th ed. New York, Academic Press, 1977, p. 545.
25. COHEN, J. M. ET AL. Taste threshold concentrations of metals in drinking water. *Journal of the American Water Works Association*, **52**: 660 (1960).

PART VI. RADIOACTIVE MATERIALS

1. INTRODUCTION

The levels of radioactivity in drinking-water recommended in the standards published by WHO in 1970 and 1971 (*1, 2*) were based on data available from the International Commission on Radiological Protection (ICRP) over the period 1959–1966 inclusive. However, since then additional information has become available (*3–5*) and has been taken into consideration in the preparation of the present guidelines.

The recommended guideline values of 0.1 Bq/litre for gross alpha activity and 1 Bq/litre for gross beta activity are based upon an adult drinking-water intake of 2 litres per day. It is recommended that levels of activity exceeding these values be reported to the competent authorities to determine what action, if any, is required. Procedures for measuring gross alpha and beta activities as well as individual radionuclides are described elsewhere (*6*).

Radioactive materials are introduced into the environment from a number of sources—naturally occurring and man-made. The naturally occurring sources include those substances produced by cosmic rays, which may find their way to water courses with rainfall and runoff, and those present in the rocks and soil, such as uranium-238 and its daughters radium-226 and radon-222. The man-made radionuclides are those resulting from fallout from nuclear tests, nuclear power production, and medical and other uses of radioactive materials.

The dose of natural radiation that a person receives depends upon a number of factors, such as the height above sea level at which he lives, the amount and type of radioactive nuclides in the soil, and the amount he takes into his body in air, food, and water. The growth in nuclear applications has led to more widespread environmental releases, which in turn add to the amount of radioactive substances in surface- and groundwater and could have a direct effect on radioactivity levels in water sources used for public water supply.

2. BASIC CONSIDERATIONS

In assessing radiation exposure, the recommendations of IAEA (7) and ICRP (4, 5) were followed. ICRP has recently introduced the concept of "detriment" in order to identify and, if possible, quantify all deleterious effects. "Detriment" is defined as the mathematical "expectation" of harm incurred from an exposure to radiation, taking into account not only the probability of each type of deleterious effect, but also the severity of the effect. The ICRP considers that knowledge of the absorbed dose is insufficient by itself to predict either the severity or the probability of deleterious effects on health resulting from irradiation under unspecified conditions. The dose-equivalent limits for radiation protection recommended by the ICRP do not apply to or include natural radiation exposure, except in so far as certain components are augmented by man's activities. Clearly, however, the ICRP recognizes that there is no sharp dividing line between levels of natural radiation that can be regarded as "normal" and those that are elevated owing to human activities or choice of environment.

The ICRP has developed a system of dose limitation that requires the following:

"(a) no practice shall be adopted unless its introduction produces a positive net benefit;

(b) all exposures shall be kept as low as reasonably achievable, economic and social factors being taken into account; and

(c) the dose equivalent to individuals shall not exceed the limits recommended for the appropriate circumstances by the Commission".

2.1 Dose-response relationship

The detrimental health effects of exposure to radiation are either somatic, i.e., those that become manifest in the exposed individual, or hereditary, i.e., those that affect the exposed individual's descendants.

For some somatic effects, such as carcinogenesis, and for the hereditary effects at dose levels involved in radiation protection, the probability of an effect occurring, rather than its severity, is regarded as a function of dose, without threshold (stochastic effects). For other somatic effects the severity varies with the dose and the effect appears above a threshold dose (non-stochastic effects).

The aim of radiation protection is to prevent detrimental non-stochastic effects and to limit the probability of stochastic effects to a level deemed acceptable. The latter aim will be achieved by the application of the system of dose limitation presented in section 2.2. The prevention of non-stochastic effects is achieved by setting up dose-equivalent limits at values sufficiently low that the threshold dose would not be reached even if the exposure persisted for the whole lifespan. The level of radioactive material in drinking-water required to reduce the incidence of stochastic effects to an acceptable level will automatically exclude the possibility of non-stochastic effects.

The dose-equivalent represents a method of quantifying dose that allows a better correlation between radiation exposure and the induced deleterious effects, more particularly the delayed stochastic effects.

A basic assumption of the ICRP is that, in the range of exposure concerned, the probability of a stochastic effect occurring is proportional to the dose received. However, different tissues in the body have different sensitivities to radiation and hence the ICRP has introduced dose-equivalent weighting factors to provide measures of equal risk. The sum of the weighted dose-equivalent factors for each tissue gives a measure of the total risk and is referred to as the effective dose-equivalent.

Moreover, in the case of long-lived radionuclides metabolized in such a way that they remain in the body for appreciable time periods, the resulting exposure may extend over many years. The committed effective dose-equivalent (H_E, 50) is defined as the total effective dose-equivalent incurred in the 50 years following intake of the nuclide.

It is this measure of exposure that is relevant to the present discussion; in what follows, the term "dose" may be used for brevity.

2.2 Dose-equivalent limits

The Commission's recommended dose limits apply to two categories of exposure—occupational and general. The limitation of the dose-equivalent refers to the sum of the annual dose-equivalents contributed by external sources and committed dose-equivalents from radioactive materials taken into the body during any year. Dose-equivalent limits established for occupational exposure are regarded as upper limits. Limitation of the dose-equivalent for members of the public is a more theoretical concept, mainly intended to ensure that it is unlikely that an individual will receive more than the specified dose-equivalent. Usually, the effectiveness is checked using sampling procedures and statistical calculations and by control of sources from which exposure is expected to arise.

The basic safety standards for radiation protection, based on ICRP's recommendations have set the effective dose-equivalent limit for the individual member of the public at 5 mSv in a year as applied to the average dose-equivalent in the critical group.[a] However, in cases where the doses to the same individuals approach this limit over many years, it would be prudent to take measures to restrict their lifetime dose-equivalent to a value corresponding to an *average* annual effective dose-equivalent of 1 mSv.

---

[a] The ICRP define "critical groups" as groups of the population with characteristics causing them to be exposed at a higher level than the rest of the exposed population from a given practice. These groups may be used as a measure of the upper limit of the individual doses from a proposed practice.

2.3 Implications for drinking-water quality

To apply the above principles to drinking-water, a measure of the potential radiation exposure from the water is necessary. While it may be possible to make some estimate on the basis of information already available—authorized discharges of man-made radioactive materials from nuclear installations, geological data on natural radioactivity, etc.—confirmation of the level of radioactivity requires direct measurement. Where it has been shown that the level is acceptably low, the necessary frequency of further measurements may be decided following a review of the particular circumstances applying and consultation with the competent authorities.

In the case of new water supplies, information concerning the activity levels in the raw water would be required. Levels in the final water after treatment should be checked.

If appropriate, the levels found in drinking-water should be related (by a competent authority) to the total exposure from all sources to which the population served by the water treatment plant is subjected.

3. SOURCES OF RADIATION EXPOSURE

The basic criterion for estimating the level of exposure to which individuals are subjected is established on the dose limitation system recommended by the ICRP (4, 5). Exposure may result from naturally occurring radionuclides at natural levels or at levels augmented by man's activities, for example, the use of phosphate fertilizers and discharge of mine waters, and from artificial radionuclides introduced into the environment, such as fallout from nuclear tests, releases from nuclear power facilities, and discharges as a result of the use of radionuclides in medicine, industry, and research (8).

Exposure levels from natural and man-made radioactivity are, in fact, regularly evaluated on a global basis, in so far as is possible, by the United Nations Scientific Committee on the Effects of Atomic Radiation (UNSCEAR), whose most recent report was published in 1977 (8). An examination of the data contained therein shows that drinking-water is a relatively minor constituent of total radiation exposure.

4. GUIDELINE VALUE FOR GROSS ALPHA AND GROSS BETA ACTIVITY

The radionuclides of interest were identified on the basis of those present in the natural environment as well as those resulting from man's activities. They are identified basically as the alpha and beta emitters, some of which have radioactive daughters. Radium-226 is typical of the naturally occurring alpha-emitting radionuclides of interest and strontium-90 is among the man-made beta emitters.

However, it is not always necessary to identify specific radionuclides present when the concentrations are low. In such cases, measurements of gross alpha and gross beta activity may serve to demonstrate that the radiotoxicity level is acceptable. Gross alpha and gross beta activity measurements are of particular interest for routine monitoring purposes.

Ideally, to use the gross activity screening procedure, the reference levels for alpha and beta activity will require to meet two criteria:

(a) they should be such as to ensure that, irrespective of the individual nuclides contributing to the gross activity, the associated exposure will be low enough not to necessitate further detailed analyses and consideration;

(b) they should be sufficiently high to ensure that the vast majority of drinking-water supplies satisfy such reference levels and hence the need for detailed analyses can be avoided.

In 1979, reference levels of 0.1 and 0.8 Bq were recommended (3) for gross alpha and gross beta activity, respectively. These levels appear to satisfy the second criterion. However, it is necessary to ensure that with the advancement of knowledge, these levels continue to satisfy the first criterion. To examine this aspect, the nuclides possibly contributing must be considered individually. Table 5 sets out the nuclides concerned and indicates, for each individual nuclide, the potential exposure from drinking two litres of water per day with contamination levels of 0.1 Bq per litre for alpha emitters and 1 Bq per litre for beta emitters.

Accurate calculations of the dose actually received by a person drinking water containing radon have not yet received general acceptance, although UNSCEAR and ICRP are at present reviewing this topic. However, it is known that the drinking-water pathway is of minor significance.

In the measurement of gross alpha and gross beta activity, it is understood that both radon and tritium (a nuclide of low radiotoxicity) are excluded. Where locally elevated levels of either of these nuclides are suspected, the competent authorities should be consulted.

It is accepted that some of the transuranium isotopes are more toxic than the nuclides listed in the table; however, the occurrence of transuranium isotopes in drinking-water at appreciable concentrations is rare. Where their presence is suspected, guidance should be sought from the competent authorities as to any special precautions that have to be taken.

Table 5.   Potential exposures to various alpha and beta emitters from drinking two litres of water per day for one year

Nuclide	$H_E,50/Bq$ (Sv)[a]	$H_E,50$ from one year's ingestion (mSv)
Alpha emitters		*73 Bq (= 0.1 Bq/litre)*
^{210}Po	4.36×10^{-7}	0.032
^{224}Ra	7.63×10^{-8}	0.006
^{226}Ra	3.05×10^{-7}	0.022
^{232}Th	7.4×10^{-7}	0.054
^{234}U	7.07×10^{-8}	0.005
^{238}U	6.32×10^{-8}	0.005
Beta emitters		*730 Bq (= 1 Bq/litre)*
^{60}Co	6.97×10^{-9}	0.005
^{89}Sr	2.17×10^{-9}	0.002
^{90}Sr	3.6×10^{-8}	0.026
^{129}I	7.4×10^{-8}	0.054
^{131}I	1.4×10^{-8}	0.010
^{134}Cs	1.98×10^{-8}	0.014
^{137}Cs	1.36×10^{-8}	0.010
^{210}Pb	1.36×10^{-6}	0.993
^{228}Ra	3.3×10^{-7}	0.214

[a] Data from the supplements to ICRP Publication 30 (*5*).

Of the alpha-emitting nuclides included in the table, thorium-232 is unlikely to be a major contributor in its own right to gross alpha activity. Hence, it would appear that attributing the alpha activity entirely to radium-226 is a conservative approach.

Considering the beta emitters, iodine-129 can be eliminated with the same proviso as for the transuranium isotopes. Radium-228 and more particularly lead-210 are the most toxic of the nuclides listed. However, these nuclides will normally represent only a small fraction of the gross beta activity, except in cases when the radon concentration in the water is high. Hence, assigning the beta activity entirely to strontium-90 would appear to be a conservative assumption, in the absence of high radon concentrations.

It can be seen that attributing 0.1 Bq of alpha activity per litre to radium-226 and 1 Bq of beta activity per litre to strontium-90 would imply an exposure of 0.048 mSv per year for a daily drinking-water intake of 2 litres.

The actual exposure associated with these concentrations will not exceed 0.048 mSv per year, and it is highly improbable that in practice even this dose level would be incurred.

According to ICRP data, this dose corresponds to a total risk in the range 10^{-7} to 10^{-6} per year, an order of magnitude less than what would be "likely to be acceptable to any individual member of the public" (*4*). This order of magnitude fully allows for the fact that drinking-water represents only one part of the general exposure.

Hence, levels of 1 Bq/litre for gross beta activity and 0.1 Bq/litre for gross alpha activity, as quoted above, are recommended as the reference levels for screening purposes.

5. RADON

Data from several countries show radon activity up to 10^3 Bq/litre in groundwater sources (deep wells) used as drinking-water supplies by some communities. Since the radon in the water is easily lost during handling, it is difficult to assess what quantity is ingested. Therefore, it is not possible to make accurate calculations of the dose actually received by a consumer drinking water containing radon, although there have been many attempts to do so (*3*).

It should be noted, however, that, as far as radon is concerned, the health risk from inhalation of the atmosphere in the room where water is drawn from a tap is greater than the risk due to ingestion of the water (*8*).

REFERENCES

1. *European standards for drinking-water*, 2nd ed. Geneva, World Health Organization, 1970.
2. *International standards for drinking-water*, 3rd ed. Geneva, World Health Organization, 1971.
3. *Radiological examination of drinking-water. Report of a WHO Working Group.* Copenhagen, WHO Regional Office for Europe, 1979 (EURO Reports and Studies No. 17).
4. INTERNATIONAL COMMISSION ON RADIOLOGICAL PROTECTION. Recommendations of the International Commission on Radiological Protection. *Annals of the ICRP*, **1** (3): 1–53 (1977) (ICRP Publication 26).
5. INTERNATIONAL COMMISSION ON RADIOLOGICAL PROTECTION. Limits for intakes of radionuclides by workers. *Annals of the ICRP*, **2–8**, (1979–1982) (ICRP Publication 30 and supplements).
6. MITCHELL, N. T. Radiological examination. In: Suess, M. J., ed. *Examination of water for pollution control*, vol. 2, Oxford, Pergamon Press, 1982, chapter 5.
7. INTERNATIONAL ATOMIC ENERGY AGENCY. *Basic safety standards for radiation protection.* Vienna, IAEA, 1982 (Safey Series No. 9).
8. UNITED NATIONS SCIENTIFIC COMMITTEE ON THE EFFECTS OF ATOMIC RADIATION. *Sources and effects of ionising radiation.* New York, United Nations, 1977.

INDEX

INDEX

WHO publications may be obtained, direct or through booksellers, from:

ALGERIA	Société Nationale d'Edition et de Diffusion, 3 bd Zirout Youcef, ALGIERS
ARGENTINA	Carlos Hirsch SRL, Florida 165, Galerias Güemes, Escritorio 453/465, BUENOS AIRES
AUSTRALIA	Hunter Publications, 58A Gipps Street, COLLINGWOOD, VIC 3066 — Australian Government Publishing Service *(Mail order sales)*, P.O. Box 84, CANBERRA A.C.T. 2600; *or over the counter from:* Australian Government Publishing Service Bookshops *at:* 70 Alinga Street, CANBERRA CITY A.C.T. 2600; 294 Adelaide Street, BRISBANE, Queensland 4000; 347 Swanston Street, MELBOURNE, VIC 3000; 309 Pitt Street, SYDNEY, N.S.W. 2000; Mt Newman House, 200 St. George's Terrace, PERTH, WA 6000; Industry House, 12 Pirie Street, ADELAIDE, SA 5000; 156–162 Macquarie Street, HOBART, TAS 7000 — R. Hill & Son Ltd., 608 St. Kilda Road, MELBOURNE, VIC 3004; Lawson House, 10–12 Clark Street, CROW'S NEST, NSW 2065
AUSTRIA	Gerold & Co., Graben 31, 1011 VIENNA I
BANGLADESH	The WHO Programme Coordinator, G.P.O. Box 250, DHAKA 5 — The Association of Voluntary Agencies, P.O. Box 5045, DHAKA 5
BELGIUM	*For books:* Office International de Librairie s.a., avenue Marnix 30, 1050 BRUSSELS. *For periodicals and subscriptions:* Office International des Périodiques, avenue Marnix 30, 1050 BRUSSELS — *Subscriptions to World Health only:* Jean de Lannoy, 202 avenue du Roi, 1060 BRUSSELS
BHUTAN	*see* India, WHO Regional Office
BOTSWANA	Botsalo Books (Pty) Ltd., P.O. Box 1532, GABORONE
BRAZIL	Biblioteca Regional de Medicina OMS/OPS, Unidade de Venda de Publicações, Caixa Postal 20.381, Vila Clementino, 04023 SÃO PAULO, S.P.
BURMA	*see* India, WHO Regional Office
CANADA	Canadian Public Health Association, 1335 Carling Avenue, Suite 210, OTTAWA, Ont. K1Z 8N8. *Subscription orders, accompanied by cheque made out to the* Royal Bank of Canada, Ottawa, Account World Health Organization, *may also be sent to the* World Health Organization, PO Box 1800, Postal Station B, OTTAWA, Ont. K1P 5R5
CHINA	China National Publications Import & Export Corporation, P.O. Box 88, BEIJING (PEKING)
CYPRUS	"MAM", P.O. Box 1722, NICOSIA
CZECHO-SLOVAKIA	Artia, Ve Smeckach 30, 111 27 PRAGUE 1
DEMOCRATIC PEOPLE'S REPUBLIC OF KOREA	*see* India, WHO Regional Office
DENMARK	Munksgaard Export and Subscription Service, Nørre Søgade 35, 1370 COPENHAGEN K (Tel: +45 1 12 85 70)
ECUADOR	Libreria Cientifica S.A., P.O. Box 362, Luque 223, GUAYAQUIL
EGYPT	Osiris Office for Books and Reviews, 50 Kasr El Nil Street, CAIRO
FIJI	The WHO Programme Coordinator, P.O. Box 113, SUVA
FINLAND	Akateeminen Kirjakauppa, Keskuskatu 2, 00101 HELSINKI 10
FRANCE	Librairie Arnette, 2 rue Casimir-Delavigne, 75006 PARIS
GABON	Librairie Universitaire du Gabon, B.P. 3881, LIBREVILLE
GERMAN DEMOCRATIC REPUBLIC	Buchhaus Leipzig, Postfach 140, 701 LEIPZIG
GERMANY, FEDERAL REPUBLIC OF	Govi-Verlag GmbH, Ginnheimerstrasse 20, Postfach 5360, 6236 ESCHBORN — W. E. Saarbach, Postfach 101 610, Follerstrasse 2, 5000 COLOGNE 1 — Alex. Horn, Spiegelgasse 9, Postfach 3340, 6200 WIESBADEN
GHANA	Fides Enterprises, P.O. Box 1628, ACCRA
GREECE	G.C. Eleftheroudakis S.A., Librairie internationale, rue Nikis 4, ATHENS (T. 126)
HAITI	Max Bouchereau, Librairie "A la Caravelle", Boîte postale 111-B, PORT-AU-PRINCE
HONG KONG	Hong Kong Government Information Services, Beaconsfield House, 6th Floor, Queen's Road, Central, VICTORIA
HUNGARY	Kultura, P.O.B. 149, BUDAPEST 62 — Akadémiai Könyvesbolt, Váci utca 22, BUDAPEST V
ICELAND	Snaebjørn Jonsson & Co., P.O. Box 1131, Hafnarstraeti 9, REYKJAVIK
INDIA	WHO Regional Office for South-East Asia, World Health House, Indraprastha Estate, Mahatma Gandhi Road, NEW DELHI 110002 — Oxford Book & Stationery Co., Scindia House, NEW DELHI 110001; 17 Park Street, CALCUTTA 700016 (*Sub-agent*)
INDONESIA	P. T. Kalman Media Pusaka, Pusat Perdagangan Senen, Block 1, 4th Floor, P.O. Box 3433/Jkt, JAKARTA
IRAN (ISLAMIC REPUBLIC OF)	Iran University Press, 85 Park Avenue, P.O. Box 54/551, TEHRAN
IRAQ	Ministry of Information, National House for Publishing, Distributing and Advertising, BAGHDAD
IRELAND	TDC Publishers, 12 North Frederick Street, DUBLIN 1 (Tel: 744835–749677)
ISRAEL	Heiliger & Co., 3 Nathan Strauss Street, JERUSALEM 94227
ITALY	Edizioni Minerva Medica, Corso Bramante 83–85, 10126 TURIN; Via Lamarmora 3, 20100 MILAN
JAPAN	Maruzen Co. Ltd., P.O. Box 5050, TOKYO International, 100–31
JORDAN, THE HASHEMITE KINGDOM OF	Jordan Book Centre Co. Ltd., University Street, P.O. Box 301 (Al-Jubeiha), AMMAN
KUWAIT	The Kuwait Bookshops Co. Ltd., Thunayan Al-Ghanem Bldg, P.O. Box 2942, KUWAIT
LAO PEOPLE'S DEMOCRATIC REPUBLIC	The WHO Programme Coordinator, P.O. Box 343, VIENTIANE
LEBANON	The Levant Distributors Co. S.A.R.L., Box 1181, Makdassi Street, Hanna Bldg, BEIRUT
LUXEMBOURG	Librairie du Centre, 49 bd Royal, LUXEMBOURG
MALAWI	Malawi Book Service, P.O. Box 30044, Chichiti, BLANTYRE 3